Bayesian Compressive Sensing for Site Characterization

Site characterization is indispensable to good geotechnical or rock engineering practice as every site is unique, but technical, budget, time, or access constraints typically result in only a tiny fraction of the underground soil and rock in a site being visually inspected, sampled, or tested. This leads to a long-lasting challenge of sparse measurements in geo-sciences and engineering. This book introduces Bayesian compressive sensing or sampling (BCS) as a highly efficient spatial data analytic and simulation method for the efficient modelling of spatial geo-data from sparse measurements, with quantified reliability and uncertainty to further optimize site characterization. It provides the necessary theory and computational tools for setting up and solving a sparse spatial data modeling problem using BCS.

This book suits graduate students, academics, researchers, and engineers interested in site characterization from sparse measurements in geotechnical and rock engineering, and also those modeling other spatially varying phenomena such as air quality data, soil or water pollution data, and meteorological data. This is supplemented with a software called Analytics of Sparse Spatial Data using Bayesian compressive sampling/sensing and illustrative examples, and enables hands-on experience of spatial data analytics and simulation using sparse measurements.

Yu Wang is Professor of Geotechnical Engineering at the Hong Kong University of Science and Technology, Hong Kong.

Tengyuan Zhao is Associate Professor of Geotechnical and Geological Engineering at Xi'an Jiaotong University, China.

Yue Hu is Humboldt Postdoctoral Fellow at Leibniz Universität Hannover, Germany.

Zheng Guan is Assistant Professor of Geotechnical Risk Assessment and Management at Delft University of Technology in the Netherlands.

Kok-Kwang Phoon is Cheng Tsang Man Chair Professor and President at Singapore University of Technology and Design (SUTD). He has twice won the American Society of Civil Engineers Norman Medal and is founding editor of the journal *Georisk.*

Challenges in Geotechnical and Rock Engineering

This series offers advanced level books focusing on state-of-the-art methods for handling problems across geotechnical engineering.

Chief Editor *Kok-Kwang Phoon, Singapore University of Technology and Design*
Assistant Editor *Dong-ming Zhang, Tongji University*

Evolutionary Process of a Steep Rocky Reservoir Bank in a Dynamic Mechanical Environment
Luqi Wang and Wengang Zhang

Uncertainty, Modelling, and Decision Making in Geotechnics
Edited by Kok-Kwang Phoon, Takayuki Shuku and Jianye Ching

Bayesian Machine Learning in Geotechnical Site Characterization
Jianye Ching

Databases for Data-Centric Geotechnics: Site Characterization (Volume 1)
Edited by Kok-Kwang Phoon and Chong Tang

Databases for Data-Centric Geotechnics: Geotechnical Structures (Volume 2)
Edited by Chong Tang and Kok-Kwang Phoon

For more information about this series, please visit: https://www.routledge.com/Challenges-in-Geotechnical-and-Rock-Engineering/book-series/CGRE?srsltid=AfmBOoriihXE0Je7bu8xQoUbZ2sAkSlhSwTOU_RxgF4La2iX-clDuDF6

Bayesian Compressive Sensing for Site Characterization

Yu Wang, Tengyuan Zhao, Yue Hu, Zheng Guan,
and Kok-Kwang Phoon

CRC Press
Taylor & Francis Group
Boca Raton London New York

CRC Press is an imprint of the
Taylor & Francis Group, an **informa** business

Designed cover image: Getty Images

First edition published 2026
by CRC Press
2385 NW Executive Center Drive, Suite 320, Boca Raton, FL 33431

and by CRC Press
4 Park Square, Milton Park, Abingdon, Oxon, OX14 4RN

CRC Press is an imprint of Taylor & Francis Group, LLC

ISBN: 9781032458090 (hbk)
ISBN: 9781032463445 (pbk)
ISBN: 9781003381235 (ebk)

DOI: 10.1201/9781003381235

Typeset in Times
by Newgen Publishing UK

Dedication

To our families
For their endless support and love

Contents

Foreword

I have the privilege of serving as Editor-in-Chief for the journal of *Georisk: Assessment and Management of Risk for Engineered Systems and Geohazards* for more than ten years. The central theme of *Georisk* is that uncertainties related to geomaterials (such as soils and rocks), geological processes, and subsequent treatments are often substantial and complex. These uncertainties play a critical role in the risk assessment and management of both engineered and natural systems. Significant theoretical and practical challenges persist in quantifying these uncertainties (such as spatial variability of soil or rock properties and statistical uncertainty due to scarce measurements), as well as in developing robust risk management methodologies that are acceptable to decision-makers and stakeholders. At the same time, rapid advancements in data-driven approaches present exciting opportunities to significantly enhance our uncertainty quantification and decision-making capabilities in the digital era.

This book *Bayesian Compressive Sensing for Site Characterization* presents a systematic exploration of a data-driven method known as Bayesian compressive sensing/sampling (BCS) for site characterization. BCS offers the distinctive ability to uncover complex spatial variation patterns from limited measurements and quantify the associated statistical uncertainty in a data-driven manner. Sparse measurements are not only a common challenge in geotechnical site characterization, but also a persistent issue across numerous research and practical domains within geo-engineering and geosciences, including risk management of geohazards (such as earthquakes, landslides, avalanches, rockfalls, and tsunamis), the safety of engineered systems (e.g., dams, buildings, offshore structures, and lifelines), air pollution assessment, and the mapping of contaminated groundwater and subsurface soils and rocks.

The development of smart cities and digital twins of geo-systems often oversimplifies subsurface environments due to a lack of extensive subsurface data and effective algorithms for interpreting sparse measurements. For instance, although subsurface stratigraphy and the spatial variability of geomaterial properties are critical for understanding landslide triggering and the large deformations of geomaterials associated with landslides, digital twins for slope safety frequently represent the subsurface as a uniform soil or rock layer with constant properties. In contrast, sophisticated three-dimensional models are created for aboveground components using photogrammetry and images captured by satellites, unmanned aerial vehicles, or ground-based instruments. This highlights a significant gap in both the research and practical implementation of smart cities and digital twins.

This book is a timely and significant contribution toward addressing the aforementioned gap. It comprehensively covers the underlying theories, algorithms, implementation procedures, software tools, and illustrative examples. I commend the authors for their outstanding work and recommend this book as essential reading and an invaluable reference for graduate students, academics, researchers, and practitioners alike.

Professor Limin Zhang
Editor-in-Chief
Georisk: Assessment and Management of Risk for Engineered Systems and Geohazards

Preface

Soils and rocks are natural geo-materials, which vary spatially because of the complicated geological processes they have undergone during their formations, making each site unique and site characterization indispensable to good geotechnical or rock engineering practice. Site characterization leads to the development of subsurface ground models and estimations of engineering properties of various types of soils or rocks encountered at a site, both of which are required for subsequent analysis, design, construction, and maintenance of geo-structures. Since subsurface soils and rocks are invisible from the ground surface, intrusive and destructive site investigation methods (e.g., drilling boreholes with a diameter of tens of centimeters to subsurface soils and rocks) are often used to obtain soil and rock samples during site characterization. Due to technical, budget, time, or access constraints, only a tiny fraction of subsurface soils and rocks within a site is visually inspected, sampled and/or tested. Such sparse measurements pose great challenges for site characterization, including (1) how to properly develop a ground model from sparse measurements, (2) how to quantify reliability or uncertainty of the developed model, (3) how to accurately depict spatial variability of the soil/rock properties, and (4) how to optimize the borehole or test layout such that as much information as possible can be obtained from the site and at the minimum expense.

This book introduces readers to an efficient spatial data analytic and simulation method called Bayesian compressive sensing/sampling (BCS) for tackling the challenges associated with sparse measurements in site characterization. Compressive sensing/sampling (CS) was originally developed for image compression and storage in digital image processing. BCS is a Bayesian extension of CS with quantified uncertainty, and it is also known as Bayesian sparse learning in machine learning literature. This book extends BCS from image processing to analytics and simulation of sparse spatial data for site characterization. It is intended to provide an easy and systematic access to the necessary theories and computational tools for using BCS to analyze and simulate sparse spatial data in the context of site characterization. The book is targeted at graduate students, academics, researchers, and practitioners interested in site characterization from sparse measurements in geotechnical and rock engineering. It is also useful for modeling other spatially varying phenomena such as the analysis and simulation of environmental data (e.g., air quality data, soil or water pollution data) and meteorological data (e.g., rain or wind data).

The book starts with an overview on CS for digital image processing, including its background and motivation, mathematical theory and associated assumptions, and potential of applications to different fields such as site characterization. Similarities and differences between site characterization and image processing are highlighted, including measurement patterns (particularly sparse measurements from site characterization versus complete measurements in digital image processing) and objectives (uncovering spatial variation patterns from sparse measurements for site characterization versus compression for efficient storage and transmission of images for digital image processing). Then, BCS algorithms are developed for analytics of sparse measurements from site characterization and random field simulation of spatial data directly from sparse measurements, facilitating the development of subsurface ground models and characterization of geotechnical property spatial variability in site characterization.

The BCS algorithms and random field generators are used to further develop quantitative methods and statistical charts for sample size determination in geotechnical site characterization, with explicit consideration of spatial variation and correlation of geotechnical properties. An uncertainty-based adaptive sampling strategy using BCS is also introduced for site characterization. The adaptive sampling strategy continuously monitors the results interpreted from already performed exploration points and automatically decides whether additional exploration points are needed and, if so, their optimal test locations. A unique feature of this book is that it is supplemented with a software package called Analytics of Sparse Spatial Data using Bayesian Compressive Sampling/Sensing (ASSD-BCS). The software ASSD-BCS allows readers to experiment with the examples in the book and get hands-on experience with spatial data analytics and simulation using sparse measurements.

Acknowledgments

The research works presented in this book have been supported by grants from the Research Grant Council of Hong Kong Special Administrative Region, the Innovation and Technology Commission of Hong Kong Special Administrative region, the Shenzhen Science and Technology Innovation Commission in China, and the City University of Hong Kong. Most works were carried out when the first four authors were affiliated with City University of Hong Kong. The support is gratefully acknowledged.

About the Authors

Yu Wang is Professor of Geotechnical Engineering at the Hong Kong University of Science and Technology. He is a registered professional engineer (Civil) in Hong Kong and an elected fellow of American Society of Civil Engineers (ASCE). His recent research efforts have focused on machine learning in geotechnical and geological engineering, geotechnical site characterization, geotechnical uncertainty, reliability and risk, geo-hazards (e.g., landslides and earthquakes), and geothermal energy. His research has earned several prestigious international awards, including the Scott Sloan Best Paper Award 2023 from *Computers and Geotechnics*, the 2023 *Thomas A. Middlebrooks Award* from ASCE, the 2022 *R.M. Quigley Award* (Honourable Mention) from Canadian Geotechnical Society, the 2020 Best Paper Award from the *ASCE-ASME Journal of Risk and Uncertainty in Engineering Systems, Part A: Civil Engineering*, the Highly Cited Research Award from the international journal of *Engineering Geology* in 2017, and the GEOSNet Award in 2025 and GEOSNet Young Researcher Award in 2025 from the Geotechnical Safety Network (GEOSNet). He has authored/co-authored more than 200 journal papers and two books in English. He served as president of ASCE Hong Kong Section from 2012 to 2013 and serves in editorial boards of several top journals in geotechnical engineering or risk and uncertainty analysis (e.g., Associate Editor for the ASCE *Journal of Geotechnical and Geoenvironmental Engineering*).

Tengyuan Zhao is Associate Professor of Geotechnical and Geological Engineering at Xi'an Jiaotong University, China. His research focuses on data analytics of geodata using random field theory and Bayesian compressive sensing, probabilistic site characterization, machine learning applications in geotechnical engineering, risk assessment, and mitigation of geo-hazards (e.g., landslides susceptibility mapping). He is a recipient of the 2020 Best Paper Award by the *ASCE-ASME Journal of Risk and Uncertainty in Engineering Systems, Part A: Civil Engineering*, and has authored/co-authored more than 50 journal papers. He currently serves on the Early Career Editorial Board of *Georisk: Assessment and Management of Risk for Engineered Systems and Geohazards*, an international journal dedicated to risk and uncertainty analysis in geotechnical and geological engineering.

Yue Hu is currently Humboldt Postdoctoral Fellow at Leibniz Universität Hannover, Germany. His research areas include digitization of geological models, machine learning in geotechnical engineering, random field simulation, and applications of compressive sampling/sensing (CS). He is a recipient of the 2020 Best Paper Award of the *ASCE-ASME Journal of Risk and Uncertainty in Engineering Systems, Part A: Civil Engineering*, the Ringo Yu Prize for Best PhD Thesis in Geotechnical Studies 2020 by the Hong Kong Institution of Engineers (HKIE), and the Humboldt Research Fellowship 2022 by the Alexander von Humboldt Foundation.

Zheng Guan is Assistant Professor of Geotechnical Risk Assessment and Management at TU Delft, the Netherlands. He obtained his PhD from City University of Hong Kong in 2021. His research focuses on integrating sparse sensing, machine learning, and Bayesian inference to advance underground digitalization and performance-/reliability-based design in geotechnical engineering.

Kok-Kwang Phoon is Cheng Tsang Man Chair Professor and President of the Singapore University of Technology and Design. He was awarded the ASCE Norman Medal twice in 2005 and 2020, the Humboldt Research Award in 2017, and the ASCE Alfredo Ang Award in 2024. He is the Founding Editor of *Georisk* and *Geodata and AI*. He is a fellow of the Academy of Engineering Singapore and Singapore National Academy of Science.

1 Introduction

Yu Wang and Kok-Kwang Phoon

ABSTRACT

Site characterization is indispensable in geotechnical and rock engineering, since subsurface soils and rocks are natural geo-materials that are spatially varying and highly heterogeneous due to the complicated geological process they have undergone. However, only a small portion of geo-materials is sampled and tested during site characterization, leading to significant uncertainty and risk in the interpreted ground conditions, subsequent geotechnical design and analysis, and construction and operation of geo-structures. An overview of the site characterization process is reviewed in this chapter. Then, a long-lasting challenge of sparse measurements in geotechnical site characterization and various disciplines in geoscience and geoengineering is highlighted. This motivates the development of Bayesian compressive sensing or sampling for geotechnical site characterization presented in this book.

1.1 SITE CHARACTERIZATION IN GEOTECHNICAL AND ROCK ENGINEERING

Geotechnical and rock engineering deals with natural geo-materials, such as soils and rocks, which often serve as the foundational elements for civil engineering projects. These materials are products of complex geological processes that have evolved over millions, or even billions, of years, resulting in highly heterogeneous and site-specific properties. The formation of soils and rocks involves a variety of geological processes such as weathering, erosion, sedimentation, tectonic activity, and mineralogical transformations, all of which contribute to their inherent variability. In some cases, the ground is affected by recent man-made activities, such as land reclamations. As a result, their mechanical and hydraulic characteristics, such as strength, permeability, and deformation properties, can differ significantly even within small spatial scales. This spatial variability poses significant challenges for engineers, as assumptions based on uniform or idealized conditions often fail to capture real-world complexities. For instance, a slope stability analysis conducted at one location may yield entirely different results just a few meters away due to changes in soil composition, stratification, or rock fracturing. Similarly, the design of foundations, tunnels, or retaining structures must account for these variations to ensure safety and performance.

DOI: 10.1201/9781003381235-1

Since subsurface soils and rocks are underground and invisible from the ground surface, it presents a fundamental challenge in geotechnical and rock engineering that necessitates a systematic investigation of project sites. These investigations are crucial for addressing three primary questions: (1) what types of soils and rocks (also anomalies in special cases) exist beneath a project site? (2) how are these materials spatially distributed? and (3) what are the engineering properties of each soil and rock type that are needed in subsequent design and analysis? There are other considerations for some sites such as the presence of harmful substances that can affect the safety of human workers. To answer these questions, engineers often employ a multi-stage investigation approach that integrates various in-situ and laboratory methods. An overview of the multi-stage investigation is provided in the following section.

1.2 OVERVIEW OF SITE CHARACTERIZATION

Site characterization aims to define subsurface stratigraphy (including identification of soil and rock types, determination of thickness and boundaries of soil and rock layers, and anomalies in some cases) and evaluate engineering properties of these materials for geotechnical analysis and design. As illustrated in Figure 1.1, it follows a general process consisting of six stages: (1) desk study, (2) site reconnaissance,

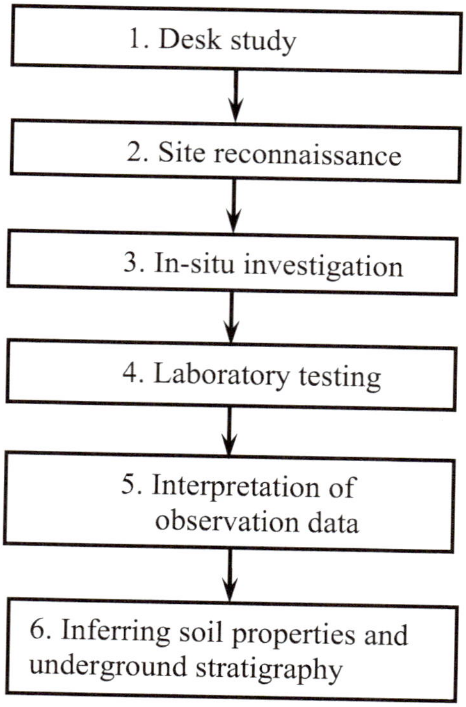

FIGURE 1.1 General process of geotechnical site characterization (after Clayton et al. 1995; Mayne et al. 2002; and Cao et al. 2017).

(3) in-situ investigation, (4) laboratory testing, (5) interpretation of field and laboratory data, and (6) derivation of soil/rock properties and subsurface stratigraphy. These stages will be briefly described in the following five subsections.

1.2.1 Desk Study

Geotechnical site characterization typically begins with a comprehensive desk study to gather general information on the region and existing site-specific information. This preliminary stage collects critical data including geological conditions, geotechnical properties and potential issues, topographic features, groundwater characteristics, meteorological records, existing infrastructure and utilities, as well as historical land use patterns (Clayton et al. 1995; Mayne et al. 2002; and Cao et al. 2017). Table 1.1

TABLE 1.1
Summary of Existing Information (after Clayton et al. 1995, Cao et al. 2017)

Type	Source
Geology	Geological maps Geological reports Geological publications Regional guides Air photographs Soil survey maps and records
Geotechnical properties and potential issues	Geotechnical reports Academic journals (e.g., Geotechnical journals, Engineering geology journals, and Civil engineering journals) Previous ground investigation reports
Site topography	Topographic maps Stereo air photographs
Groundwater conditions	Topographical maps Air photographs Well records Previous ground investigation reports
Meteorological conditions	Meteorological records
Existing construction and services	Topographical maps Plans held by utilities Mining records Construction presses
Previous land use and past construction activities	Out-of-print topographical maps Out-of-print geological maps Air photographs Airborne remote sensing Archaeological society records Mining records Past construction records

systematically summarizes the types of available information and their potential sources.

Geological information, including geological history, can be obtained from existing geological records such as published geological maps, regional guidebooks, technical reports, soil survey maps, aerial photographs, and academic textbooks. Geotechnical parameters and documented ground-related issues, including adverse ground conditions, soil classification, material properties, and stratigraphic profiles, are typically available in existing geotechnical reports (Kulhawy and Mayne 1990), peer-reviewed journals (particularly in geotechnical, engineering geology, and civil engineering publications), and historical ground investigation records from the site or adjacent areas.

Topographic information can be acquired through topographic maps and stereoscopic aerial photography. Groundwater conditions, including water table levels, are typically documented in well records, previous investigation reports, topographic maps, and aerial photographs. Meteorological data sources provide relevant climatic information. Furthermore, details regarding existing infrastructure, utilities, historical land use, and past construction activities, including mining activities, can be extracted from topographic maps, aerial photographs, and mining records.

1.2.2 Site Reconnaissance

After desk study, site reconnaissance is carried out to confirm and supplement the information that is previously collected during desk study. Site reconnaissance involves site visits and visual examination and documentation of surface conditions at the project site. It provides a first-hand understanding of the site's physical characteristics and potential geotechnical challenges before more invasive (and costly) subsurface investigations begin. Its primary objectives include identifying visible geohazards such as slope instability, erosion, or evidence of past ground failures, which could impact construction safety and design. The reconnaissance also assesses site accessibility, logistics, and constraints, such as existing structures, utilities, or environmental features that may influence investigation methods or project planning. By documenting surface geology, vegetation patterns, drainage conditions, and man-made features, engineers can optimize the locations for subsequent boreholes, field tests, and instrumentation. Additionally, this stage helps develop a preliminary geotechnical conceptual model, guiding cost-effective and risk-informed investigation strategies. The findings ensure compliance with regulatory requirements while minimizing unforeseen challenges during later stages. Ultimately, site reconnaissance enhances investigation efficiency, reduces project risks, and supports informed decision-making by bridging desk studies with field-based data collection.

For large or important sites, emerging technologies, such as unmanned aerial vehicle (UAV, also known as drone) and other remote sensing methods, are often used. Remote sensing provides large-scale, high-resolution data on terrain, soil composition, and subsurface conditions without direct contact. It utilizes technologies such as satellite imagery, UAV, aerial photography, light detection and ranging (LiDAR), and ground-penetrating radar to assess site characteristics efficiently. One key advantage is its ability to detect geological hazards like landslides, soil erosion, and fault lines, enabling engineers to identify risk-prone areas before ground exploration begins. For example, LiDAR generates detailed topographic maps, revealing subtle elevation changes. Thermal imaging can be used to detect moisture variations, highlighting potential drainage or groundwater issues. Additionally, remote sensing enhances geotechnical site characterization in inaccessible or hazardous locations, reducing the need for extensive manual surveys. By integrating remote sensing data with traditional geotechnical testing, engineers can develop more comprehensive site models, improving foundation design, slope stability analysis, and construction planning. As technology advances, remote sensing continues to refine the accuracy and efficiency of geotechnical site characterization, while reducing the time and cost required.

1.2.3 In-situ Investigation

In-situ investigation work generally includes borehole drilling, soil and rock sampling, in-situ testing, and groundwater investigation (Clayton et al. 1995; Mayne et al. 2002; Cao et al. 2017). Boreholes are essential in geotechnical site investigations to obtain subsurface soil/rock samples and assess subsurface conditions. They provide critical data on stratigraphy, groundwater levels, and material properties, informing foundation design, slope stability analysis, and construction planning. The drilling process involves selecting appropriate methods (rotary, percussion, or auger drilling) based on ground conditions. As drilling progresses, sampling (disturbed or undisturbed) is conducted at specified depths, followed by in-situ tests like standard penetration tests (SPT). Borehole logs are created documenting soil/rock types, moisture content, and strength. Borehole data is later analyzed in labs to determine geotechnical parameters, ensuring safe and cost-effective engineering solutions.

Real-time data about subsurface conditions and drilling performance may also be collected during drilling operation. This is referred to as measurement while drilling (MWD), which involves sensors embedded in the drill string that continuously measure parameters such as wellbore trajectory, pressure, temperature, and formation properties. This data is transmitted to the surface and allows practitioners to make immediate adjustments for optimizing drilling efficiency and safety. One key advantage of MWD is its ability to provide accurate wellbore positioning, ensuring that directional drilling follows the planned path.

SPT is commonly used in-situ test worldwide. It involves driving a thick-walled sampler into the borehole base using repeated blows from a 63.5 kg hammer dropped from a height of 760 mm. The test measures the blow count (SPT N-value) required to advance the sampler 300 mm after an initial 150 mm seating drive (Clayton 1995; Mayne et al. 2002). While providing soil samples for classification and laboratory testing, the SPT offers a relatively simple and robust method for various soil types. However, its results exhibit significant variability due to operational and equipment inconsistencies, limiting their reliability for precise engineering analysis (e.g., Kulhawy and Trautmann 1996; Mayne et al. 2002).

Another widely used in-situ testing method is cone penetration test (CPT). CPT involves advancing a cylindrical steel probe into the ground at a constant rate (typically 20 mm/s) while measuring penetration resistance (Lunne et al. 1997; Mayne et al., 2002). The test measures three key parameters: cone tip resistance, sleeve friction, and pore water pressure when equipped with a piezometer (piezocone test). Over the past several decades, CPT has gained global recognition due to its rapid execution, minimal operator dependence, and ability to generate near-continuous subsurface profiles. However, the cone cannot penetrate gravel or boulder layers effectively and does not recover physical soil samples (Mayne et al. 2002).

1.2.4 Laboratory Testing

Following in-situ investigations, retrieved soil and rock samples are sent to laboratory and undergo comprehensive laboratory testing to determine their engineering properties. Commonly used laboratory tests on soil samples include (1) index property tests (moisture content, specific gravity, unit weight, particle size distribution, Atterberg limits, and compaction characteristics); (2) strength and stiffness evaluations (unconfined compression, direct shear, and triaxial tests); (3) permeability assessments (constant-head and falling-head tests); and (4) consolidation testing (oedometer tests) (Clayton et al. 1995; Mayne et al. 2002). These tests yield critical soil parameters including Atterberg limits (liquid, plastic, and shrinkage limits), strength characteristics (undrained shear strength, effective friction angle, etc.), stiffness moduli (Young's modulus and shear modulus), hydraulic conductivity, compressibility coefficients, and pre-consolidation pressure.

Common tests of rock samples in laboratory include uniaxial compressive strength (UCS) tests, which measure the maximum axial load a rock sample can withstand before failure, providing critical data for stability analysis in tunnels and slopes. Brazilian tensile strength tests determine indirect tensile strength by applying diametral compression to disc-shaped samples, important for assessing fracturing potential. Triaxial compression tests evaluate rock strength under confining pressures, simulating in-situ stress conditions for deep excavations or reservoir engineering. Point load tests offer a rapid estimate of compressive strength using irregular samples, useful for preliminary assessments. Direct shear tests measure shear strength along discontinuities, crucial for analyzing jointed rock masses. Other key tests include Young's modulus and Poisson's ratio determination through stress–strain analysis, slake durability tests for weathering resistance, and permeability tests to assess fluid flow through fractures. These tests provide data for numerical modeling, design optimization, and risk mitigation in rock engineering projects. While laboratory testing has high accuracy, sample preparation and testing protocols must account for scale effects and natural heterogeneity in rock masses. It is useful to point out that laboratory tests on intact rock samples are generally less informative for design and construction than those on soil samples, because the behavior of a rock mass at an engineering scale (slope, tunnel, etc.) is controlled by discontinuities that cannot be characterized easily using small rock cores.

While laboratory testing measures geotechnical properties under more controlled boundary conditions compared to in-situ tests, this advantage comes with significant trade-offs. The process demands substantial time for sample extraction and preparation, specialized equipment operation, and detailed analysis. Additionally, laboratory tests incur higher costs due to equipment maintenance, skilled personnel requirements, and controlled environmental needs. These constraints necessitate a strategic approach to test selection, where engineers must balance data quality with practical considerations. For critical parameters like shear strength or compressibility, laboratory testing remains indispensable, but for preliminary assessments, in-situ methods may suffice. Ultimately, the choice depends on project complexity, accuracy requirements, and budget constraints, ensuring optimal resource allocation without compromising essential data quality.

1.2.5 Interpretation of Observation Data and Inferring the Site Subsurface Conditions

Measured properties obtained from in-situ or laboratory tests often cannot be directly applied in geotechnical analysis and design. These raw measurements typically require conversion into design properties through appropriate transformation models. Geotechnical literature provides numerous such models, which may incorporate empirical correlations, theoretical relationships, or a combination of both, to derive the necessary design parameters from test measurements (e.g., Kulhawy and Mayne 1990).

Geotechnical engineers synthesize multiple data sources to evaluate site conditions for analysis and design purposes. This comprehensive assessment integrates (1) preexisting information from various sources obtained during desk study, (2) field observation data collected through in-situ testing, including remote sensing and geophysical testing, (3) laboratory test results, (4) appropriate transformation models, and (5) professional engineering judgment. Through this process, engineers develop reliable estimates of subsurface conditions, including soil/rock properties and stratigraphic profiles. The final deliverables include a detailed geotechnical site characterization report documenting all findings and providing design recommendations based on the interpreted ground conditions (Mayne et al. 2002).

A survey of geodata including site characterization data is recently conducted by Phoon and Tang (2025). An extensive compilation of site characterization data is given in Phoon and Tang (2024) and Phoon et al. (2024a).

1.3 CHALLENGES IN SITE CHARACTERIZATION

One of the long-lasting challenges in site characterization is a lack of measurements or data. Figure 1.2a illustrates the subsurface geological profile of a site in Hong Kong, depicting distinct soil and rock layers, or zones, represented by color-coded units. The stratigraphy includes fill, soft marine deposits, heterogeneous diamict (a mixed sedimentary unit), and bedrock formations of marble and granite. This profile reveals significant spatial variability, with abrupt transitions between materials reflecting local complex geological history in Hong Kong. Such complexity necessitates a careful characterization and interpretation of the site for geotechnical analysis and design, as the irregular interfaces between units may influence slope stability and foundation performance.

As outlined in Section 1.2, borehole drilling serves as a fundamental method for subsurface exploration during geotechnical site characterization. While these investigations provide direct observations on soil and rock stratigraphy, groundwater

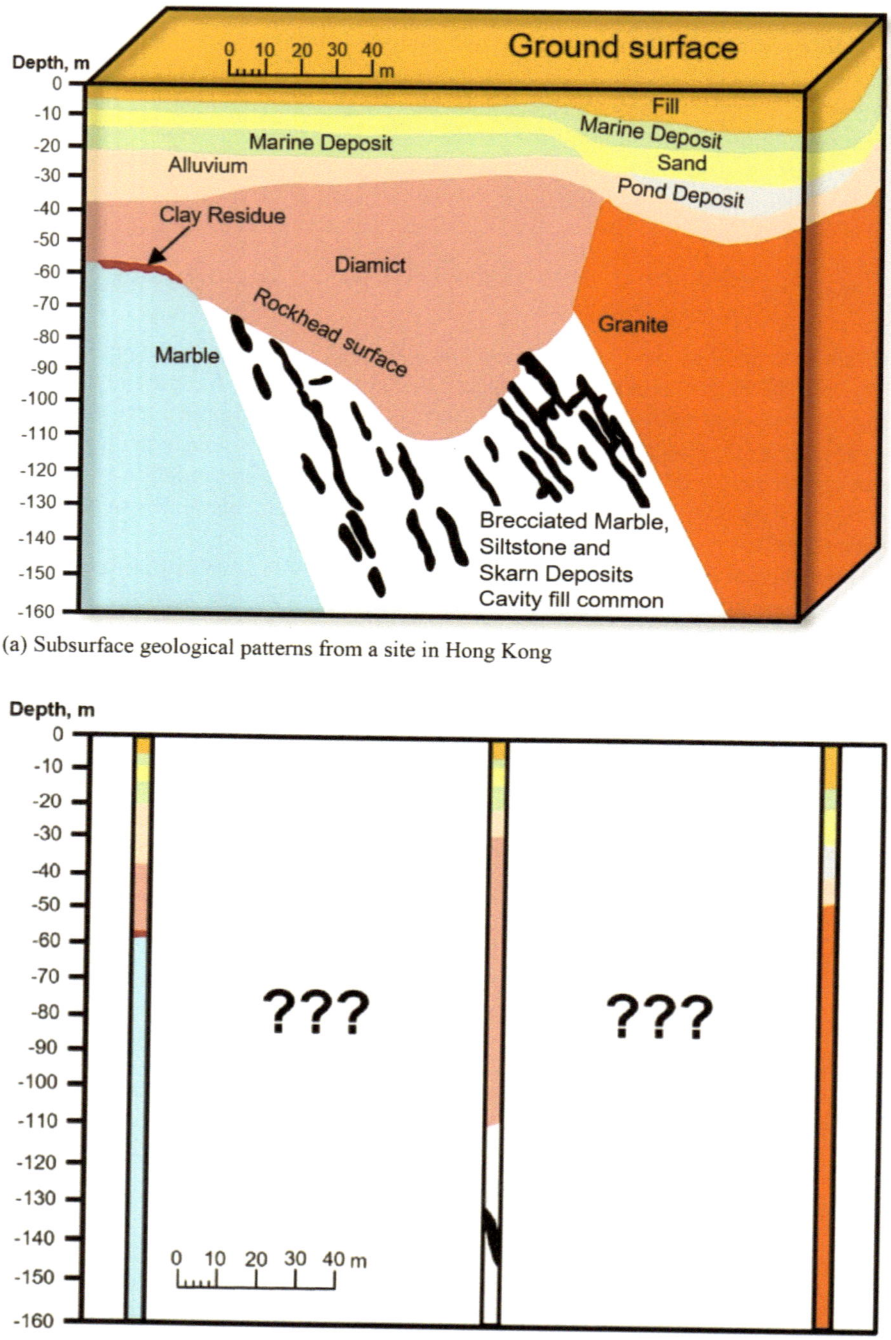

(a) Subsurface geological patterns from a site in Hong Kong

(b) Drilled boreholes and borehole logs

FIGURE 1.2 An illustration of a sparse measurement in site characterization.

conditions, and material properties, their implementation faces significant practical constraints. First, the intrusive nature of drilling, which disturbs or destroys the surrounding ground, limits the number of boreholes that can be feasibly executed. Moreover, logistical and economic factors further restrict borehole deployment. Drilling operations incur substantial costs, including equipment mobilization and labor. Time constraints also play a role, particularly in large-scale projects where rapid progress is essential. Additionally, physical accessibility often poses challenges; urban sites may lack sufficient space for drilling rigs, while remote or rugged terrains complicate equipment transportation and setup.

Due to these limitations, borehole spacing typically ranges from tens to hundreds of meters in most geotechnical applications. Phoon (2017) called this the "curse of small sample size". In regional-scale projects, such as highway or railway alignments, spacing may even extend to thousands of meters. This sparse sampling necessitates interpolation between adjacent boreholes, as shown by questions marks in Figure 1.2b.

A commonly used method for interpolation of stratigraphic boundaries between adjacent boreholes involves connecting them with straight lines to construct geological profiles. While this linear interpolation method works for consistent subsurface layers, it fails to accurately represent complex geological conditions where soil/rock types or sequences differ significantly between boreholes, as illustrated by the contrasting color-coded units in Figure 1.2b. Such discrepancies highlight the inherent limitations of sparse borehole data in capturing subsurface variability. The challenge of interpreting spatially limited measurements, where key stratigraphic features or data patterns may be missed between widely spaced measurements, remains a persistent issue in geotechnical site characterization. This book introduces an imaged-based data analysis method called BCS to systematically tackle this important challenge of sparse measurements in geotechnical site characterization. Other challenges in data-driven site characterization are presented in Phoon et al. (2022) and Phoon et al. (2024b).

1.4 CHALLENGES OF SPARSE MEASUREMENTS IN GEOSCIENCE AND ENGINEERING

The challenge of sparse measurements is also frequently encountered across various disciplines of geoscience and geoengineering. Natural systems display continuous variations across space and time, yet observational capabilities often remain restricted to isolated sampling points and discrete moments. For example, rain gauges measure precipitation at specific locations, and deploying multiple gauges across a region can capture spatial rainfall variability. However, rain gauges are often widely spaced and may miss localized extremes. Similarly, anemometers measure wind speed and direction at fixed locations. During typhoons, strategically placed anemometers record crucial real-time data to track storm intensity and wind field evolution. However, sparse networks struggle to capture the storm's full spatiotemporal variability due to: (1) inability to resolve small-scale turbulence and sudden wind shifts between stations, (2) missed localized extreme gusts from convective bands, and (3) insufficient density to map the typhoon's asymmetric structure accurately. These gaps may lead to underestimation of peak winds in unmonitored zones, affecting early warnings and engineering wind load assessments.

The BCS presented in this book offers a powerful framework to address these sparse measurement challenges. By integrating probability principles with compressive sensing, BCS enables reconstruction of complete spatial or temporal fields from limited observations with quantified uncertainty. It is particularly valuable for interpolation between sparse sensor measurements in environmental monitoring networks and reconstruction of time-continuous processes from intermittent measurements. Its ability to rigorously quantify uncertainty makes it particularly appealing across geoscience and geoengineering applications, where optimal data acquisition strategies must balance information gain with practical constraints.

1.5 ORGANIZATION OF THE BOOK

This book is organized into eight chapters. After the introduction in this chapter, Chapter 2 gives an overview on compressive sensing for digital image processing, including its background and motivation, mathematical theory and associated assumptions, and potential of applications to different fields. Chapter 3 highlights similarities and differences between geotechnical site characterization and image or signal processing, such as measurement patterns (particularly sparse measurements versus complete measurements) and objectives, leading to development of BCS for sparse measurements in geotechnical site characterization. Chapter 4 extends the BCS algorithms developed in Chapter 3 to random field simulation of spatial or temporal data directly from sparse measurements, facilitating subsurface stratification and characterization of three-dimensional (3D) soil variability in geotechnical site characterization.

Chapter 5 applies the BCS algorithms and random field generators described in Chapters 3 and 4 to further develop quantitative methods and statistical charts for sample size determination in geotechnical site characterization with explicit consideration of spatial variation and correlation of geotechnical properties. In Chapter 6, an uncertainty-based adaptive sampling strategy using BCS is introduced for geotechnical site characterization. The adaptive sampling strategy continuously monitors the results interpreted from already performed exploration points and automatically decides whether additional exploration points are needed and, if so, their optimal test locations. Chapter 7 presents a computer software package for implementing BCS using the Python programming language. It contains step-by-step procedures that allow readers to gain hands-on experience with BCS. This will hopefully help the readers to properly use and apply BCS in geotechnical site characterization. Some additional comments and future research directions are provided in Chapter 8.

REFERENCES

Cao, Z., Y. Wang, and D. Li. 2017. *Probabilistic approaches for geotechnical site characterization and slope stability analysis.* Springer-Verlag Berlin Heidelberg, ISBN 978-3-662-52912-6.

Clayton, C. R. I. 1995. *The standard penetration test (SPT): Methods and Use*, Report 143, Construction Industry Research and Information Association, UK.

Clayton, C. R. I., M. C. Matthews, and N. E. Simons. 1995. *Site Investigation*. Cambridge, MA: Blackwell Science.

Kulhawy, F. H. and P. W. Mayne. 1990. *Manual on estimating soil properties for foundation design*, Report EL 6800. Electric Power Research Inst., Palo Alto.

Kulhawy, F. H. and C. H. Trautmann. 1996. Estimation of in-situ test uncertainty. *Uncertainty in the Geologic Environment: From Theory to Practice. Geotechnical Special Publication* 58 (I): 269–286.

Lunne, T., P. K. Robertson, and J. J. M. Powell. 1997. *Cone Penetration Testing in Geotechnical Practice*. London: Blackie-Academic Publishing,

Mayne, P. W., B. R. Christopher, and J. DeJong. 2002. *Subsurface Investigations – Geotechnical Site Characterization*, No. FHWA NHI-01-031, Federal Highway Administration, U. S. Department of Transportation, Washington D C.

Phoon, K. K. 2017. Role of reliability calculations in geotechnical design. *Georisk: Assessment and Management of Risk for Engineered Systems and Geohazards* 11 (1): 4–21.

Phoon, K. K., J. Ching, and T. Shuku. 2022. Challenges in data-driven site characterization. *Georisk: Assessment and Management of Risk for Engineered Systems and Geohazards* 16 (1): 114–126.

Phoon, K. K. and C. Tang. 2024. *Databases for Data-Centric Geotechnics: Site Characterization*. Boca Raton, FL: CRC Press.

Phoon, K. K., J. Ching, and Y. Tao. 2024a. Soil and rock parametric uncertainties. *Uncertainty, Modelling, and Decision Making in Geotechnics* (Chapter 2, 37–114). Boca Raton, FL: CRC Press.

Phoon, K. K., J. Ching, and C. Tang. 2024b. Role of site characterization information in data-centric geotechnics. *Databases for Data-Centric Geotechnics: Site Characterization* (Chapter 1, 1–49). Boca Raton, FL: CRC Press.

Phoon, K. K. and C. Tang. 2025. A short survey of geodata – Geotechnical engineering. *Georisk: Assessment and Management of Risk for Engineered Systems and Geohazards*, under review.

2 Compressive Sensing (CS)

Yue Hu and Yu Wang

ABSTRACT

This chapter provides a brief overview of compressive sensing (CS) for digital image processing, covering its background, motivation, mathematical theory, assumptions, and potential applications across various fields. The discussion begins with image processing and compression and later expands to more general signals, including spatial and temporal data, such as geotechnical data examined in this book. Technically, CS relies on three key prerequisites: data sparsity, a properly designed sensing matrix, and reconstruction algorithms. The first one pertains to the intrinsic characteristics of the signal or image, the second one to the measurement modality, and the third one to the optimization required for solving a system of linear equations. These three fundamental aspects are explained in detail in this chapter. Additionally, the chapter highlights the challenges of achieving reliable and robust CS solutions, describes quantitative criteria for evaluating CS performance, and reviews two widely used and effective reconstruction algorithms. Several real-data examples, featuring images and geotechnical spatial data, are presented for illustration.

2.1 MOTIVATION OF COMPRESSIVE SENSING/SAMPLING IN IMAGE PROCESSING

Nowadays, with the rapid development and innovation of both hardware and software in imaging systems, there has been a growing trend of using digital cameras, including smartphones, to take photos or images with high resolutions (e.g., more than five or even ten megapixels (MP) for a single image). For example, with just a simple tap of the shutter button, a life moment can be instantly captured using a smartphone. Quite often, this high-resolution digital image will be immediately saved as a Joint Photographic Experts Group (JPEG) format file in a smartphone and vividly displayed on the screen. Figure 2.1a gives an example of a grayscale digital image showing a campus view of the City University of Hong Kong (CityU). By zooming into the logo of CityU in the middle, some fine-scale details of this image can be observed, as shown in Figure 2.1b. At this local scale, it is clearly seen that objects in this image are represented by a series of discrete "picture elements" with different brightness. Each "picture element", or "pixel" in short, represents a single

DOI: 10.1201/9781003381235-2

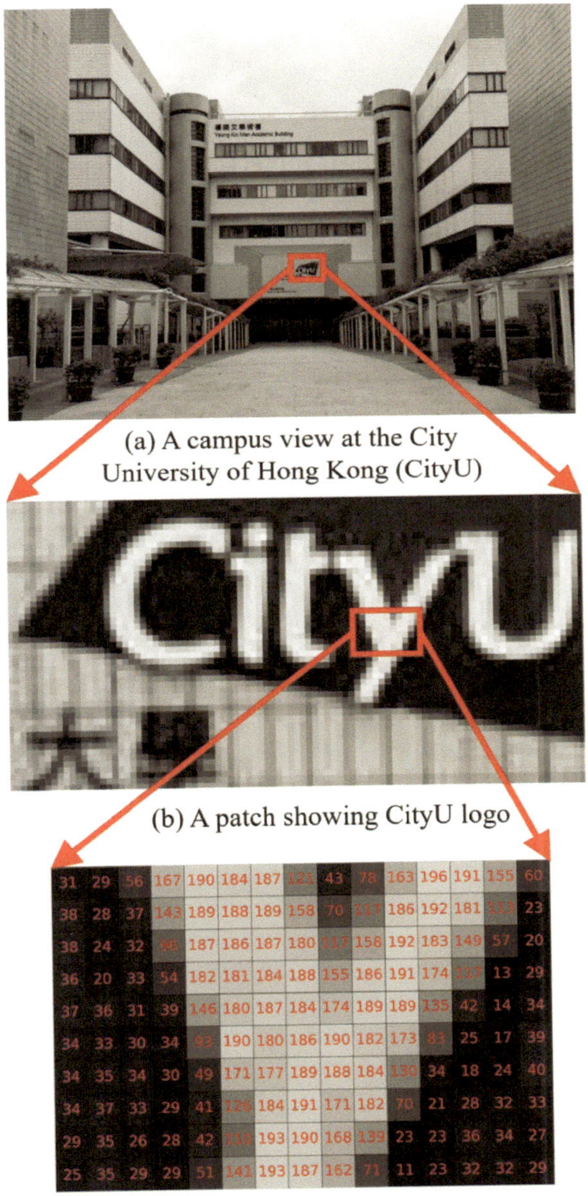

(a) A campus view at the City University of Hong Kong (CityU)

(b) A patch showing CityU logo

(c) Local pixels with brightness information

FIGURE 2.1 A grayscale digital image showing a campus view at City University of Hong Kong consists of a series of pixels with different intensity values.

point in the image and holds specific brightness information (for a grayscale image), or both the color and brightness information (for a colored image). When these pixels are arranged in such a structured grid, they collectively form a complete image that a person can see on screen.

The resolution of the image, defined by the number of pixels in width and height, determines its clarity and details. The original image shown in Figure 2.1a has a resolution of 2,736×2,052 = 5,614,272 million pixels (i.e., over 5 MP). By further zooming into the local part of the alphabet "y" as denoted by a red frame in Figure 2.1b, the associated pixels are enlarged and shown in Figure 2.1c. Figure 2.1c illustrates how the brightness information can be represented numerically by intensity values. The associated intensity values, or raw data, for each pixel in this small image patch are shown by red integers. Note that the intensity value in each pixel is quantitatively represented by a non-negative integer, with "0" for pure black and a specific maximum integer for pure white. For an 8-bit grayscale image, the maximum integer is $2^8 - 1 = 256 - 1 = 255$, leading to 256 intensity possibilities represented by an integer from 0 (black) to 255 (white). In this way, a grayscale image can be digitally represented by a matrix of integers in a smartphone or a computer. To display a grayscale digital image on a screen, a smartphone or computer decides the brightness, or greascale intensity, at each pixel based on the integer value at the corresponding pixel and shows all pixels collectively. Technically for a graphics processing unit (GPU), an image is a matrix of integers. This enables many fantastic analyses of images, including graphics rendering, machine learning, and face recognition.

While high-resolution images offer great visual fidelity and details, they also come with obvious demands on computational resources and present processing challenges. In general, a high-resolution image has large file size and requires long acquiring time, large storage space, and high data transmission capability. Processing such large high-resolution images can be computationally intensive and resource-consuming, particularly when handling a huge number of images. To improve the efficiency of processing high-resolution images, extensive theories and algorithms have been developed under the topic of image compression. Consider, for example, modern JPEG and JPEG-2000 lossy image compression standards developed by the Joint Photographic Experts Group. The standard JPEG compression routinely begins with a transformation of a high-resolution image from a spatial domain to a frequency domain using discrete cosine transform (DCT), which represents an image as a sum of sinusoids with varying magnitudes and frequencies. DCT is able to effectively separate and prioritize image details based on their visual importance. In general, relatively low frequency components are usually more important than those high frequency components because subtle changes in color and brightness are less perceptible to human eyes. As a result, only those sinusoids with relatively low frequency possess significant magnitudes and are non-trivial. Therefore, only the corresponding significant DCT coefficients are encoded, and the rest are just discarded. This process is called quantization or compression, and it leads to a significant reduction in file size of the image and facilitates data storage and transmission. When necessary, the encoded DCT coefficients can be used to decompress or reconstruct the original image. JPEG compression is highly effective for complex images, making it a popular choice for digital imaging, web usage, and multimedia applications (e.g., Ahmed et al. 1974; Wallace 1992).

This transform-and-compress traditional paradigm in DCT, however, suffers from several bottlenecks and sometimes appears inefficient (e.g., Duarte et al. 2008; Romberg 2008). Firstly, the compression needs to repeatedly compute the associated transform coefficients for a range of frequencies, but finally only limited ones are kept and coded. Secondly, both the locations and magnitudes of coded coefficients are necessary for compression and image reconstruction, while they apparently vary for different images with different spatial patterns. The unknown locations or magnitudes of the significant coefficients coded for a specific image introduce non-linearity to image compression and reconstruction. This leads to an important question: is there an efficient and unified way to incorporate the compression process directly into the transformation process? The answer to this question is compressive sensing/sampling (CS).

CS is a new sampling paradigm for efficiently sampling and reconstructing a signal (e.g., Candès and Romberg 2007; Candès and Wakin 2008; Romberg 2008). It goes against the conventional wisdom of the Nyquist–Shannon sampling theorem, which states that the sampling rate must be at least twice the maximum frequency present in a signal (e.g., Shannon 1948). Essentially, CS focuses on the limited "intrinsic" information embedded in a signal, rather than the actual length of the signal (or the actual resolution of an image). In contrast to conventional signal processing techniques, it aims to acquire directly the "intrinsic" information, or the compressed version of the signal, from far fewer measured data than the original signal length. CS has been successfully applied to signal processing, imaging processing, seismology, among others, mainly for data compression purposes. For example, in the context of image processing, an appealing application is the implementation of single-pixel camera, which is to sense a real-world image using the inner products between the image itself and a series of independent random sensing vectors. It has been demonstrated that an image can be properly reconstructed from a limited number of such inner products via sparse recovery algorithms. Technically, a successful CS application shall satisfy three prerequisites: (1) data sparsity, (2) properly designed sensing matrix, and (3) reconstruction algorithm. The first one focuses on the nature of the signal/image itself, while the second one pertains to the measurement modality, and the third one is essentially an optimization for solving a group of linear equations. The following sections will briefly introduce these three aspects of CS, respectively.

2.2 SPARSE REPRESENTATION

2.2.1 Definition of a Sparse Signal

Sparse representation is also known as sparse approximation theory. It refers to the practice of expressing a signal or image as a linear combination of a small number of basis elements from an overcomplete dictionary. Consider, for example, the Fourier series expansion in undergraduate mathematics. Fourier series is a mathematical tool used to analyze and represent a function through a sum of a series of trigonometric functions (e.g., sine and cosine functions). An interesting and noteworthy fact is that many natural signals (e.g., audios, images, and videos) exhibit quite concise patterns in terms of a limited number of trigonometric function coefficients when using the Fourier series expansion. To illustrate this idea, Figure 2.2 shows a simple signal in spatial domain by a red line. Although it seems that the signal in spatial domain

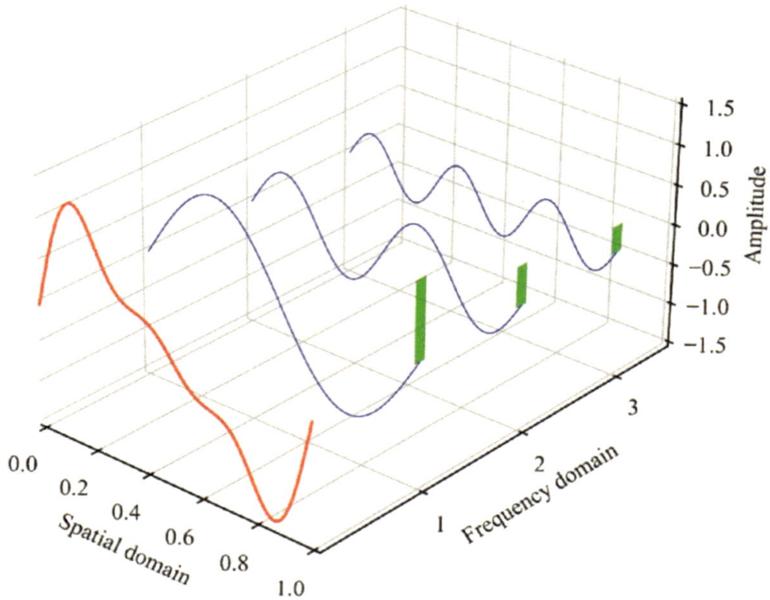

FIGURE 2.2 Fourier series expansion of a continuous signal in spatial domain.

may not be directly described by a common parametric function, it is actually just a summation of three different sine function components, as shown by three blue lines in Figure 2.2. In the frequency domain shown in Figure 2.2, the amplitudes or weights of the three sine functions are shown by green bars. The idea of sparse representation is clearly illustrated in Figure 2.2: although it may require many data points to accurately characterize the signal in the spatial domain, it may only need three weight coefficients in the frequency domain to achieve equivalent characterization.

Conceptually, sparsity of a signal refers to a phenomenon that a natural signal itself contains a limited number of non-zero elements when projected onto a properly selected feature space (e.g., a group of basis functions obtained from discrete cosine transform, discrete wavelet transform, or Fourier transform). Mathematically, a digital signal is denoted as an N-length real-valued column vector $\boldsymbol{f}$. Note that a digital image, or a matrix shown in Figure 2.1c, may be reshaped into a vector $\boldsymbol{f}$ as well, and the formulation below is equally applicable to digital images. Under a context of sparse representation, the vector $\boldsymbol{f}$ can be considered as a weighted summation of N orthonormal basis functions (e.g., cosine function or wavelet function) and expressed as:

$$\boldsymbol{f} = \mathbf{B}\boldsymbol{\omega} \tag{2.1}$$

where $\mathbf{B}$ is an $N \times N$ matrix, with each column representing a prescribed orthonormal basis function, and $\boldsymbol{\omega}$ is an N-length column vector, with its elements representing the weight coefficients corresponding to columns of $\mathbf{B}$. Given an appropriate basis function **B** for f, most elements in $\boldsymbol{\omega}$ are very small or virtually zero, except for S (S<<N) non-trivial ones with significantly large magnitudes (e.g., Tipping 2001; Candès and Wakin 2008). Therefore, the vector $\boldsymbol{\omega}$ is regarded as S-sparse if it satisfies the following condition:

$$\|\boldsymbol{\omega}\|_0 \leq S \tag{2.2}$$

where $\|\boldsymbol{\omega}\|_0$ denotes the l_0 norm of $\boldsymbol{\omega}$ or simply the number of non-zero elements. A general form of l_p norm when p is greater than zero is defined as:

$$\|\boldsymbol{\omega}\|_p = \left(\sum_{i=1}^{N} |\omega_i|^p\right)^{1/p} \quad p > 0 \tag{2.3}$$

For example, when p=1, the l_1 norm is simply the sum of the absolute values of all elements in $\boldsymbol{\omega}$; when p=2, the l_2 norm represents the Euclidean norm of vector $\boldsymbol{\omega}$. In practice, the $\boldsymbol{\omega}$ vector is usually regarded as nearly sparse or "compressible" if the errors between $\|\boldsymbol{\omega}\|_0$ and $\|\boldsymbol{\omega}_s\|_0$ decay quickly, where $\|\boldsymbol{\omega}_s\|_0$ is a copy of $\|\boldsymbol{\omega}\|_0$ with the remaining N-S trivial elements as zeros. Four real examples of sparse signals, including digital image, meteorological time series, and geotechnical spatial data, are illustrated below.

2.2.2 2D Image Example

Figure 2.3a reshows an example of CityU campus photo with a resolution of 2,736×2,052. Based on Eq. (2.1), for those around 5.61 million pixels in this image, there shall exist about 5.61 million DCT coefficients. For example, when a compression is implemented for this image using the largest 1% of those DCT coefficients, or equivalently about the largest 56,100 DCT coefficients, the image reconstructed after such a compression is shown in Figure 2.3b. The main patterns and many details in Figure 2.3a have been well preserved in Figure 2.3b. The difference between the original image and the compressed image is minor although Figure 2.3b appears vaguer than Figure 2.3a. The absolute values of associated DCT coefficients expressed as a coefficient matrix are shown in Figure 2.4a, which seems purely white. This DCT coefficient matrix has the same dimension (i.e., 2,736×2,052) as the resolution of the image. Each coefficient represents the combination of a specific vertical frequency and a specific horizontal frequency, and the darkness indicates its weight. It can be seen from Figure 2.4a that most coefficients are purely white and almost zero because the portion of non-trivial coefficients are too small to be seen clearly. A zoom-in view of the DCT coefficients is shown in Figure 2.4b. It can be observed that a quite limited number of DCT coefficients exhibit significant values, and they generally present around the top-left corner (e.g., relatively low frequency along both directions) of the DCT coefficient matrix. The 1% coefficients used for reconstructing the image in Figure 2.3b are further denoted by red dots in Figure 2.4c. It shows that the limited number of DCT coefficients can reconstruct the photo reasonably well. Therefore, the CityU campus photo shown in Figure 2.3a can

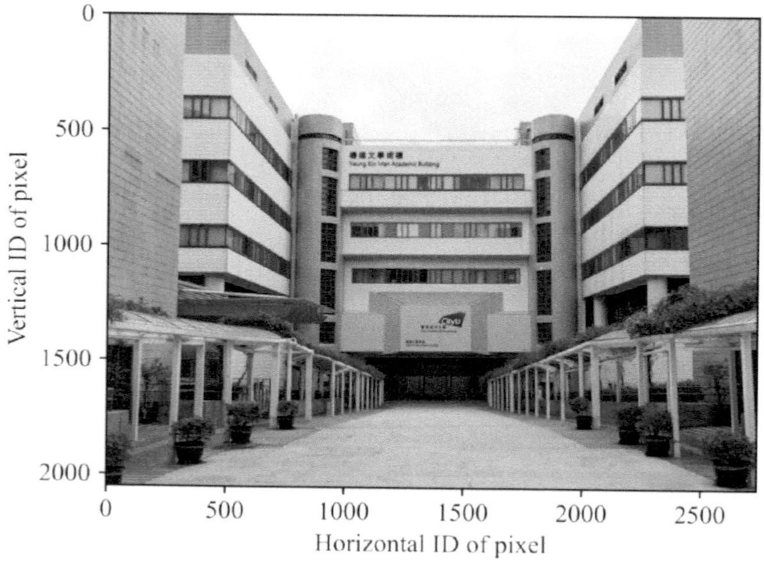

(a) Original image

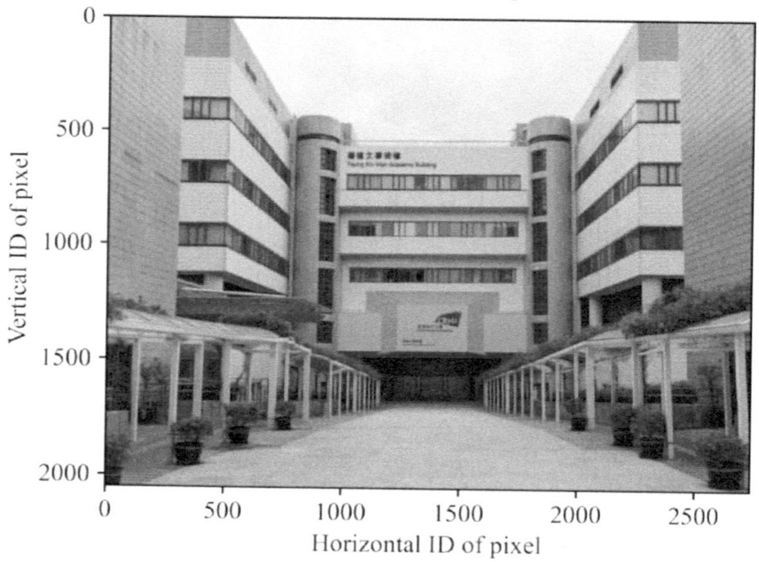

(b) Reconstructed images using 1% percent of DCT coefficients

FIGURE 2.3 Illustration of image compression using an image showing a campus view at City University of Hong Kong.

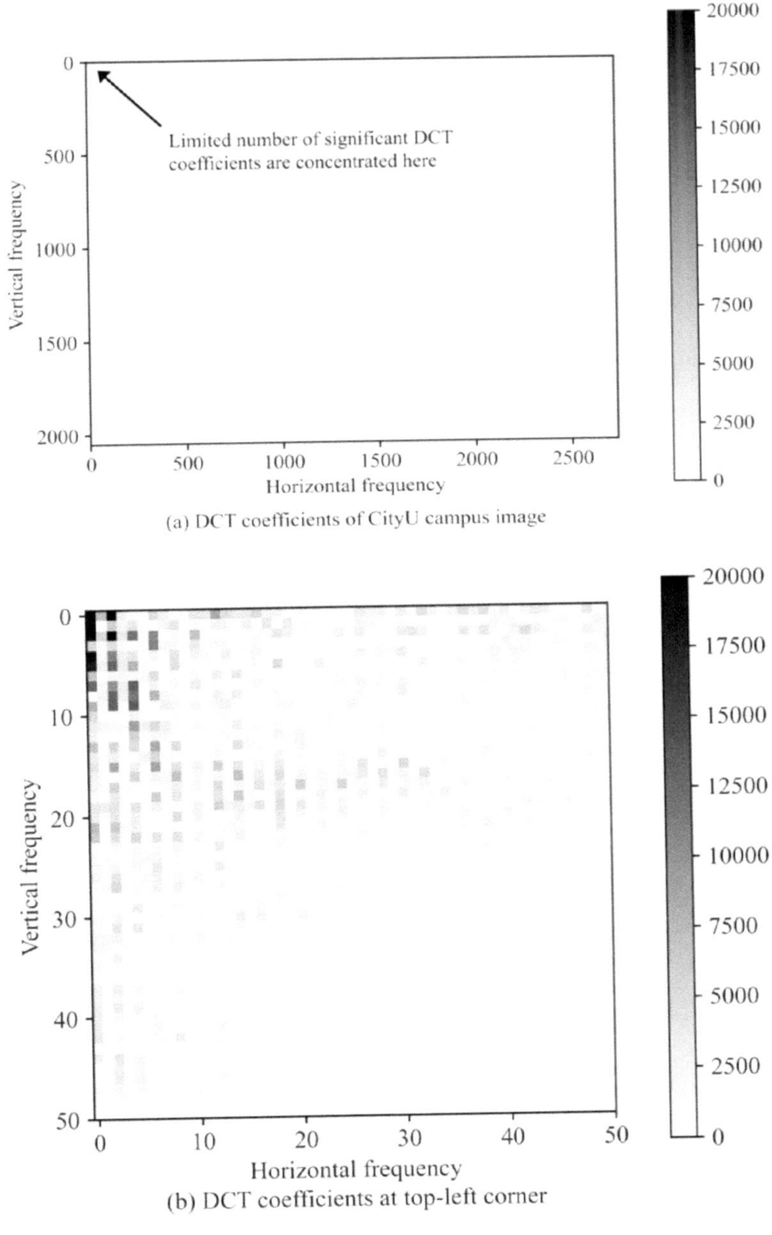

FIGURE 2.4 DCT coefficients of CityU campus photo.

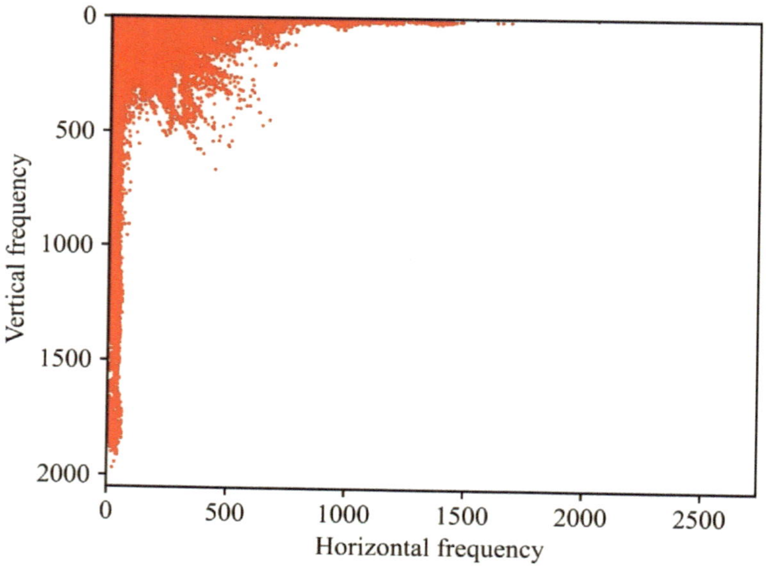

(c) The largest 1% of DCT coefficients

FIGURE 2.4 DCT coefficients of CityU campus photo. (Continued)

be regarded as a sparse or "compressible" signal and is used to illustrate the sparse representation of 2D images.

2.2.3 1D Wind Speed Time Series Example

Figure 2.5 shows a set of real wind speed time series analyzed using DCT. This 1D example documented a wind speed data series recorded at a monitoring station in Shenzhen, China, from 20 to 21 October 2016, during the typhoon Haima. The recorded data series covers a duration of 2,000 minutes, with a measurement interval of 1 minute, leading to a vector with a length of $N = 2{,}000$. This data series in Figure 2.5a clearly shows the variability of wind speed during typhoon Haima, particularly when it passed by Shenzhen area, and exhibits highly non-stationary patterns. In addition, significant local-scale random fluctuations are observed. Nonetheless, this highly variable wind speed time series has a concise pattern after a DCT. In Figure 2.5b, DCT coefficients of this wind speed time series are denoted by black crosses. The majority of the obtained DCT coefficients are close to zeros except for a limited number of non-trivial ones with small DCT coefficients ID (i.e., low frequency components). Based on the sparse representation, a reconstructed wind speed series (see the red dashed line in Figure 2.5a) from the first 50 largest DCT coefficients (see the red circles in Figure 2.5b) has an overall trend that is quite consistent with the original one. In other words, the intrinsic information contained in this wind speed time series is usually much smaller

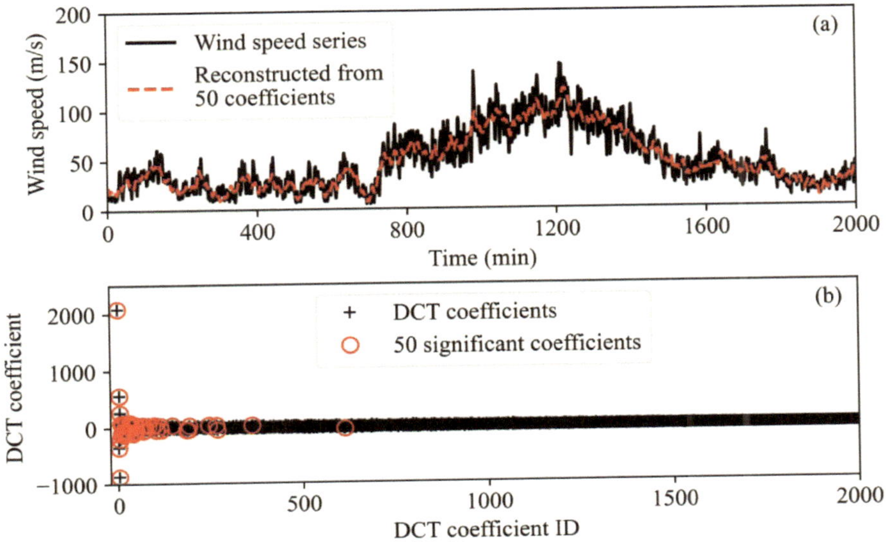

FIGURE 2.5 A real wind speed profile example: (a) wind speed profile and (b) DCT coefficients.

than its temporal length, and hence, a large number of the trivial coefficients may be discarded without losing much important information when reconstructing the signal.

2.2.4 1D Soil Data Example

Sparse representation of signals is applicable to not only digital images and temporal data, but also to many other naturally occurring signals or processes. For example, cone penetration test (CPT) is a widely used in-situ testing method in geotechnical engineering, and it provides detailed information about the subsurface conditions of geological deposits along depth. By pushing a cone-shaped penetrometer into the ground, CPT records the ground responses and mechanical properties of soils along the depth, offering a nearly continuous profile of soils. The CPT data profile can also be regarded as a signal. Figure 2.6 provides an example of CPT tip resistance q_t data profile of the sounding number MSC014 obtained from the United States Geological Survey (USGS) database (https://earthquake.usgs.gov/research/cpt/data/). Due to the spatially correlated soil depositional process at geological time scale, CPT data profile usually exhibits spatial patterns and therefore can be analyzed under the sparse representation framework. The q_t data profile of the sounding number MSC014 was sampled over a 12.75m-thick sand layer with a sampling interval of 0.05m. As a result, the q_t data profile has N=256 points and is shown by a black solid line in Figure 2.6a. It is observed that the q_t profile exhibits complex variability over

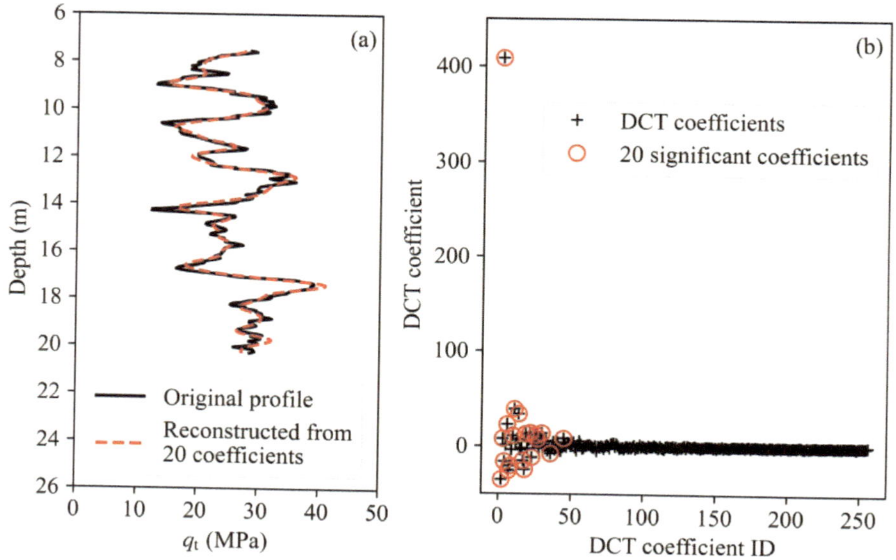

FIGURE 2.6 A cone penetration test (CPT) q_t example: (a) q_t data profile and (b) DCT coefficients.

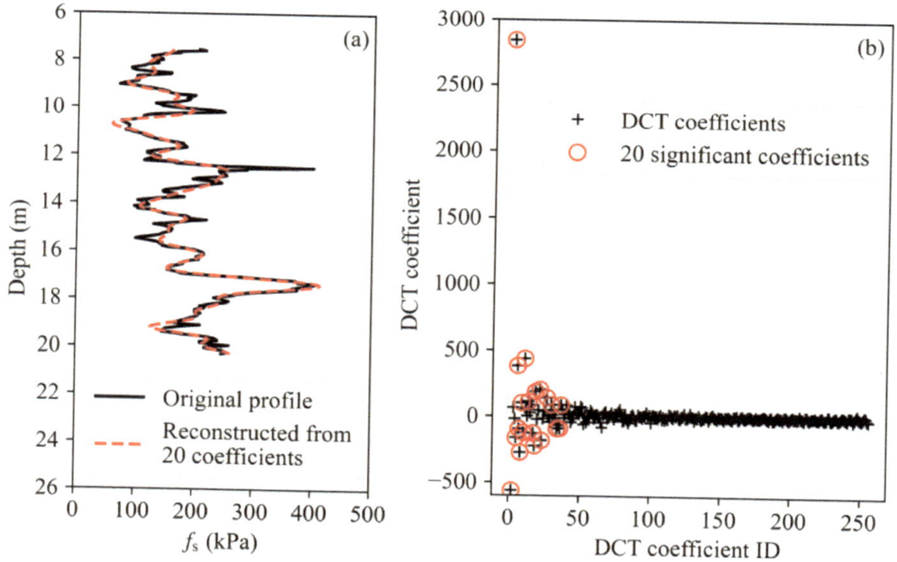

FIGURE 2.7 A cone penetration test (CPT) f_s example: (a) f_s data profile and (b) DCT coefficients.

depth. However, the associated 256 DCT coefficients are quite sparse, and only a limited number of DCT coefficients are non-zero, as shown by a series of black crosses in Figure 2.6b. A q_t data profile is reconstructed from the 20 largest DCT coefficients (around 7.8% of N=256) denoted by red circles in Figure 2.6b. The reconstructed q_t data profile, as shown in Figure 2.6a by a red dashed line, is almost overlapped with the original one. Similar observations are obtained for the sleeve friction f_s profile of the same CPT, as shown in Figure 2.7. The above examples demonstrate that many images or natural signals have sparse representation, and they are compressible.

2.3 DESIGN OF MEASUREMENT MATRIX

2.3.1 CS Measurements

Since all the four examples shown previously in Section 2.2 have demonstrated signal sparsity, they can be reconstructed by a simple summation of those selected DCT basis functions with significant or large weight (or transformation) coefficients. However, as the signal of interest grows in size and complexity, this conventional compression paradigm appears inefficient. The computational cost of associated transform coefficients for a large number of frequencies would become extensive, but finally only limited ones are kept. More importantly, both locations and magnitudes of the coded coefficients are necessary for compression and reconstruction, while they apparently vary for different signals with different patterns. Automatic and fast identification of those significant coefficients for any given signal is challenging. To tackle this problem, a properly designed sensing matrix is used in CS by taking linear measurements of the signal. These measurements effectively reduce the dimension of the original signal, without compromising quality of the signal reconstruction. Consider, for example, a signal of interest expressed as an N-length vector $\boldsymbol{f}$. The original signal $\boldsymbol{f}$ may not be a sparse signal but it may support sparse representation after a carefully selected transformation (e.g., DCT), and hence the $\boldsymbol{\omega}$ vector of DCT coefficients is usually sparse. The associated measuring process can be formulated as:

$$\boldsymbol{y} = \boldsymbol{\Psi} \boldsymbol{f} = \boldsymbol{\Psi} \mathbf{B} \boldsymbol{\omega} = \mathbf{A} \boldsymbol{\omega} \tag{2.4}$$

where $\boldsymbol{y}$ is an M-length vector representing measurement data; $\boldsymbol{\Psi}$ is the sensing matrix of size $M \times N$ for $\boldsymbol{f}$; $\mathbf{A} = \boldsymbol{\Psi}\mathbf{B}$ is the modified sensing matrix with a size of $M \times N$, and it is also known as measurement matrix. It is worth noting that the measurement process in CS is different from the common measurement process of a specific physical quantity. Instead of collecting point information of the signal at specific locations, CS requires and directly operates on a full complete signal. It formulates the measurement as an inner product of the signal $\boldsymbol{f}$ and a series of sensing functions

$\boldsymbol{\Psi}_k$, or equivalently the inner product of the sparse vector $\boldsymbol{\omega}$ and the modified sensing function $\boldsymbol{A}_k$:

$$y_k = <\boldsymbol{f}, \boldsymbol{\Psi}_k> = \sum_{i=1}^{N} f_i \Psi_{k,i} = <\boldsymbol{\omega}, \boldsymbol{A}_k> = \sum_{i=1}^{N} \omega_i A_{k,i} \quad k \in \{1, 2, \ldots, M\} \tag{2.5}$$

where y_k is the kth measurement out of M measurement data ($M<<N$); $\langle \cdot, \cdot \rangle$ is the inner product operation; $\boldsymbol{\Psi}_k$ is the kth sensing function and the kth row of matrix $\boldsymbol{\Psi}$; $\boldsymbol{A}_k$ is the kth modified sensing function and the kth row of the modified sensing matrix, or interchangeably measurement matrix $\mathbf{A}$. Note that $\boldsymbol{A}_k$ has the same length as the sparse vector $\boldsymbol{\omega}$. For a digital signal commonly expressed as a vector, the inner product is equivalent to the dot product in the Euclidean vector space. It is calculated as the sum of the element-wise product of two vectors. Eq. (2.5) is a typical and fundamental formulation in CS and is also the essential mathematical foundation in various relevant fields, e.g., single-pixel camera, magnetic resonance imaging, sampling theory, machine learning, and low-rank matrix recovery. The measurement process is essentially a dimensionality reduction. An illustration of the measurement process is shown in Figure 2.8. The measuring process is formulated as a 4×10 sensing matrix $\boldsymbol{\Psi}$ multiplying a signal vector $\boldsymbol{f}$ with a length of 10. The signal $\boldsymbol{f}$ is simply a number sequence from 1 to 10. For example, the first element in measurement data vector $\boldsymbol{y}$ is obtained by the dot product of the first row of $\boldsymbol{\Psi}$ and the $\boldsymbol{f}$. This process hence leads to a measurement data vector with a length of 4 on the right. In Figure 2.8, both a Gaussian random sensing matrix (Figure 2.8a) and a deterministic spike measurement matrix (Figure 2.8b) are adopted for illustration, with the values at each element denoted. The deterministic spike sensing matrix, which uses a Dirac delta function, contains only one nonzero element, i.e., one, in each sensing function $\boldsymbol{\psi}_k$ (k=1, 2, 3, 4). This is equivalent to the traditional measurement performed at four specific locations, e.g., the f_1, f_4, f_7, f_{10}. In contrast, the Gaussian measurement matrix may seem unnatural. Each sensing function consists of a series of random numbers. In this scenario, each of the measurement data on the right can be regarded as a linear summation of all elements in the original signal $\boldsymbol{f}$ with random weights. It has been suggested that, under DCT framework, the Gaussian random sensing matrix in Figure 2.8a effectively facilitates sparse signal recovery (e.g., Candès and Wakin 2008; Foucart and Rauhut 2013).

Traditional wisdom on signal processing, e.g., analog-to-digital conversion and medical imaging, suggests that the number M of measurements shall be as large as the total dimension N of the signal of interest. Although this generally guarantees quality of the reconstruction, it is a waste of effort as substantial efforts are required to conduct the measurements at each entry of the signal while most of them are discarded in a compressed version. In contrast, it has been proved theoretically that the minimal number of measurements M=2S might be sufficient

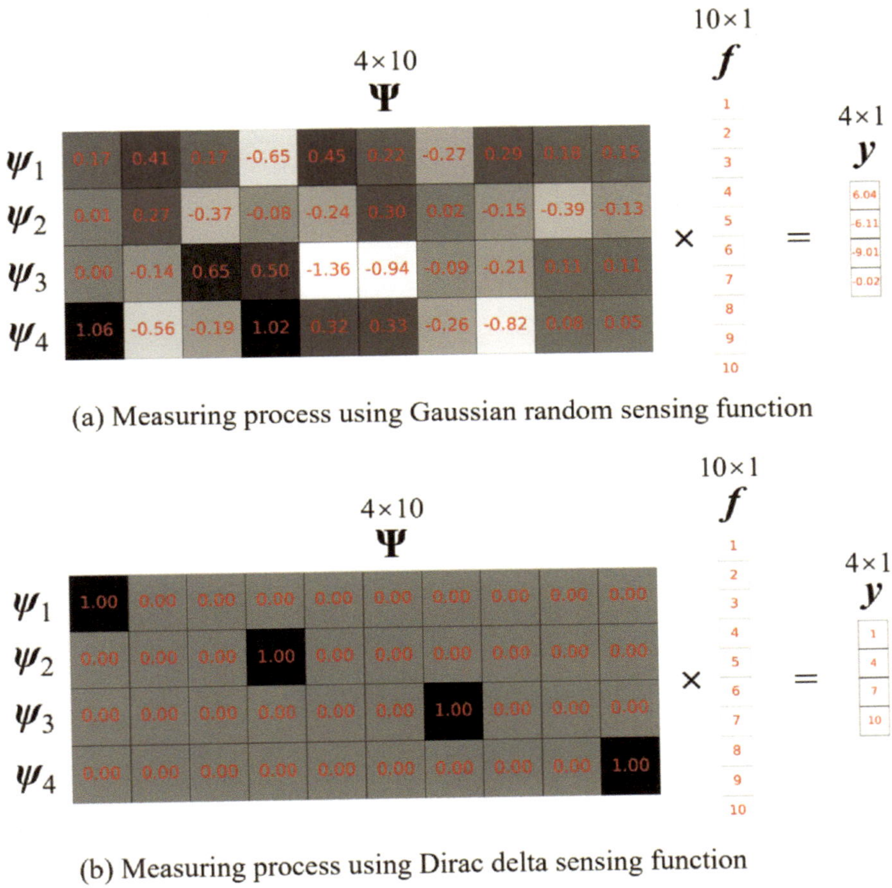

FIGURE 2.8 Comparison of measuring processes using two different sensing functions.

to reconstruct any S-sparse signal and even $M=S+1$ might be sufficient to reconstruct a specific S-sparse signal (e.g., Foucart and Rauhut 2013). It is worth noting that when the number of measurement M is smaller than N, Eq. (2.4) essentially establishes an underdetermined system, resulting in the exact solution to Eq. (2.4) being not directly available. According to Eq. (2.4), the measurement matrix $\mathbf{A}$ is determined by both the adopted basis function $\mathbf{B}$ and the sensing functions $\mathbf{\Psi}$, simultaneously formulating the signal sparsity and measurement modality. Design of a proper measurement matrix $\mathbf{A}$ is the core of CS applications. A noteworthy fact of the underdetermined system in Eq. (2.4) is that the measurement process requires the availability of the full data in sparse vector $\boldsymbol{\omega}$, or alternatively the full data of signal/image f itself, as the main purpose of CS is to simultaneously compress and sample the signal or image of interest. Therefore, the design of a specific measurement matrix $\mathbf{A}$ or sensing matrix $\mathbf{\Psi}$ is possible.

2.3.2 Conditions of Measurement Matrix

A significant challenge of solving Eq. (2.4) is the associated computational complexity, as both the locations and magnitudes of non-trivial weight coefficients may vary for different signals. Solving the S-sparse weight coefficients $\boldsymbol{\omega}$ from measurement $\boldsymbol{y}$ can be regarded as solving an l_0-minimization problem:

$$\min_{\boldsymbol{\omega} \in C^N} \|\boldsymbol{\omega}\|_0 \text{ subject to } \boldsymbol{y} = \mathbf{A}\boldsymbol{\omega} \tag{2.6}$$

A straightforward brute-force way to solve the S-sparse $\boldsymbol{\omega}$ is to try out all possible S-subset of N, leading to a number, $C(N, S)$, of systems like Eq. (2.4). The associated computational effort is generally prohibitively large and impractical. It has been proved that solving the l_0-minimization problem in Eq. (2.5) is NP-hard, which means the potential solutions cannot be found in polynomial time associated with the signal length N. Although it seems intractable to directly solve the underdetermined system in Eq. (2.4), it will be demonstrated later that a properly designed $\mathbf{A}$ matrix in CS may facilitate a successful recovery of the original signal.

It is usually non-trivial to verify quantitatively the suitability of a measurement matrix $\mathbf{A}$ for CS, particularly when the sparsity level and measurement precision of a real signal or image are often unknown beforehand. However, some performance metrics, such as coherence and restricted isometry property (RIP), have been developed to conveniently evaluate the quality of a measurement matrix $\mathbf{A}$. The coherence μ quantifies the correlation between any two columns of the measurement matrix $\mathbf{A}$ and is defined as:

$$\mu = \max_{1 \le i \le j \le N} \left| < \boldsymbol{a}_i, \boldsymbol{a}_j > \right| \tag{2.7}$$

where $\boldsymbol{a}_i$ and $\boldsymbol{a}_j$ are the ith and the jth l_2-normlized columns of $\mathbf{A}$, respectively. Essentially, coherence describes the correlation or interdependence between any two columns, and by definition, it has the maximum value of 1. Naturally, for a measurement matrix $\mathbf{A}$ with $M \ge N$, μ=0 if and only if the columns of $\mathbf{A}$ are orthonormal. In the CS framework, it is common to have $\mathbf{A}$ matrix with $M<N$. In this case, it is of concern that how small the coherence μ can be and how it guarantees the quality of measurement matrix for successful sparse signal recovery. A lower bound of coherence known as Welch bound is defined as (e.g., Strohmer and Heath 2003; Foucart and Rauhut 2013):

$$\mu \ge \sqrt{\frac{N-M}{M(N-1)}} \tag{2.8}$$

For a large N, the coherence approaches to $1/\sqrt{M}$. Based on the coherence, a sufficient condition has been verified for the recovery of an S-sparse signal using basis pursuit algorithm (e.g., Foucart and Rauhut 2013):

$$\mu < \frac{1}{2S-1} \tag{2.9}$$

With a measurement matrix satisfying the small coherence condition, recovery of S-sparse signal is achievable using basis pursuit when the number of measurement data satisfies:

$$M \geq C\,S^2 \tag{2.10}$$

where C is the positive constant adaptive to different signals. This indicates that the required number of measurement data is proportional to a square of the signal's sparsity level. Eq. (2.10) suggests that many measurement data are required for a signal with a high sparsity level S. On the other hand, CS is particularly appealing for highly compressible signals with a small sparsity level S. Details of basis pursuit algorithm will be provided in Section 2.4.

In CS, a generalized criterion called restricted isometry property (RIP) has been developed to ensure the suitability of a measurement matrix. As a convenient alternative of coherence property, RIP is an important concept in CS to guarantee successful recovery of sparse signal. In contrast to the coherence which quantifies the correlation between any two columns in a measurement matrix **A**, the RIP evaluates all possible S columns in **A** directly, and therefore, the quality of **A** efficiently. The detailed mathematical proofs of the RIP under different scenarios are provided in the literature on CS and signal processing (e.g., Candès and Wakin 2008; Baraniuk et al. 2008; Foucart and Rauhut 2013; Boche et al. 2015). Mathematically, the RIP condition is defined by a restricted isometry constant δ:

$$(1-\delta)\|\boldsymbol{\omega}\|_2^2 \leq \|\boldsymbol{A}\,\boldsymbol{\omega}\|_2^2 \leq (1+\delta)\|\boldsymbol{\omega}\|_2^2 \tag{2.11}$$

The measurement matrix **A** is considered to satisfy the RIP condition if the constant δ is small for sufficiently large S. δ is a pre-specified constant to quantify the strictness of RIP condition. If the RIP condition is satisfied, the measurement data $\boldsymbol{y}$ in Eq. (2.4) preserves the l_2-norm of S-sparse $\boldsymbol{\omega}$. From a linear algebra viewpoint, this indicates that the columns in measurement matrix **A** are highly orthogonal to a specific extent, and $\boldsymbol{y}$ is effectively within the column space of **A**. From a spectral perspective, RIP also indicates the energy present in the sparse $\boldsymbol{\omega}$ shall be effectively captured by measurement data $\boldsymbol{y}$. The RIP condition guarantees that the exact solutions for the S-sparse $\boldsymbol{\omega}$ can be obtained from measurements $\boldsymbol{y}$ (e.g., Foucart and Rauhut 2013).

It has been proved that, instead of matching the measurement matrix **A** to the specific patterns exhibited in S-sparse $\boldsymbol{\omega}$, the ideal configuration is the exact opposite. Interestingly, the modified sensing function $\boldsymbol{A}_k$ (i.e., the kth row of measurement matrix **A**) should be unstructured and random noise-like (e.g., Candès and Romberg 2007; Candès and Wakin 2008; Romberg 2008). For example, the Gaussian random numbers with a zero mean and a variance of $1/M$ (e.g., as shown in Figure 2.9a) or a symmetric Bernoulli process (e.g., as shown in Figure 2.9b) can be used to

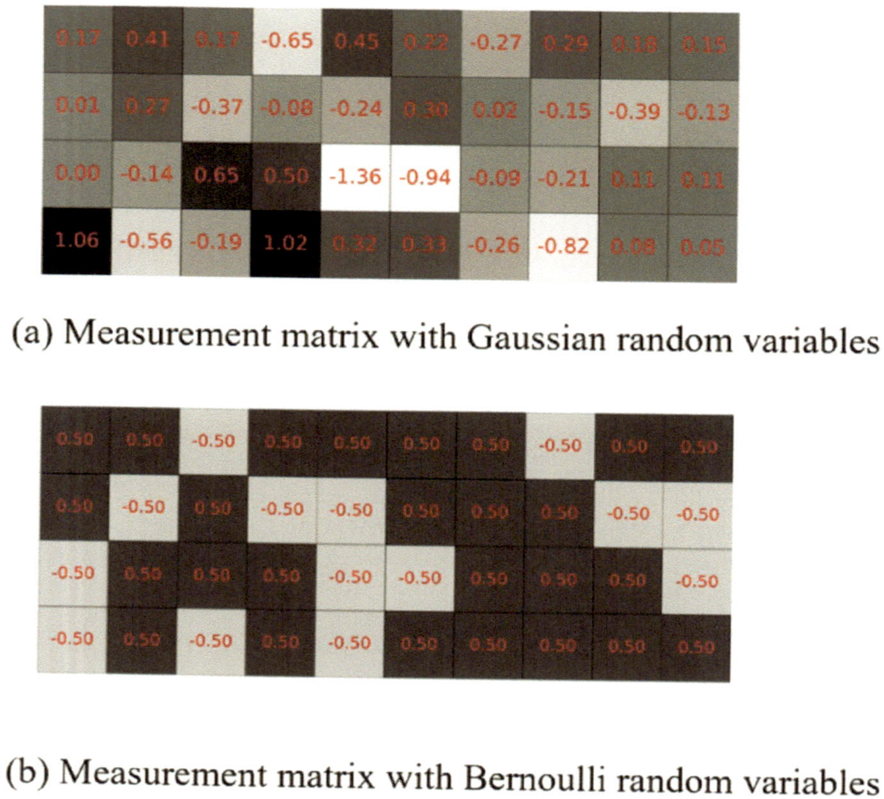

FIGURE 2.9 Two different measurement matrix **A** with random functions.

generate the modified sensing function $\boldsymbol{A}_k$ for efficient sparse signal recovery. Under this setting, all columns of **A** are also random vectors. As the dot product of two zero-mean random vectors is expected to be close to zero, the coherence of measurement matrix **A** in Eq. (2.7) is well controlled to a low value as well. This subsequently leads to good performance in reconstructing N-length $\boldsymbol{\omega}$ from M-length measurement data $\boldsymbol{y}$. In other words, this setting renders each measurement data highly independent and efficient for acquiring the "intrinsic" information embedded in S-sparse $\boldsymbol{\omega}$. An important conclusion in CS states that, all S-sparse signal can be reconstructed with overwhelmingly high probability (e.g., a probability of successful recovery exceeding $1 - \delta$) using random sensing functions if the number M of measurements satisfies the following condition (e.g., Candès and Wakin 2008; Foucart and Rauhut 2013):

$$M > 2S\ln\left(N / S\right) \tag{2.12}$$

Note that the RIP condition guarantees effective signal recovery in a statistical manner. However, to find a deterministic measurement matrix satisfying RIP for a specific signal is still an open question (e.g., Candès and Romberg 2007; Foucart and Rauhut 2013).

2.4 BASIS PURSUIT ALGORITHM

The mathematical algorithms for solving Eq. (2.4) may be generally divided into three categories, i.e., optimization, greedy algorithms, and threshold-based algorithms (e.g., Baraniuk 2007; Foucart and Rauhut 2013). The basis pursuit is introduced in this section, which is a common optimization algorithm exploiting the sparsity for solving the $\boldsymbol{\omega}$. Note that solving the l_0-minimization in Eq. (2.6) is essentially a nonconvex problem, which is NP-hard. On the other hand, norm minimizations with orders higher than 1 fail to recover sparse signals. However, the l_1-minimization happens to become a convex problem:

$$\min_{\boldsymbol{\omega} \in C^N} \|\boldsymbol{\omega}\|_1 \text{ subject to } \boldsymbol{y} = \mathbf{A}\boldsymbol{\omega} \tag{2.13}$$

The above l_1-minimization is also called basis pursuit in CS. Although the l_1-norm may not be differentiable everywhere and may not always yield unique solution, it has been proved to be an appropriate choice for the sparse signal recovery (e.g., Candès and Wakin 2008; Foucart and Rauhut 2013).

In basis pursuit, the null space property is a necessary and sufficient condition for exact recovery of sparse signals. A matrix $\mathbf{A} \in \mathbf{R}^{M\times N}$ is said to satisfy the null space property relative to sparsity S if:

$$\|\mathbf{v}_S\|_1 < \|\mathbf{v}_{\bar{S}}\|_1 \quad \textit{for all}\, \mathbf{v} \in ker\mathbf{A} \setminus \{0\} \tag{2.14}$$

where $\mathbf{v} = \mathbf{v}_S + \mathbf{v}_{\bar{S}}$ is a vector in the kernel space $ker\mathbf{A}$ or null space of $\mathbf{A}$; $\mathbf{v}_S$ is the N-length and S-sparse vector, and $\mathbf{v}_{\bar{S}}$ is its complement. It can then be derived based on Eq. (2.14) that given a matrix $\mathbf{A} \in \mathbf{R}^{M\times N}$, every S-sparse vector $\boldsymbol{\omega} \in \mathbf{R}^N$ is the unique solution of Eq. (2.6) using measurement data in Eq. (2.4), if and only if matrix $\mathbf{A} \in \mathbf{R}^{M\times N}$ satisfies the null space property relative to sparsity S. This condition essentially aligns with the Nyquist–Shannon sampling theorem and is sufficient to guarantee the exact recovery of a sparse signal. The basis pursuit algorithm also features measurement data $\boldsymbol{y}$ which might be rescaled, reshuffled, or added, provided that the measurement matrix $\mathbf{A}$ is adapted accordingly.

Another two important issues of basis pursuit are the sparsity defect and the measurement precision. The sparsity of a signal following Eq. (2.2) is usually idealized. In practice, it is often necessary to assume that the signal of interest is approximately S-sparse or just compressible. Therefore, the stability of sparse signal recovery with a control of the sparsity defect is also important. A strengthened null space property, i.e., stable null space property, is described as:

$$\|\mathbf{v}_S\|_1 < \rho \|\mathbf{v}_{\bar{S}}\|_1 \quad for\ all\ \mathbf{v} \in ker\mathbf{A} \setminus \{0\} \tag{2.15}$$

where the constant 0< ρ <1 characterizes the proximity of sparsity. Then the difference between the N-length sparse $\boldsymbol{\omega}$ and the approximated N-length sparse vector $\hat{\boldsymbol{\omega}}$ can be controlled by:

$$\|\boldsymbol{\omega}-\hat{\boldsymbol{\omega}}\|_1 \le \frac{2(1+\rho)}{1-\rho}\sigma_s(\boldsymbol{\omega})_1 \tag{2.16}$$

where $\hat{\boldsymbol{\omega}}$ is the sparse approximation of $\boldsymbol{\omega}$ and $\sigma_s(\boldsymbol{\omega})_1$ is the l_1 actual error of the S-sparse vector ω. On the other hand, real measurement data obtained inevitably involves measurement noise. This indicates that the measurement data in Eq. (2.4) is only an approximation, and the associated l_2-norm errors are usually expressed as:

$$\|y - \mathbf{A}\boldsymbol{\omega}\|_2 < \eta \tag{2.17}$$

where $\eta > 0$ is a constant. The robustness refers to the quality of an algorithm that can yield approximation with errors controlled by the constant η. The basis pursuit optimization in Eq. (2.6) is then reformulated as:

$$\min_{\boldsymbol{\omega}\in\mathbf{C}^N} \|\boldsymbol{\omega}\|_1 \ \text{subject to} \ \|y - \mathbf{A}\boldsymbol{\omega}\|_2 < \eta \tag{2.18}$$

Then a robust null space property is derived:

$$\|\boldsymbol{\omega}-\hat{\boldsymbol{\omega}}\|_1 \le \frac{2(1+\rho)}{1-\rho}\sigma_s(\boldsymbol{\omega})_1 + \frac{4\tau}{1-\rho}\eta \tag{2.19}$$

in which $\tau > 0$ is another constant affecting the approximation quality.

To illustrate the basis pursuit algorithm, the CPT q_t data profile example shown in Figure 2.6 is re-analyzed in this section using limited measurement under the CS framework. As mentioned above, the q_t data profile is obtained over a 12.75 m-thick sand layer with a sampling interval of 0.05 m and consists of N=256 data points along depth. As demonstrated in Figure 2.6, the profile supports sparse representation with DCT basis functions. Therefore, many data points within the profile might be redundant from an “intrinsic” information perspective. Under CS framework, it is possible to efficiently compress and sample the q_t data profile simultaneously while preserving most of the intrinsic information of the spatial variability of the sand layer property. To evaluate the effect of the design of measurement matrix, two different types of measurement matrix, i.e., Gaussian random sensing matrix and Bernoulli random sensing matrix, are evaluated under three different measurement scenarios, i.e., M=10, 40, and 80. The measuring process follows the illustration in Figure 2.8. For example, when Gaussian random sensing matrix is adopted to obtain M=10 measurement data from the q_t data profile $\boldsymbol{f}$, a 10×256 random sensing matrix $\boldsymbol{\Psi}$ is generated by realizing repeatedly independent and identically distributed Gaussian variables with a zero-mean and a variance of $1/M$. Based on Eq. (2.4), the measurement matrix **A** is obtained by multiplication $\boldsymbol{\Psi}\mathbf{B}$, where **B** is the DCT basis function matrix adopted. Each data point in measurement $\boldsymbol{y}$ is formulated as the inner product of a row in $\boldsymbol{\Psi}$ and the profile $\boldsymbol{f}$. Alternatively, each measurement data is also the inner product of a row in **A** and the DCT coefficient vector $\boldsymbol{\omega}$. After the measurement process, an underdetermined system (e.g., see Eq. (2.4)) is established. In this example, measurement error is not considered. By using the basis pursuit algorithm with l_1-minimization, solving Eq. (2.4) is reformulated as solving Eq. (2.13) (e.g., Rao and Kreutz-Delgado 1999; Candès and Wakin 2008; Foucart and Rauhut 2013).

After the approximated sparse coefficient vector $\hat{\boldsymbol{\omega}}$ is solved, the original q_t data profile is reconstructed and denoted as $\hat{\boldsymbol{f}}$. The associated reconstruction results under different scenarios are shown in Figure 2.10. The original q_t data profile and the reconstructed profile are shown by a black solid line and a red dashed line, respectively. Figures 2.10a–c show the results by adopting Gaussian random sensing matrix, under the three measurement scenarios. When M=10, quite noisy patterns are observed in the reconstructed profile, and it suggests that the number of measurement data is not sufficient, and the corresponding measurement matrix might not preserve RIP condition to guarantee recovery. As the number of measurement data increases to M=40 and M=80, the associated measurement matrix evolves and gradually satisfies the RIP condition, and therefore the reconstructed profiles plot gradually close to the original profile. Similar observations are obtained when Bernoulli random sensing matrix is adopted, as shown in Figures 2.10d–f. Note that the DCT coefficients of the q_t data profile shown in Figure 2.6b indicate many negligibly small but non-zero coefficients. Therefore, the q_t data profile can only be regarded as a nearly sparse signal and a perfect reconstruction is usually intractable.

2.5 ORTHOGONAL MATCHING PURSUIT

This section introduces another commonly used algorithm, called orthogonal matching pursuit (OMP), for solving the underdetermined system in Eq. (2.4) in the context of sparse representation. OMP is an iterative greedy algorithm that identifies columns that are most correlated with the measurement data $\boldsymbol{y}$ (e.g., Davis et al. 1997; Tropp 2004). The idea of the OMP algorithm is to iteratively find out the columns in **A** that can approximate the measurements $\boldsymbol{y}$. At the first iteration, the column that is the most correlated with $\boldsymbol{y}$ is selected. The associated non-trivial weight coefficient vector is estimated as $\hat{\boldsymbol{\omega}}$. Then, subtract the contribution of the selected column from $\boldsymbol{y}$ and iterate the process using the residual obtained. The iteration is terminated when the overall number of iterations reaches a specified limit, or when the difference

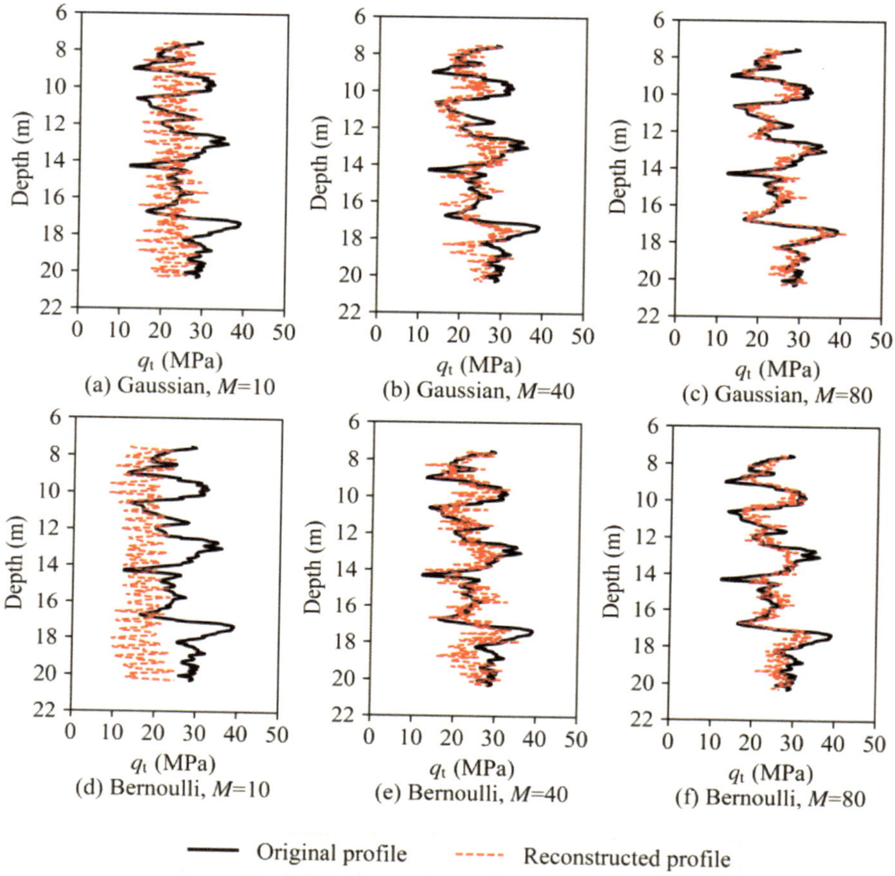

FIGURE 2.10 Recovery of the q_t data profile of CPT MSC014 using basis pursuit algorithm with two types of measurement matrix under three measurement scenarios.

between the approximated $\boldsymbol{y}$ using all selected columns and the original $\boldsymbol{y}$ is smaller than an acceptable stopping criterion, which can be formulated as the ratio between the l_2 norms of residual and $\boldsymbol{y}$ (e.g., Tropp and Gilbert 2007; Wang and Zhao 2016). The stopping criteria can be prescribed based on the magnitude of the measurement error in practice. The OMP algorithm can be summarized as follows:

Step 1: Set initial values to the variables in the iteration, i.e., set iteration counter t = 1; the residual $\boldsymbol{r}_t = \boldsymbol{y}$, the predicted measurement data, $\boldsymbol{y}' = \boldsymbol{0}$ ($\boldsymbol{0}$ is an M-length zero vector). Initiate an empty matrix $\mathbf{A}_s$ to store the columns that will be selected from $\mathbf{A}$.

Step 2: Find the column $\boldsymbol{a}_j$ that solves the maximization problem $\max_j \left| < \boldsymbol{r}_t, \boldsymbol{a}_j > / \sqrt{< \boldsymbol{r}_t, \boldsymbol{r}_t >< \boldsymbol{a}_j, \boldsymbol{a}_j >} \right|$. Append column $\boldsymbol{a}_j$ to $\mathbf{A}_s$.

Step 3: Estimate the non-trivial coefficient vector $\hat{\omega} = \left[\mathbf{A}_s^{\mathrm{T}}\mathbf{A}_s\right]^{-1}\mathbf{A}_s^{\mathrm{T}}\boldsymbol{y}$; re-estimate the measurement data $\boldsymbol{y}' = \mathbf{A}_s\hat{\boldsymbol{\omega}}$; update the residual $\boldsymbol{r}_t = \boldsymbol{y} - \boldsymbol{y}'$; and update the iteration counter $t = t + 1$.

Step 4: Repeat the above process until the relative error $\|\boldsymbol{r}_t\| / \|\boldsymbol{y}\|$ is smaller than a specified threshold value; and finally reconstruct the $\boldsymbol{f}$ using the estimated non-trivial coefficients and the associated basis functions.

To illustrate the OMP algorithm, the CPT q_t data profile example shown in Figure 2.6 is reanalyzed in this section using limited measurement under the CS framework. Similar to the discussion in Section 2.4, two different types of measurement matrix, i.e., Gaussian random sensing matrix and Bernoulli random sensing matrix, are evaluated under three different measurement scenarios, i.e., M=10,

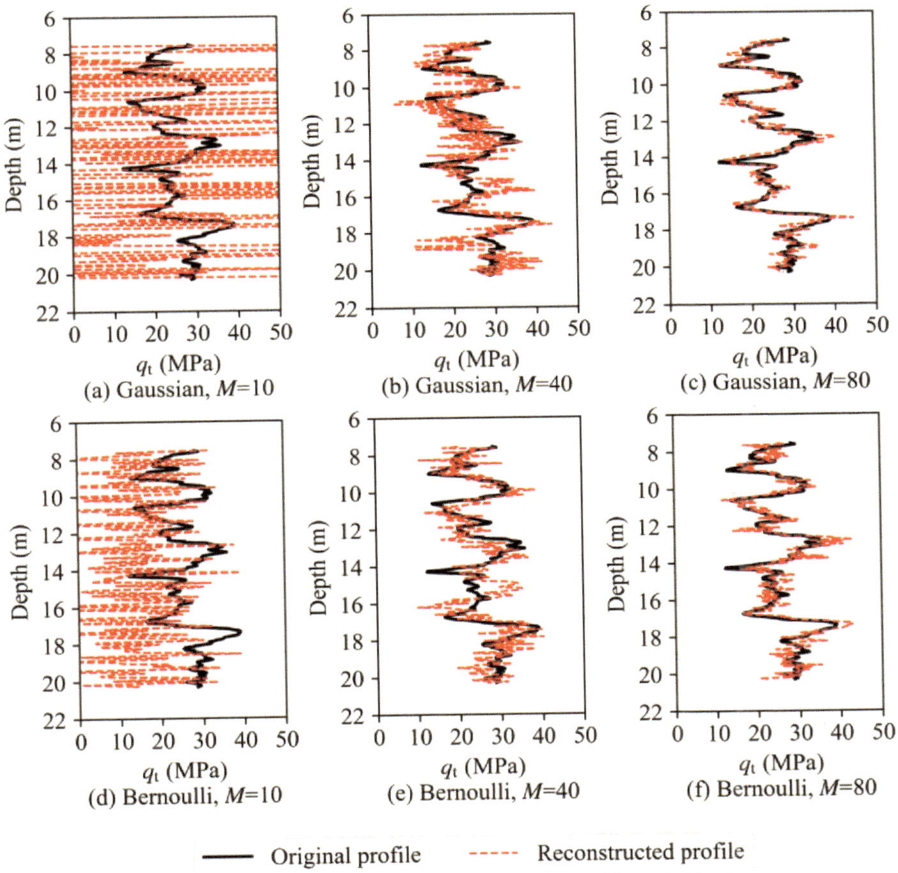

FIGURE 2.11 Recovery of the q_t data profile of CPT MSC014 using orthogonal matching pursuit algorithm with two types of measurement matrix under three measurement scenarios.

40, and 80. The associated reconstruction results under different scenarios are shown in Figure 2.11. Figure 2.11 uses the same legends as Figure 2.10. When M= 10, reconstructed profiles using either of these two sensing matrices yield significant noisy patterns. Particularly, the reconstructed profiles in Figures 2.11a and d exhibit much more unrealistic noisy patterns than the counterparts in Figures 2.10a and d. When the number of measurement data increases to M=40 and M=80, the recovery performances of OMP are comparable to basis pursuit algorithm. One of the limitations of OMP algorithm is that an incorrect or suboptimal basis function might be selected during iteration, but it cannot be removed until the end of OMP implementation (e.g., Boche et al. 2015).

REFERENCES

Ahmed, N., T. Natarajan, and K. R. Rao. 1974. Discrete cosine transform. *IEEE Transactions on Computers* 100 (1): 90–93.

Baraniuk, R. G. 2007. Compressive sensing [lecture notes]. *IEEE Signal Processing Magazine* 24 (4): 118–121.

Baraniuk, R., M. Davenport, R. DeVore, and M. Wakin. 2008. A simple proof of the restricted isometry property for random matrices. *Constructive Approximation* 28: 253–263.

Boche, H., R. Calderbank, G. Kutyniok, and J. Vybíral. 2015. A survey of compressed sensing. In *Compressed Sensing and Its Applications: MATHEON Workshop 2013*, 1–39. Cham: Springer International Publishing.

Candès, E., and J. Romberg. 2007. Sparsity and incoherence in compressive sampling. *Inverse Problems* 23 (3): 969.

Candès, E. J., and M. B. Wakin. 2008. An introduction to compressive sampling. *IEEE Signal Processing Magazine* 25 (2): 21–30.

Davis, G., S. Mallat, and M. Avellaneda. 1997. Adaptive greedy approximations. *Constructive Approximation* 13: 57–98.

Duarte, M. F., M. A. Davenport, D. Takhar, J. N. Laska, T. Sun, K. F. Kelly, and R. G. Baraniuk. 2008. Single-pixel imaging via compressive sampling. *IEEE Signal Processing Magazine* 25 (2): 83–91.

Foucart, S., and H. Rauhut. 2013. Restricted isometry property. In *A Mathematical Introduction to Compressive Sensing*, 133–163. Applied and Numerical Harmonic Analysis. New York: Birkhäuser.

Rao, B. D., and K. Kreutz-Delgado. 1999. An affine scaling methodology for best basis selection. *IEEE Transactions on Signal Processing* 47 (1): 187–200.

Romberg, J. 2008. Imaging via compressive sampling. *IEEE Signal Processing Magazine* 25 (2): 14–20.

Shannon, C. E. 1948. A mathematical theory of communication. *The Bell System Technical Journal* 27 (3): 379–423.

Strohmer, T., and R. W. Heath Jr. 2003. Grassmannian frames with applications to coding and communication. *Applied and Computational Harmonic Analysis* 14 (3): 257–275.

Tipping, M. E. 2001. Sparse Bayesian learning and the relevance vector machine. *Journal of Machine Learning Research* 1 (June): 211–244.

Tropp, J. A. 2004. Greed is good: Algorithmic results for sparse approximation. *IEEE Transactions on Information Theory* 50 (10): 2231–2242.

Tropp, J. A., and A. C. Gilbert. 2007. Signal recovery from random measurements via orthogonal matching pursuit. *IEEE Transactions on Information Theory* 53 (12): 4655–4666.

Wallace, G. K. 1992. The JPEG still picture compression standard. *IEEE Transactions on Consumer Electronics* 38 (1): xviii–xxxiv.

Wang, Y., and T. Zhao. 2016. Interpretation of soil property profile from limited measurement data: A compressive sampling perspective. *Canadian Geotechnical Journal* 53 (9): 1547–1559.

3 Analytics of Sparse Measurements on Spatial Data by Bayesian Compressive Sensing (BCS)

Tengyuan Zhao and Yu Wang

ABSTRACT

This chapter addresses the challenges and complexities associated with analysis of sparse spatial measurements in geotechnical site characterization using compressive sensing (CS), particularly Bayesian CS (BCS). It highlights key differences in measurement patterns, objectives, and sensing matrices between geotechnical site characterization and image or signal processing. While geotechnical or geological data exhibit spatial variability, they are typically measured at only a limited number of locations during site characterization, resulting in sparse measurements, unlike image or signal processing, where nearly complete datasets are collected. To effectively analyze sparse spatial measurements using BCS, careful modifications to the sensing matrix and algorithms are required, including (1) adapting the original orthogonal matching pursuit (OMP) algorithm to identify nontrivial coefficients and corresponding basis functions from sparse measurements; and (2) quantifying interpretation uncertainties due to sparse measurements within a Bayesian framework. This chapter develops and applies BCS methods for one-dimensional (1D), two-dimensional (2D), and three-dimensional (3D) geotechnical site characterization. Both simulated and real-world data are used to illustrate and validate these methods. Results demonstrate that BCS effectively characterizes 1D, 2D, and 3D geotechnical property profiles with quantified statistical uncertainty, providing valuable insights for geotechnical site characterization.

3.1 CHALLENGES ASSOCIATED WITH ANALYTICS OF SPARSE MEASUREMENTS ON SPATIAL DATA

As discussed and illustrated in Chapter 2, compressive sensing or sampling (CS) is able to reconstruct a spatially or temporally varying signal of interest from compressed

DOI: 10.1201/9781003381235-3

measurements, if the signal itself is sparse or has patterns. In image or signal processing, almost full measurements on the signal $\boldsymbol{f}$ of interest are often acquired, e.g., by a high-resolution camera in a smartphone. Then, a random sensing matrix $\boldsymbol{\Psi}$ is used for dimension reduction of the measurement data obtained to facilitate data storage and transmission. This, however, is not the case in geotechnical site characterization, where the objective is to reconstruct, or predict, a complete but unknown profile $\boldsymbol{f}$ from its sparse measurements $\boldsymbol{y}$, instead of compressed measurements with dimension reduction by the random sensing matrix. Because the objective in geotechnical site characterization is different from that in image processing, the original CS/BCS algorithms developed for image or signal processing are NOT directly applicable to analytics of sparse measurements on spatial data in site characterization. Some careful modifications to the original CS/BCS algorithms are therefore needed to deal with sparse measurements, rather than compressed measurements, in site characterization, as discussed in this chapter.

3.1.1 Objectives and Measurements in Geotechnical Site Characterization

The objectives and measurements in geotechnical site characterization are quite different from those in image or signal processing. The objective in geotechnical site characterization is to accurately depict or predict spatially varying patterns of underground geomaterials (e.g., soils or rocks) from in-situ or laboratory tests that are often intrusive and destructive. Because of the time, resource, access or technical constraints, the tests are typically performed at a limited number of locations in the field or in the laboratory using a small number of soil or rock samples taken from the field (e.g., Mayne et al. 2002). From a signal or image processing perspective, a spatially varying pattern of underground geomaterials may also be denoted as $\boldsymbol{f}$ that is to be estimated from limited measurements $\boldsymbol{y}$ obtained in situ or from laboratory. The measurements $\boldsymbol{y}$ are sparse and limited in geotechnical site characterization. This is a major difference between geotechnical site characterization and image or signal processing, where almost full measurements on the signal $\boldsymbol{f}$ of interest are often available. The objective in image or signal processing is for efficient data compression, transmission, and reconstruction, while the objective in geotechnical site characterization is to recover, or predict, the spatially varying patterns of interest $\boldsymbol{f}$ from limited measurements.

Consider, for example, a spatially varying signal of interest $\boldsymbol{f}$ with a total length of $N = 1{,}000$, as shown in Figure 3.1a. Then, sparse measurements $\boldsymbol{y} = \boldsymbol{\Psi f}$ is obtained using a sensing matrix $\boldsymbol{\Psi}$ with a dimension of M by N, where M is much less than N. In the context of original CS/BCS algorithms, $\boldsymbol{\Psi}$ is a random sensing matrix simulated, e.g., from a standard Gaussian distribution with a zero mean and a unit variance (e.g., Candès and Wakin 2008). Figure 3.1b shows, by open circles, some examples of sparse or compressed measurements $\boldsymbol{y}$ with a length of $M = 50$ obtained after applying such a random sensing matrix. As discussed in Chapter 2, using a random sensing matrix ensures that the measurement data points in $\boldsymbol{y}$ are independent among each other and that each measurement data point in $\boldsymbol{y}$ contains information from each data point of $\boldsymbol{f}$. Once compressed measurements $\boldsymbol{y}$ with a length much less than $\boldsymbol{f}$ are available, efficient and stable transmission, e.g., over the Internet, using

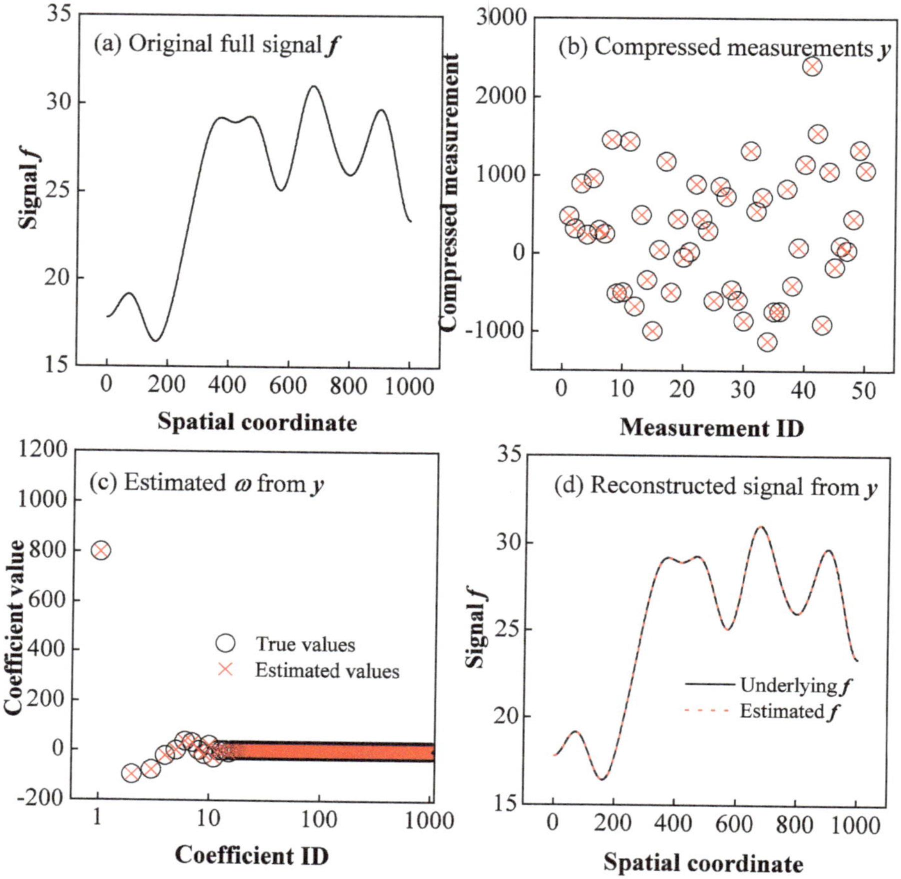

FIGURE 3.1 Illustration of compression and reconstruction by original CS algorithms: (a) original full signal $\boldsymbol{f}$; (b) compressed measurements $\boldsymbol{y}$; (c) estimated $\boldsymbol{\omega}$ from $\boldsymbol{y}$; and (d) reconstructed signal from $\boldsymbol{y}$.

limited bandwidth is possible (e.g., Alan et al. 1989; Salomon 2007). When $\boldsymbol{y}$ is received after transmission, the signal $\boldsymbol{f}$ can be reconstructed from $\boldsymbol{y}$ by signal decomposition $\boldsymbol{f} = \mathbf{B}\boldsymbol{\omega}$, where $\mathbf{B}$ is an orthonormal matrix, representing N basis functions, e.g., discrete cosine functions; and $\boldsymbol{\omega}$ represents the weight vector with a length of N corresponding to N basis functions. Under the CS framework, most elements of $\boldsymbol{\omega}$ are negligibly small except only a few non-trivial ones, as shown in Figure 3.1c. This enables an estimation of $\boldsymbol{\omega}$ from $\boldsymbol{y}$, because $\boldsymbol{y} = \boldsymbol{\Psi f} = \boldsymbol{\Psi}\mathbf{B}\boldsymbol{\omega} = \mathbf{A}\boldsymbol{\omega}$, where $\mathbf{A} = \boldsymbol{\Psi}\mathbf{B}$. Using, e.g., OMP (e.g., Donoho 2006; Donoho et al. 2012), $\boldsymbol{\omega}$ can be estimated and denoted as $\hat{\boldsymbol{\omega}}$ (see the crosses in Figure 3.1c). Then, the signal $\boldsymbol{f}$ can be reconstructed as $\boldsymbol{f} \approx \mathbf{B}\hat{\boldsymbol{\omega}}$, as shown in Figure 3.1d by dotted lines. When compared with the original signal $\boldsymbol{f}$ which has a length of $N = 1{,}000$, $\boldsymbol{y}$ only contains $M = 50$ data points, which greatly reduces the demand for storage and improves transmission efficiency.

This, however, is not the case in geotechnical site characterization, where the objective is to characterize the spatially varying pattern or signal $\boldsymbol{f}$ (see Figure 3.2a) from sparse measurements on $\boldsymbol{f}$, as shown in Figure 3.2b. In this case, $\boldsymbol{y}$ is a sub-vector of $\boldsymbol{f}$, rather than compressed measurements of $\boldsymbol{f}$. Specifically, "$\boldsymbol{\Psi}$" here cannot be designed as random sensing matrix but serves as a matrix to strictly reflect corresponding sampling locations of $\boldsymbol{y}$ from $\boldsymbol{f}$. It is worth noting that satisfaction of the RIP criterion discussed in Chapter 2 cannot be guaranteed in this case. Besides, the number of measurements in geotechnical site characterization is often much smaller than those in signal or image processing. As a result, it is difficult to directly use the original CS/BCS algorithms developed for image or signal processing to estimate $\boldsymbol{\omega}$ from $\boldsymbol{y}$ in geotechnical site characterization. For example, using the data $\boldsymbol{y}$ shown in Figure 3.2b as input, OMP (i.e., the one used in Figure 3.1d) is

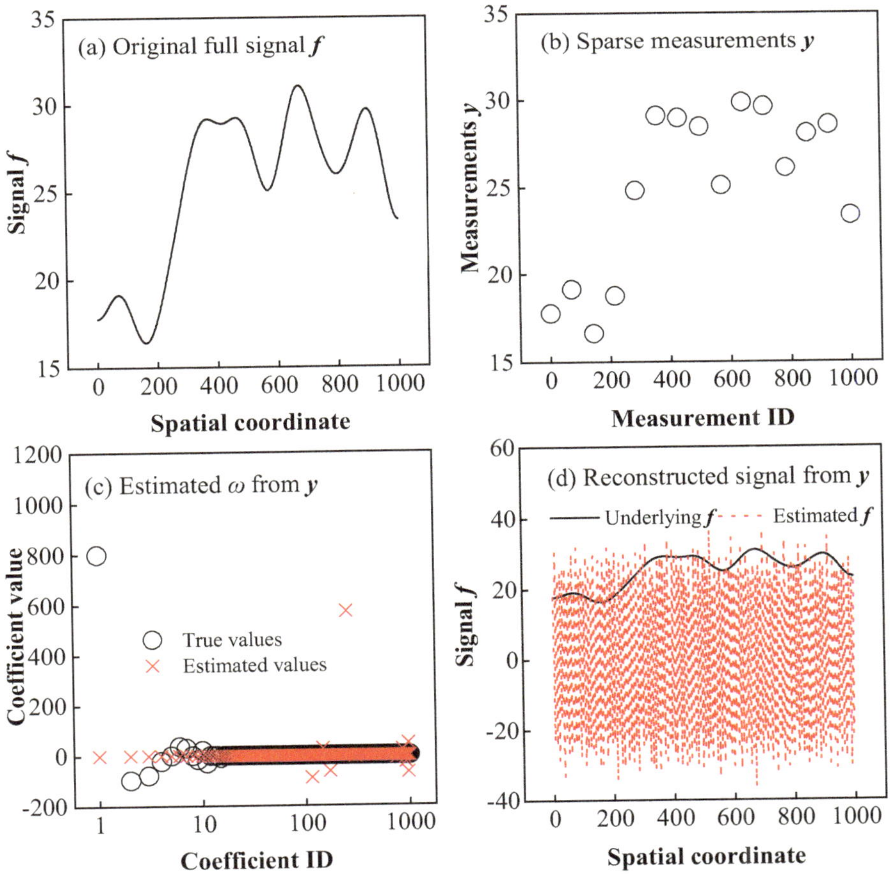

FIGURE 3.2 Illustration of direct application of original CS algorithms in geotechnical site characterization: (a) original full signal $\boldsymbol{f}$; (b) sparse measurements $\boldsymbol{y}$; (c) estimated $\boldsymbol{\omega}$ from $\boldsymbol{y}$; and (d) reconstructed signal from $\boldsymbol{y}$.

directly adopted to estimate $\hat{\boldsymbol{\omega}}$. Figure 3.2c shows the estimated $\hat{\boldsymbol{\omega}}$ by crosses. For comparison, the underlying true $\boldsymbol{\omega}$ is shown in open circles. It is obvious that the crosses do not match the open circles. More importantly, some trivial coefficients are identified as non-trivial ones with relatively large magnitudes. Incorrect identification of nontrivial coefficients and unsuitable contribution from the trivial coefficients lead to significantly large fluctuations in the reconstructed $\boldsymbol{f}$, as shown in Figure 3.2d by dotted lines. In general, these trivial coefficients correspond to basis functions (i.e., columns of matrix $\mathbf{B}$) with high frequencies, as discussed in Section 2.2. Therefore, they should be removed before further analysis in geotechnical site characterization, as discussed in Section 3.1.3.

3.1.2 Sensing Matrix for Sparse Measurements in Site Characterization

As pointed out in Chapter 2 and the previous Section 3.1.1, sensing matrix $\boldsymbol{\Psi}$ in the original CS/BCS algorithms for signal or image processing is a randomized matrix for data compression, storage, and transmission. In contrast, $\boldsymbol{y}$ is a sub-vector of $\boldsymbol{f}$ in geotechnical site characterization. In this case, sensing matrix $\boldsymbol{\Psi}$ reflects the positions of the measurement data points of $\boldsymbol{y}$ in $\boldsymbol{f}$, and it is a matrix with binary elements of either zero or one. Generally speaking, $\boldsymbol{\Psi}$ in geotechnical site characterization can be constructed from an identity matrix $\mathbf{I}_N$ with a dimension of N by N. Specifically, when the ith ($i = 1, 2,\ldots, M$) component of $\boldsymbol{y}$ is the nth ($n =1, 2,\ldots, N$) component of $\boldsymbol{f}$, the nth row of the identity matrix $\mathbf{I}_N$ is adopted as the ith row of $\boldsymbol{\Psi}$. For example, if the first component of $\boldsymbol{y}$ is the 10th component of $\boldsymbol{f}$, the 10th row of identity matrix $\mathbf{I}_N$ is taken as the first row of $\boldsymbol{\Psi}$. Each row of the matrix $\boldsymbol{\Psi}$, corresponding to a position of element of $\boldsymbol{y}$, can be obtained in a similar manner. Figures 3.3a and b compare the difference of $\boldsymbol{\Psi}$ in geotechnical site characterization and signal or image processing.

3.1.3 Removal of High-Frequency Noises

Given the sensing matrix $\boldsymbol{\Psi}$ and measurements $\boldsymbol{y}$ in geotechnical site characterization, $\boldsymbol{\omega}$ can be estimated theoretically using CS or BCS algorithms, as demonstrated in Section 3.1.1. However, the predicted or reconstructed $\boldsymbol{f}$ might be unrealistic because of the sparse measurements $\boldsymbol{y}$, unsatisfied RIP criterion, and high-frequency noises corresponding to trivial components of the basis functions. To improve the performance of CS or BCS on sparse measurements, high-frequency components in the spatially varying quantity should be removed first before using CS or BCS algorithms for reconstruction or prediction in geotechnical site characterization. Removal of high-frequency noises is the second key difference between the original CS algorithms for image or signal processing and the CS or BCS algorithms for geotechnical site characterization. As discussed in Chapter 2, $\boldsymbol{f}$ can be considered as a summation of different basis functions with different frequencies and weights, i.e., $\boldsymbol{f} = \mathbf{B}\boldsymbol{\omega} = \sum_{j=1}^{N} \boldsymbol{B}_j \cdot \omega_j$. Each component in $\boldsymbol{\omega}$, i.e., ω_j, corresponds to a basis function with a corresponding

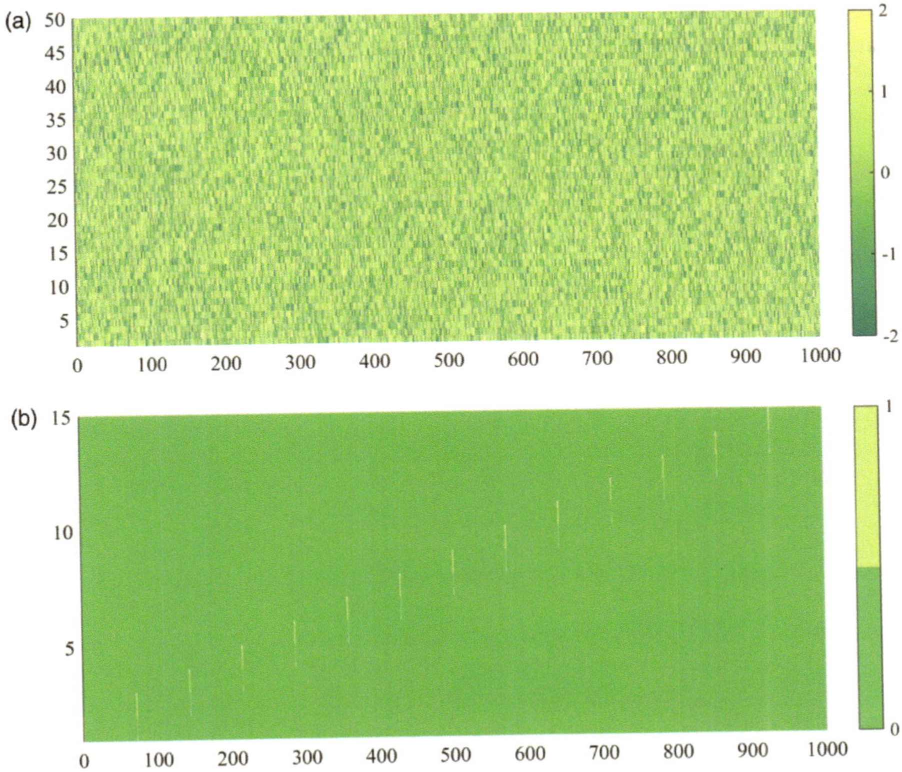

FIGURE 3.3 Comparison of sensing matrix $\boldsymbol{\Psi}$ in the field of (a) signal or image processing and (b) geotechnical site characterization.

frequency, v_j (j=1, 2, ..., N). v_j can be expressed as (e.g., Rao and Yip 1990; Salomon 2007; Wang and Zhao 2016)

$$v_j = \frac{v_s}{2} \times \frac{j-1}{N} = \frac{1}{2}\left(\frac{j-1}{N} v_s\right) \tag{3.1}$$

where v_s is the target sampling frequency for $\boldsymbol{f}$, and it represents the number of data points per unit length (e.g., 1 m) that the reconstructed soil property profile is expected to have. It is equal to the reciprocal of the target resolution η of the reconstructed $\boldsymbol{f}$. Each frequency v_j corresponds to a fluctuation period T_j of a cosine function, and $T_j = v_j^{-1}$. Eq. (3.1) shows that a high value of j corresponds to a high-frequency v_j, and thus corresponds to a short fluctuation period T_j. In geotechnical site investigation, due to the long period of geological formation process, soil property varies relatively slow with depths or horizontally, and the fluctuation period is relatively large. Therefore, a relatively small fluctuation period (e.g., 0.1m) can be taken as a lower bound fluctuation period, T_l, of soil properties. Subsequently, $T_j = v_j^{-1} \geq T_l$, and correspondingly, basis function ID j should satisfy (e.g., Wang and Zhao 2016):

$$j \leq j^* = 2N/(T_l\, v_s) + 1 \tag{3.2}$$

where j^* is the maximum basis function ID, corresponding to the maximum frequency v_{j^*} of a cosine function. Cosine functions with frequencies greater than v_{j^*}, or equivalently, with basis function ID $j \geq j^*$ are excluded by Eq. (3.2). Correspondingly, the components with $j \geq j^*$ in $\boldsymbol{\omega}$ are removed. Subsequently, it filters out high-frequency noises when reconstructing $\boldsymbol{f}$ from sparse data (e.g., Wang and Zhao 2016).

In geotechnical engineering practice, soil properties are usually tested with a spatial interval d. For a given sampling interval d^* in a unit of meter in a specific site, the minimum fluctuation period of soil property that can be estimated is $2d^*$ (e.g., Fenton and Griffiths 2008; Nyquist 1928; Shannon 1948). In the context of geotechnical site investigation in this book, sampling period $T_j = v_j^{-1}$ for a cosine function that corresponds to component ω_j should satisfy $T_j = v_j^{-1} \geq 2d^*$ (Wang and Zhao 2016). Combing $T_j = v_j^{-1} \geq 2d^*$ and Eq. (3.1) leads to:

$$j \leq j^* = N/(d^* v_s) + 1 \tag{3.3}$$

Similar to Eq. (3.2), Eq. (3.3) provides a threshold for removing cosine basis functions with high frequencies during the reconstruction of $\boldsymbol{f}$. Basis function ID j should satisfy both Eqs. (3.2) and (3.3) simultaneously. Since each element of $\boldsymbol{\omega}$, i.e., ω_j, corresponds to one column in $\mathbf{A}$, and j is limited to a range of $1 \leq j \leq j^*$, $\mathbf{A}$ may be simplified as $\mathbf{A}_s$, which is the first j^* columns of $\mathbf{A}$, with j satisfying both Eqs. (3.2) and (3.3). Removal of high-frequency noise greatly improves the possibility of correctly identifying columns $\mathbf{A}_s$ by an appropriate algorithm such as OMP (e.g., Tropp and Gilbert 2007; Needell and Vershynin 2010; Cai and Wang 2011), even when the number of measurement data points is quite small in geotechnical site characterization. In addition, because the OMP algorithm identifies one column in $\mathbf{A}_s$ in each iteration, the j^* from Eqs. (3.2) and (3.3) also acts as an upper bound for the iteration counter.

In addition to the theoretical derivation above for determining the upper bound of high-frequency component used in the reconstruction of spatially varying profile $\boldsymbol{f}$, a rule of thumb maximum number of j^* as M~$1.4M$ might also be used in accordance with previous experiences on geotechnical site investigation data. After the removal of high-frequency components, nontrivial elements of $\boldsymbol{\omega}$ can be readily and reasonably identified and estimated in two steps, i.e., identification of nontrivial elements in $\boldsymbol{\omega}$ by OMP, and estimation of the identified nontrivial $\boldsymbol{\omega}$ values under a Bayesian framework, as discussed in detail in Sections 3.2 and 3.3.

3.2 MODIFIED OMP FOR IDENTIFICATION OF NONTRIVIAL ELEMENTS OF ω

In this section, OMP with a deterministic and binary sensing matrix $\boldsymbol{\Psi}$ with entries of 1 or 0 is used to identify the nontrivial elements of $\boldsymbol{\omega}$ after removal of high-frequency components in Section 3.1. In this case, matrix **A** in $\boldsymbol{y} = \mathbf{A}\boldsymbol{\omega}$ is reduced to $\mathbf{A}_s$, which contains the first j^* columns of **A**. Based on the discussions in Section 3.1, the upper bound of column ID j^* is determined via either Eq. (3.3) or $j^* = 1M\sim1.4M$. Then, the columns of $\mathbf{A}_s$ significantly contributing to $\boldsymbol{y}$, corresponding to nontrivial elements of $\boldsymbol{\omega}$ (i.e., $\boldsymbol{\omega}_s$), can be determined. Implementation of modified OMP is summarized as:

I Let a column vector $\hat{\boldsymbol{y}}$ represent $\boldsymbol{y}$ re-estimated from the nontrivial basis functions and corresponding weight. $\boldsymbol{y}_r$ represents the residuals between $\boldsymbol{y}$ and $\hat{\boldsymbol{y}}$, i.e., $\boldsymbol{y}_r = \boldsymbol{y}-\hat{\boldsymbol{y}}$. Similarly, let $\hat{\mathbf{A}}$ represent the matrix, with columns corresponding to the nontrivial elements of $\boldsymbol{\omega}$ that are selected from $\mathbf{A}_s$. Initially, $\hat{\boldsymbol{y}}$ is a zero-column vector, which has the same length as $\boldsymbol{y}$; $\hat{\mathbf{A}}$ is an empty set, containing nothing.

II Compute the linear correlation between the jth column of $\mathbf{A}_s$, i.e., $\boldsymbol{A}_{s,j}$ (j=1, 2, …, j^*), and the current residual $\boldsymbol{y}_r$ using a dot product. Namely, $\lambda_j = \dfrac{(\boldsymbol{y}_r)^{\mathrm{T}} \cdot \boldsymbol{A}_{s,j}}{\sqrt{(\boldsymbol{y}_r)^{\mathrm{T}} \cdot \boldsymbol{A}_{s,j}}\, g \sqrt{(\boldsymbol{A}_{s,j})^{\mathrm{T}} \cdot \boldsymbol{A}_{s,j}}}$. Compare the absolute results of $|\lambda_j|$, j=1, 2, …, j^*, and find out the one with the largest value. The largest value among $|\lambda_j|$ for j=1, 2, …, j^* indicates greatest contribution of the corresponding column of $\mathbf{A}_s$ to $\boldsymbol{y}$. The corresponding column is then selected as the most important one, denoted as $\boldsymbol{A}_c$.

III Update the matrix $\hat{\mathbf{A}}$ by appending the newly selected column $\boldsymbol{A}_c$ to $\hat{\mathbf{A}}$. Subtract the contribution of the selected column $\boldsymbol{A}_c$ in $\mathbf{A}_s$, i.e., set the corresponding column in $\mathbf{A}_s$ to a zero vector. This thus avoids a potential problem of repeated selection of the same column in $\mathbf{A}_s$.

IV Estimate the coefficients $\boldsymbol{\omega}_s$ corresponding to $\hat{\mathbf{A}}$ using the least-squares method as $\boldsymbol{\omega}_s = [(\mathbf{A}_s)^{\mathrm{T}}(\mathbf{A}_s)]^{-1}(\mathbf{A}_s)^{\mathrm{T}}\boldsymbol{y}$. Re-estimate the measurement data as $\hat{\boldsymbol{y}} \approx \hat{\mathbf{A}}\boldsymbol{\omega}_s$; update the residual as $\boldsymbol{y}_r = \boldsymbol{y} - \hat{\boldsymbol{y}}$ and calculate the relative error between $\boldsymbol{y}$ and $\hat{\boldsymbol{y}}$,

i.e., $\varsigma = \sqrt{\sum_{i=1}^{M} |\hat{y}_i - y_i|^2 \Big/ \left(\sum_{i=1}^{M} |y(i)|^2\right)^{\frac{1}{2}}}$.

V Repeat substeps (II)–(IV) until a termination criterion is satisfied, e.g., the relative error ς less than a given threshold ζ, i.e., $\varsigma \leq \zeta$.

Consider, for example, the sparse measurements in Figure 3.2b with a number, $M = 15$, of measurements. The maximum column ID may be determined as $j^* = 1.4M = 21$. Matrix **A** reduces $\mathbf{A}_s$ containing only the first 21 columns from **A**. After that, the OMP algorithm mentioned in this section is implemented, the nontrivial coefficients $\boldsymbol{\omega}_s$ can be identified and estimated, as shown in Figure 3.4b. For comparison, Figure 3.4a shows the estimated $\boldsymbol{\omega}_s$ by crosses using the original OMP without removing high-frequency components of the signal. Figure 3.4b shows that the crosses are much more consistent with those of the true frequency components shown by open circles than those shown in Figure 3.4a. Such observations indicate

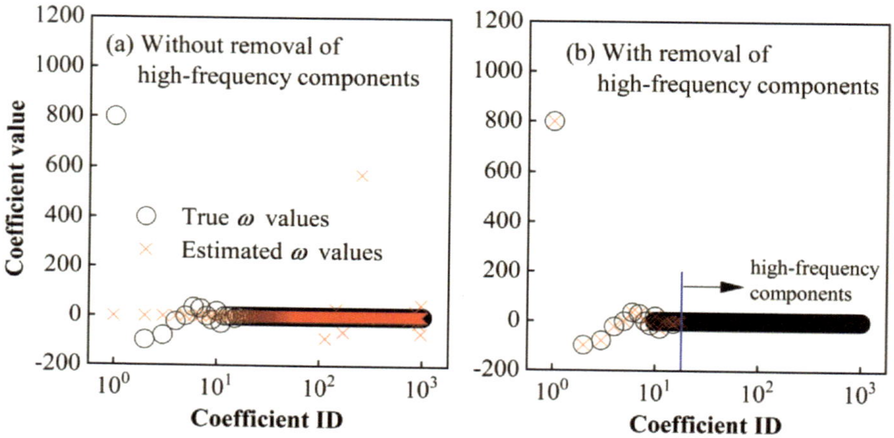

FIGURE 3.4 Illustration of the effect of removal of high-frequency components in the estimation of nontrivial coefficients in $\boldsymbol{\omega}$.

that removal of high-frequency components significantly improves both the identification and estimation accuracy of nontrivial coefficients in $\boldsymbol{\omega}$, even when the number of measurements is quite limited in geotechnical site characterization. Note that, although the nontrivial elements of $\boldsymbol{\omega}$ have been estimated as $\boldsymbol{\omega}_s$ by a least-squares method in OMP, uncertainty associated with $\boldsymbol{\omega}_s$ cannot be quantitatively evaluated. The quantified uncertainty not only provides a measure of reliability for the estimated $\boldsymbol{\omega}_s$, but also paves a way for quantifying the uncertainty associated with the profile of geotechnical property in site characterization. A Bayesian framework is used to quantify the uncertainty on the estimated $\boldsymbol{\omega}_s$, as discussed in Section 3.3.

3.3 BAYESIAN ESTIMATION OF ω_S

The estimation of $\boldsymbol{\omega}_s$ by OMP as described in Section 3.2 cannot deal with the uncertainties associated with measurements, such as measurement error, statistical uncertainty, or model uncertainty (e.g., residuals). These uncertainties are important and affect reliability of geotechnical design in practice (e.g., Phoon and Kulhawy 1999; Wang et al. 2016). In this section, after identification of nontrivial elements in $\boldsymbol{\omega}$ by the modified OMP in Section 3.2, $\boldsymbol{\omega}_s$ is re-estimated probabilistically under a Bayesian framework to consider various uncertainties, as formulated below:

$$p(\boldsymbol{\omega}_s \mid \boldsymbol{y}) = \frac{p(\boldsymbol{y} \mid \boldsymbol{\omega}_s)p(\boldsymbol{\omega}_s)}{p(\boldsymbol{\omega}_s)} \tag{3.4}$$

where $p(\boldsymbol{\omega}_s|\boldsymbol{y})$ = the posterior probability density function (PDF) of $\boldsymbol{\omega}_s$, representing the updated knowledge on $\boldsymbol{\omega}_s$ given the sparse measurement data $\boldsymbol{y}$ and prior knowledge; $p(\boldsymbol{y}|\boldsymbol{\omega}_s)$ = the likelihood function of observing $\boldsymbol{y}$ given $\boldsymbol{\omega}_s$, reflecting a goodness of fit between $\boldsymbol{y}$ given $\boldsymbol{\omega}_s$; $p(\boldsymbol{\omega}_s)$ = the prior distribution of $\boldsymbol{\omega}_s$ reflects the existing knowledge on $\boldsymbol{\omega}_s$ in the absence of measurement data $\boldsymbol{y}$; $p(\boldsymbol{y})$ = a normalizing constant. The likelihood function $p(\boldsymbol{y}|\boldsymbol{\omega}_s)$ and the prior distribution $p(\boldsymbol{\omega}_s)$ are two essential ingredients in the Bayesian framework, which are discussed in detail next.

3.3.1 Likelihood Function

To facilitate the derivation of the likelihood function $p(\boldsymbol{\omega}_s|\boldsymbol{y})$, the sparsely measured geotechnical property $\boldsymbol{y}$ is re-written as follows:

$$\boldsymbol{y} = \mathbf{A}_s\boldsymbol{\omega}_s + \boldsymbol{\varepsilon} \tag{3.5}$$

where $\mathbf{A}_s$ represents the selected columns corresponding to the nontrivial elements, i.e., $\boldsymbol{\omega}_s$, as defined in Section 3.2; $\boldsymbol{\varepsilon}$ is a column vector with the same length as $\boldsymbol{y}$, and it represents the lumped uncertainty of measurement errors and residuals between $\boldsymbol{y}$ and $\mathbf{A}_s\boldsymbol{\omega}_s$. For derivation convenience, $\boldsymbol{\varepsilon}$ is considered as M realizations of a Gaussian random variable with a mean of zero and unknown variance. Moreover, the reciprocal of the unknown variance of ε is denoted as τ, which will be updated together with $\boldsymbol{\omega}_s$. Based on the discussions above and Eq. (3.5), the likelihood function is expressed as

$$p(\boldsymbol{y}|\boldsymbol{\omega}_s,\tau) = \frac{\tau^{M/2}}{\left(\sqrt{2\pi}\right)^{M/2}}\exp\left(-\frac{\tau(\boldsymbol{y}-\mathbf{A}_s\boldsymbol{\omega}_s)^{\mathrm{T}}(\boldsymbol{y}-\mathbf{A}_s\hat{\boldsymbol{\omega}}_s)}{2}\right) \tag{3.6}$$

Eq. (3.6) implies that the lumped uncertainty $\boldsymbol{\varepsilon}$ is considered weakly correlated or independent.

3.3.2 Prior Distribution

Both $\boldsymbol{\omega}_s$ and τ are unknown and taken as random variables. In this case, a joint prior PDF of $\boldsymbol{\omega}_s$ and τ is needed, which is expressed as:

$$p(\boldsymbol{\omega}_s,\tau) = p(\boldsymbol{\omega}_s)p(\tau) \tag{3.7}$$

$p(\boldsymbol{\omega}_s)$ and $p(\tau)$ are the prior PDFs of $\boldsymbol{\omega}_s$ and τ, respectively, and they are constructed one by one below.

In this chapter, a three-level hierarchical model is used to model the prior PDF of $\boldsymbol{\omega}_s$. First, each element of $\boldsymbol{\omega}_s$, denoted as $\omega_{s,t}$ (t = 1, 2, …, s), is taken to follow a Gaussian distribution with a mean of zero and unknown variance of σ_t^2. The subscript "t" serves as an index to represent the tth element of the vector $\boldsymbol{\omega}_s$. Note that $\omega_{s,t}$ can be either negative or positive. For modeling convenience, the reciprocal of σ_t^2, i.e.,

($\alpha_t = 1/\sigma_t^2$, $\alpha_t > 0$), is used as a parameter in the prior PDF of $\omega_{s,t}$. Then, the prior distribution of $\boldsymbol{\omega}_s$ is expressed as (e.g., Zhao and Wang 2020):

$$p(\omega_s \mid \boldsymbol{\alpha}) = \prod_{t=1}^{s} \frac{\alpha_t^{1/2}}{\sqrt{2\pi}} \exp\left(-\frac{\alpha_t(\omega_{s,t})^2}{2}\right) = \frac{[\det(\mathbf{D}_s^{\alpha})]^{1/2}}{(2\pi)^{s/2}} \exp\left(-\frac{(\omega_s)^{\mathrm{T}} \mathbf{D}_s^{\alpha} (\omega_s)}{2}\right) \tag{3.8}$$

where $\boldsymbol{\alpha}$ represents α_t expressed in a vector form, i.e., $\boldsymbol{\alpha} = [\alpha_1, \alpha_2, \ldots, \alpha_s]$; $\mathbf{D}_s^{\alpha}$ is a diagonal matrix with diagonal element $D_s^{\alpha}(t,t) = \alpha_t$ ($t = 1, 2, \ldots, s$). Note that the prior distribution given in Eq. (3.8) is conjugated to the likelihood function in Eq. (3.6). This suggests that the conditional posterior PDF to be derived later follows a Gaussian distribution too.

Secondly, because $\boldsymbol{\alpha} = [\alpha_1, \alpha_2, \ldots, \alpha_s]$ are unknown, they are modeled as random variables in this study. Inspired by Zhao et al. (2015), each element of $\boldsymbol{\alpha}$, i.e., α_t ($t = 1, 2, \ldots, s$) is taken to follow an inverse Gamma distribution with a parameter $\gamma/2$ ($\gamma > 0$) and a constant parameter 1, and $p(\alpha_t \mid \gamma) = \gamma/2\alpha_t^{-2}\exp[-\gamma/2\alpha_t^{-1}]$. Then, the prior PDF of $\boldsymbol{\alpha}$ is expressed as:

$$p(\boldsymbol{\alpha}|\gamma) = \prod_{t=1}^{s} p(\alpha_t \mid \gamma) = \prod_{t=1}^{s} \frac{\gamma}{2} \alpha_t^{-2} \exp[-\frac{\gamma}{2} \alpha_t^{-1}] \tag{3.9}$$

Thirdly, to avoid manual selection of γ values, γ is also treated as a random variable and taken to follow a Gamma distribution, which is expressed as

$$p(\gamma) = b_0^{a_0} \gamma^{a_0 - 1} \exp(-b_0 \gamma) / \Gamma(a_0) \tag{3.10}$$

where a_0 and b_0 are nonnegative constants. In this chapter, a_0 and b_0 are taken as small values, e.g., $a_0 = b_0 = 10^{-4}$. In such a case, γ has a possibility to vary across a very wide range, with a tendency to zero. This implies that α_t tends to take small values, i.e., $1/\alpha_t$ (i.e., the variance of $\omega_{s,t}$) tends to be significant, and the prior PDF of $\omega_{s,t}$ is uninformative. Such a hierarchical formulation explicitly models the uncertainty in both $\boldsymbol{\alpha}$ and γ and allows $\omega_{s,t}$ to be updated efficiently by measurements. For illustration, Figures 3.5a to 3.5c, respectively, show the prior PDF of γ, α_t, and $\omega_{s,t}$. It is seen that γ varies from almost zero to a value of at least 10^3, but has a larger possibility to be small values. α_t varies from 10^{-3} to 10^6, which further facilitates an un-informative prior distribution of $\omega_{s,t}$ in Figure 3.5c. $\omega_{s,t}$ peaks sharply at zero, and varies flexibly from almost negative infinity to positive infinity. This indicates that an un-informative prior PDF of $\omega_{s,t}$ has been achieved by the three-level hierarchical model used herein. Note that the horizontal axis in Figure 3.5c is truncated to a range of [-400, 400] for display purpose.

In addition, the reciprocal of the variance of ε, i.e., τ, is taken to follow a Gamma distribution, and the prior PDF of τ is expressed as (e.g., Babacan et al. 2010; Zhao et al. 2015):

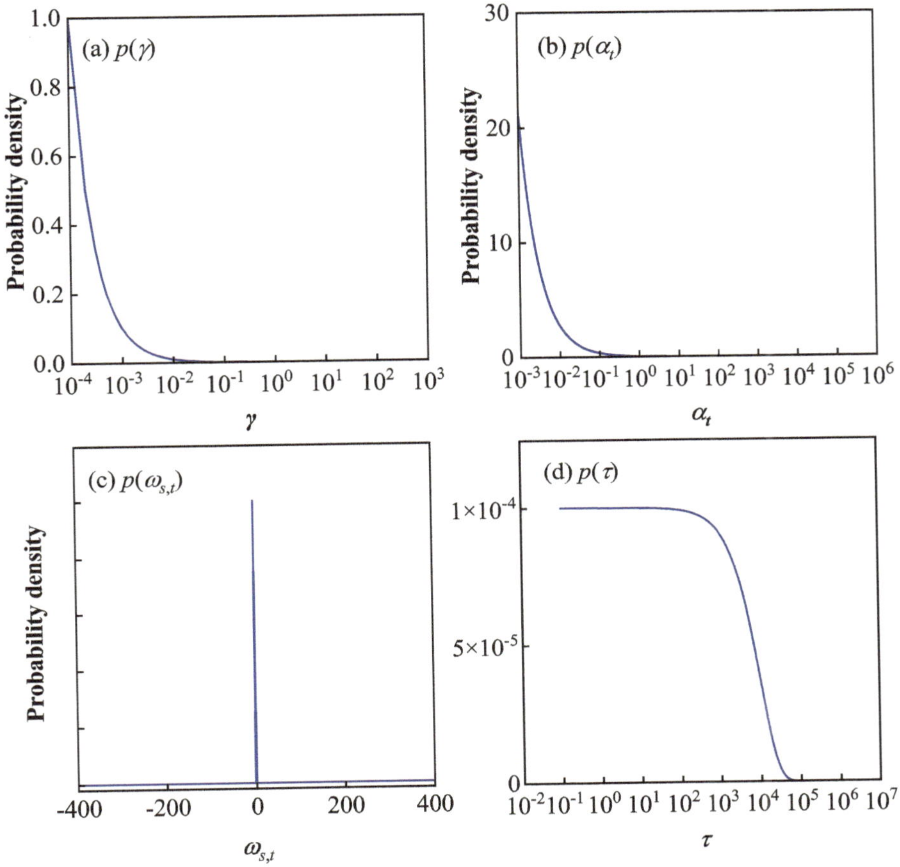

FIGURE 3.5 Visualization of prior PDFs of (a) γ, i.e., $p(\gamma)$; (b) α_t ($t = 1, 2, \ldots, s$), i.e., $p(\alpha_t)$; (c) $\omega_{s,t}$, i.e., $p(\omega_{s,t})$; and (d) τ, i.e., $p(\tau)$.

$$p(\tau) = \frac{d_0^{\,c_0}\,\tau^{c_0-1}}{\Gamma(c_0)}\exp\left(-d_0\tau\right) \tag{3.11}$$

Similar to a_0 and b_0, c_0 and d_0 are also non-negative parameters, and they are taken as $c_0 = 1$ and $d_0 = 10^{-4}$ in this chapter. This means that τ has a mean of $1/10^{-4} = 10^4$ and a significant variance of $1/(10^{-4})^2 = 10^8$. The prior PDF of τ, therefore, is uninformative, as shown in Figure 3.5d.

Combining Eqs. (3.8) and (3.11) leads to the joint prior PDF of $\boldsymbol{\omega}_s$, $\boldsymbol{\alpha}$, γ, and τ, which is expressed as:

$$p(\boldsymbol{\omega}_s, \boldsymbol{\alpha}, \gamma, \tau) = p(\boldsymbol{\omega}_s \mid \boldsymbol{\alpha})\,p(\boldsymbol{\alpha}|\gamma)\,p(\gamma)\,p(\tau) \tag{3.12}$$

where τ is taken as independent of $(\boldsymbol{\omega}_s, \boldsymbol{\alpha}, \gamma)$. Note that Eq. (3.12), instead of Eq. (3.7), shall be used for the prior PDF since $\boldsymbol{\omega}_s$, $\boldsymbol{\alpha}$, γ, and τ are all taken as random variables, which will be updated together by sparse measurements.

3.3.3 Posterior Distribution

In accordance with the Bayes' theorem, the posterior PDF of $\boldsymbol{\omega}_s$, $\boldsymbol{\alpha}$, γ, and τ can be obtained by combining the likelihood function in Eq. (3.6) and the joint prior PDF in Eq. (3.12) and expressed as follows (e.g., Zhao and Wang 2020):

$$p(\boldsymbol{\omega}_s,\boldsymbol{\alpha},\gamma,\tau|\mathbf{y}) = \frac{p(\mathbf{y}|\boldsymbol{\omega}_s,\tau)p(\boldsymbol{\omega}_s|\boldsymbol{\alpha})p(\boldsymbol{\alpha}|\gamma)p(\gamma)p(\tau)}{p(\mathbf{y})}$$
$$= \frac{\tau^{M/2}}{(2\pi)^{M/2}}\exp\left(-\frac{\tau(\mathbf{y}-\mathbf{A}_s\boldsymbol{\omega}_s)^{\mathrm{T}}(\mathbf{y}-\mathbf{A}_s\boldsymbol{\omega}_s^{1D})}{2}\right)\frac{[\det(\mathbf{D}_s^{\alpha})]^{1/2}}{(2\pi)^{s/2}}\exp\left(-\frac{(\boldsymbol{\omega}_s)^{\mathrm{T}}\mathbf{D}_s^{\alpha}(\boldsymbol{\omega}_s)}{2}\right)$$
$$\times\prod_{t=1}^{s}\frac{\gamma}{2}\alpha_t^{-2}\exp[-\frac{\gamma}{2}\alpha_t^{-1}]\times\frac{d_0^{c_0}\tau^{c_0-1}}{\Gamma(c_0)}\exp\left(-d_0\tau\right)\times\frac{1}{p(\mathbf{y})} \tag{3.13}$$

Note that there is no analytical expression for the posterior PDF in Eq. (3.13). In addition, only $\boldsymbol{\omega}_s$ is of interest in the formulation. Therefore, the marginal posterior PDF of $\boldsymbol{\omega}_s$ can be obtained theoretically by marginalization, i.e., integrating other parameters (i.e., $\boldsymbol{\alpha}$, γ, and τ) out of Eq. (3.13), $p(\boldsymbol{\omega}_s|\mathbf{y}) = \int p(\boldsymbol{\omega}_s,\boldsymbol{\alpha},\gamma,\tau|\mathbf{y})d\boldsymbol{\alpha}d\gamma d\tau$ (Sivia and Skilling 2006; Gelman et al. 2004). In such a case, the influence of $\boldsymbol{\alpha}$, γ, and τ is explicitly considered. However, such a procedure is often infeasible in practice due to the high-dimensional integration. Consider, for example, the estimated $\boldsymbol{\omega}_s$ in Figure 3.4b. In this case, 17 variables in total should be marginalized, including α_t (t = 1, 2, …, 15), γ, and τ. Note that if each variable is discretized as 100 data points in its domain for integration, the above marginalization involves 10^{17} times of computation, which is computationally expensive and might be out of memory when using a personal computer. To address this issue, several methods have been developed, including variational Bayesian inference (e.g., Bishop and Tipping 2000; Beal 2003; Zhao and Wang 2021), as discussed in Section 3.4.

3.4 VARIATIONAL BAYESIAN INFERENCE

3.4.1 Mathematical Derivations

Variational Bayesian inference (VBI) aims at finding a tractable distribution for properly approximating the true joint posterior PDF of interest (e.g., Beal 2003). In this section, VBI is used to find a joint PDF $q(\boldsymbol{\omega}_s,\boldsymbol{\alpha},\gamma,\tau)$ for approximating the posterior PDF $p(\boldsymbol{\omega}_s, \boldsymbol{\alpha}, \gamma, \tau|\mathbf{y})$. Note that the approximate PDF obtained from the VBI is denoted as "$q(\cdot)$" to make a distinction from the true PDF "$p(\cdot)$". In the VBI, the approximate PDF $q(\boldsymbol{\omega}_s, \boldsymbol{\alpha}, \gamma, \tau)$ is obtained by minimizing a metric of difference, such as Kullback–Leibler divergence (KLD), between $q(\boldsymbol{\omega}_s, \boldsymbol{\alpha}, \gamma, \tau)$ and $p(\boldsymbol{\omega}_s, \boldsymbol{\alpha}, \gamma, \tau|\mathbf{y})$. The KLD between $q(\boldsymbol{\omega}_s, \boldsymbol{\alpha}, \gamma, \tau)$ and $p(\boldsymbol{\omega}_s, \boldsymbol{\alpha}, \gamma, \tau|\mathbf{y})$, denoted as KLD ($q||p$), is expressed as (e.g., Bishop and Tipping 2000; Murphy 2012):

$$\mathrm{KLD}(q \,\|\, p) = -\int q(\boldsymbol{\omega}_s, \boldsymbol{\alpha}, \gamma, \tau) \ln \frac{p(\boldsymbol{\omega}_s, \boldsymbol{\alpha}, \gamma, \tau \mid \mathbf{y})}{q(\boldsymbol{\omega}_s, \boldsymbol{\alpha}, \gamma, \tau)} d\boldsymbol{\omega}_s \, d\boldsymbol{\alpha} \, d\gamma \, d\tau \tag{3.14}$$

Note that KLD($q||p$) is non-negative. The smaller the KLD($q||p$) is, the better $q(\boldsymbol{\omega}_s, \boldsymbol{\alpha}, \gamma, \tau)$ approximates $p(\boldsymbol{\omega}_s, \boldsymbol{\alpha}, \gamma, \tau|\mathbf{y})$. In accordance with the rules of conditional probability, $p(\boldsymbol{\omega}_s, \boldsymbol{\alpha}, \gamma, \tau \mid \mathbf{y}) = p(\boldsymbol{\omega}_s, \boldsymbol{\alpha}, \gamma, \tau, \mathbf{y}) / p(\mathbf{y})$. Substituting this expression into Eq. (3.14), and rearranging the terms leads to

$$\mathrm{KLD}(q \,\|\, p) = -\int q(\boldsymbol{\omega}_s, \boldsymbol{\alpha}, \gamma, \tau)[\ln p(\boldsymbol{\omega}_s, \boldsymbol{\alpha}, \gamma, \tau, \mathbf{y}) - \ln q(\boldsymbol{\omega}_s, \boldsymbol{\alpha}, \gamma, \tau) - \ln p(\mathbf{y})] d\boldsymbol{\omega}_s \, d\boldsymbol{\alpha} \, d\gamma \, d\tau \tag{3.15}$$

Note that $\ln[p(\mathbf{y})]$ is independent of the distribution $q(\boldsymbol{\omega}_s, \boldsymbol{\alpha}, \gamma, \tau)$. Therefore, $\int q(\boldsymbol{\omega}_s, \boldsymbol{\alpha}, \gamma, \tau) \ln p(\mathbf{y}) d\boldsymbol{\omega}_s \, d\boldsymbol{\alpha} \, d\gamma \, d\tau = \ln p(\mathbf{y})$. As a result, Eq. (3.14) is simplified as:

$$\mathrm{KL\,D}(q \,\|\, p) = -\int q(\boldsymbol{\omega}_s, \boldsymbol{\alpha}, \gamma, \tau)[\ln \frac{p(\boldsymbol{\omega}_s, \boldsymbol{\alpha}, \gamma, \tau, \mathbf{y})}{q(\boldsymbol{\omega}_s, \boldsymbol{\alpha}, \gamma, \tau)}] d\boldsymbol{\omega}_s \, d\boldsymbol{\alpha} \, d\gamma \, d\tau + \ln p(\mathbf{y}) \tag{3.16}$$

Let $L(q) = \int q(\boldsymbol{\omega}_s, \boldsymbol{\alpha}, \gamma, \tau) \ln\left(\frac{q(\boldsymbol{\omega}_s, \boldsymbol{\alpha}, \gamma, \tau, \mathbf{y})}{q(\boldsymbol{\omega}_s, \boldsymbol{\alpha}, \gamma, \tau)}\right) d\boldsymbol{\omega}_s \, d\boldsymbol{\alpha} \, d\gamma \, d\tau$. Then, Eq. (3.16) is rewritten as KLD($q||p$) = -$L(q)$+ln[$p(\mathbf{y})$], from which ln[$p(\mathbf{y})$] is obtained as

$$\ln[p(\mathbf{y})] = L(q) + \mathrm{KL\,D}(q \,\|\, p) \tag{3.17}$$

Because $p(\mathbf{y})$ is a constant, ln[$p(\mathbf{y})$] is a constant too. Eq. (3.17) also indicates that minimization of KLD($q||p$) is equivalent to maximization of $L(q)$. Furthermore, KLD($q||p$) is non-negative by definition (see Murphy 2012 for more information), and $L(q)$ therefore represents the lower bound of ln[$p(\mathbf{y})$].

In order to find out $q(\boldsymbol{\omega}_s, \boldsymbol{\alpha}, \gamma, \tau)$ that maximizes $L(q)$ in Eq. (3.17), $q(\boldsymbol{\omega}_s, \boldsymbol{\alpha}, \gamma, \tau)$ may be factorized as (e.g., Bishop 2006):

$$q(\boldsymbol{\omega}_s, \boldsymbol{\alpha}, \gamma, \tau) = q(\boldsymbol{\omega}_s) q(\boldsymbol{\alpha}) q(\gamma) q(\tau) \tag{3.18}$$

where $q(\boldsymbol{\omega}_s)$, $q(\boldsymbol{\alpha})$, $q(\gamma)$, and $q(\tau)$, respectively, represent the tractable PDFs of $\boldsymbol{\omega}_s$, $\boldsymbol{\alpha}$, γ, and τ from the VBI, which approximates, respectively, the true posterior PDF of $\boldsymbol{\omega}_s$, $\boldsymbol{\alpha}$, γ, and τ. Then, in accordance with the factorization in Eq. (3.18) and Bishop and Tipping (2000), the approximate PDF of a random variable of interest (e.g., $q(\boldsymbol{\omega}_s)$) or its logarithm version (e.g., ln $[q(\boldsymbol{\omega}_s)]$) that maximizes $L(q)$ can be expressed as the expectation of the true joint PDF (e.g., ln $p(\boldsymbol{\omega}_s, \boldsymbol{\alpha}, \gamma, \tau, \mathbf{y})$) under the approximate posterior PDF of other random variables (e.g., $q(\boldsymbol{\alpha})$, $q(\gamma)$, $q(\tau)$). Mathematically, $q(\boldsymbol{\omega}_s)$ that maximizes $L(q)$ is expressed as:

$$\ln[q(\boldsymbol{\omega}_s)] = \int q(\boldsymbol{\alpha})q(\gamma)q(\tau)\ln p(\boldsymbol{\omega}_s, \boldsymbol{\alpha}, \gamma, \tau, \mathbf{y})d\boldsymbol{\alpha}d\gamma d\tau + const_1 \tag{3.19}$$

where "$const_1$" in Eq. (3.19) is a constant that ensures integration of $q(\boldsymbol{\omega}_s)$ is one. Based on the rules of conditional probability, ln $p(\boldsymbol{\omega}_s, \boldsymbol{\alpha}, \gamma, \tau, \mathbf{y})$ is expressed as:

$$\ln p(\boldsymbol{\omega}_s, \boldsymbol{\alpha}, \gamma, \tau, \mathbf{y}) = \ln p(\mathbf{y} | \boldsymbol{\omega}_s, \tau) + \ln p(\boldsymbol{\omega}_s | \boldsymbol{\alpha}) + \ln p(\boldsymbol{\alpha} | \gamma) + \ln p(\gamma) + \ln p(\tau) \tag{3.20}$$

Then, substituting Eq. (3.20) into Eq. (3.19), and rearranging the terms, leads to

$$\ln[q(\boldsymbol{\omega}_s)] = \int q(\tau)\ln p(\mathbf{y} | \boldsymbol{\omega}_s, \tau)d\tau + \int q(\boldsymbol{\alpha})\ln p(\boldsymbol{\omega}_s | \boldsymbol{\alpha})d\boldsymbol{\alpha} + const_2 \tag{3.21}$$

where "$const_2$" represents the summation of the term "$const_1$" and other terms that do not involve $\boldsymbol{\omega}_s$. Note that $q(\boldsymbol{\alpha})$ and $q(\gamma)$ are independent of ln $p(\mathbf{y} | \boldsymbol{\omega}_s, \tau)$. Therefore, $\int q(\boldsymbol{\alpha})q(\gamma)q(\tau)\ln p(\mathbf{y} | \boldsymbol{\omega}_s, \tau)d\boldsymbol{\alpha}d\gamma d\tau$ may be simplified as $\int q(\tau)\ln p(\mathbf{y} | \boldsymbol{\omega}_s, \tau)d\tau$. Similarly, $\int q(\boldsymbol{\alpha})q(\gamma)q(\tau)\ln p(\boldsymbol{\omega}_s | \boldsymbol{\alpha})d\boldsymbol{\alpha}d\gamma d\tau$ may be simplified as $\int q(\boldsymbol{\alpha})\ln p(\boldsymbol{\omega}_s | \boldsymbol{\alpha})d\boldsymbol{\alpha}$. Subsequently, substituting Eqs. (3.6) and (3.8) into Eq. (3.21), and rearranging the terms, leads to:

$$\begin{aligned}\ln[q(\boldsymbol{\omega}_s)] &= \int q(\tau)\ln p(\mathbf{y} | \boldsymbol{\omega}_s, \tau)d\tau + \int q(\boldsymbol{\omega})\ln p(\boldsymbol{\omega}_s | \boldsymbol{\alpha})d\boldsymbol{\alpha} + const_2 \\ &= -\frac{E(\tau)(\mathbf{y} - \mathbf{A}_s\boldsymbol{\omega}_s)^{\mathrm{T}}(\mathbf{y} - \mathbf{A}_s\boldsymbol{\omega}_s)}{2} - \frac{(\boldsymbol{\omega}_s)^{\mathrm{T}}E(\mathbf{D}_s^{\alpha})^{\mathrm{T}}\boldsymbol{\omega}_s}{2} + const_3\end{aligned} \tag{3.22}$$

where $E(\tau) = \int \tau q(\tau)d\tau$ represents the posterior mean of τ. Similarly, $E(\mathbf{D}_s^{\alpha})$ represents the posterior mean of α_t in a matrix format, i.e., $E(\mathbf{D}_s^{\alpha}(t,t)) = E(\alpha_t) = \int \alpha_t q(\alpha_t)d\alpha_t$, where $q(\alpha_t)$ represents the tractable distribution that can properly approximate the true posterior PDF of α_t. "$const_3$" in Eq. (3.22) represents the summation of "$const_2$" and new terms that do not involve $\boldsymbol{\omega}_s$. Completing the square for $\boldsymbol{\omega}_s$ in Eq. (3.22), and rearranging the terms, leads to:

$$\ln[q(\boldsymbol{\omega}_s)] = -\frac{(\boldsymbol{\omega}_s)^{\mathrm{T}}[\mathbf{A}_s^{\mathrm{T}}\mathbf{A}_s E(\tau) + E(\mathbf{D}_s^{\alpha})]\boldsymbol{\omega}_s - 2(\boldsymbol{\omega}_s)^{\mathrm{T}}\mathbf{A}_s^{\mathrm{T}}\mathrm{y}E(\tau) + (\mathbf{y})^{\mathrm{T}}\mathbf{y}E(\tau)}{2} + const_3 \tag{3.23}$$

Let $\mathbf{COV}_{\omega_s} = [\mathbf{A}_s^{\mathrm{T}}\mathbf{A}_s E(\tau) + E(\mathbf{D}_s^{\alpha})]^{-1}$. $(\boldsymbol{\omega}_s)^{\mathrm{T}}[\mathbf{A}_s^{\mathrm{T}}\mathbf{A}_s E(\tau) + E(\mathbf{D}_s^{\alpha})]\boldsymbol{\omega}_s$ is then rewritten as $(\boldsymbol{\omega}_s)^{\mathrm{T}}(\mathbf{COV}_{\omega_s})^{-1}\boldsymbol{\omega}_s$. Besides, $(\boldsymbol{\omega}_s)^{\mathrm{T}}\mathbf{A}_s^{\mathrm{T}} yE(\tau)$ is rearranged as $(\boldsymbol{\omega}_s)^{\mathrm{T}}(\mathbf{COV}_{\omega_s})^{-1}\mathbf{COV}_{\omega_s}\mathbf{A}_s^{\mathrm{T}} yE(\tau)$. Subsequently, Eq. (3.23) is rewritten as

$$\ln[q(\boldsymbol{\omega}_s)] = -\frac{(\boldsymbol{\omega}_s)^{\mathrm{T}}(\mathbf{COV}_{\omega_s})^{-1}\boldsymbol{\omega}_s - 2(\boldsymbol{\omega}_s)^{\mathrm{T}}(\mathbf{COV}_{\omega_s})^{-1}\boldsymbol{\mu}_{\omega_s} + (\boldsymbol{\mu}_{\omega_s})^{\mathrm{T}}(\mathbf{COV}_{\omega_s})^{-1}\boldsymbol{\mu}_{\omega_s}}{2} - \frac{(\boldsymbol{\mu}_{\omega_s})^{\mathrm{T}}(\mathbf{COV}_{\omega_s})^{-1}\boldsymbol{\mu}_{\omega_s}}{2} + \frac{(\boldsymbol{y})^{\mathrm{T}} \boldsymbol{y}E(\tau)}{2} + const_3 \quad (3.24)$$

where $\boldsymbol{\mu}_{\omega_s} = \mathbf{COV}_{\omega_s}\mathbf{A}_s yE(\tau)$. Let a term "$const_4$" incorporate "$const_3$" and new terms which do not involve $\boldsymbol{\omega}_s$. Eq. (3.24) may be further simplified as

$$\ln[q(\boldsymbol{\omega}_s)] = -\frac{(\boldsymbol{\omega}_s)^{\mathrm{T}}(\mathbf{COV}_{\omega_s})^{-1}\boldsymbol{\omega}_s - 2(\boldsymbol{\omega}_s)^{\mathrm{T}}(\mathbf{COV}_{\omega_s})^{-1}\boldsymbol{\mu}_{\omega_s} + (\boldsymbol{\mu}_{\omega_s})^{\mathrm{T}}(\mathbf{COV}_{\omega_s})^{-1}\boldsymbol{\mu}_{\omega_s}}{2} + const_4 \quad (3.25)$$

Take the exponential operation for both sides of Eq. (3.25), and rearrange the terms, leading to

$$q(\boldsymbol{\omega}_s) = \frac{1}{\sqrt{(2\pi)^s \det(\mathbf{COV}_{\omega_s})}} \exp\left[-\frac{(\boldsymbol{\omega}_s - \boldsymbol{\mu}_{\omega_s})^{\mathrm{T}}(\mathbf{COV}_{\omega_s})^{-1}(\boldsymbol{\omega}_s - \boldsymbol{\mu}_{\omega_s})}{2}\right] \times \sqrt{(2\pi)^s \det(\mathbf{COV}_{\omega_s})}\exp(const_5) \quad (3.26)$$

A close examination of Eq. (3.26) shows that $\frac{1}{\sqrt{(2\pi)^s \det(\mathbf{COV}_{\omega_s})}} \exp\left[-\frac{(\boldsymbol{\omega}_s - \boldsymbol{\mu}_{\omega_s})^{\mathrm{T}}(\mathbf{COV}_{\omega_s})^{-1}(\boldsymbol{\omega}_s - \boldsymbol{\mu}_{\omega_s})}{2}\right]$ is the multivariate normal distribution. Therefore, its integration with respect to $\boldsymbol{\omega}_s$ is one. In addition, note that $q(\boldsymbol{\omega}_s)$ is a PDF, therefore $\int q(\boldsymbol{\omega}_s) d\boldsymbol{\omega}_s = 1$. As a result, the term "$\sqrt{(2\pi)^s \det(\mathbf{COV}_{\omega_s})}\exp(const_5)$" in Eq. (3.26) is equal to 1. In such cases, Eq. (3.26) may be simplified as

$$q(\boldsymbol{\omega}_s) = \frac{1}{\sqrt{(2\pi)^s \det(\mathbf{COV}_{\omega_s})}} \exp\left[-\frac{(\boldsymbol{\omega}_s - \boldsymbol{\mu}_{\omega_s})^{\mathrm{T}}(\mathbf{COV}_{\omega_s})^{-1}(\boldsymbol{\omega}_s - \boldsymbol{\mu}_{\omega_s})}{2}\right] \quad (3.27)$$

with a mean of $\boldsymbol{\mu}_{\omega_s}$ and covariance matrix $\mathbf{COV}_{\omega_s}$, which are expressed as

$$\begin{aligned}\boldsymbol{\mu}_{\omega_s} &= \mathbf{COV}_{\omega_s}\mathbf{A}_s^{\mathrm{T}}\boldsymbol{y}E(\tau)\\ \mathbf{COV}_{\omega_s} &= [\mathbf{A}_s^{\mathrm{T}}\mathbf{A}_s^{\mathrm{T}}E(\tau)+E(\mathbf{D}_s^{\alpha})]^{-1}\end{aligned} \tag{3.28}$$

where $E(\tau)$ represents the expectation of τ; $E(\mathbf{D}_s^{\alpha})$ represents the expectation of α_t, i.e., $E(\alpha_t)$ (t = 1,2, …, s) in a diagonal matrix. Both $E(\tau)$ and $E(\alpha_t)$ will be derived in the following paragraphs.

Following a derivation procedure similar to that of $q(\boldsymbol{\omega}_s)$ discussed above, tractable PDF of $\boldsymbol{\alpha}$, i.e., $q(\boldsymbol{\alpha})$, is derived as

$$q(\boldsymbol{\alpha}) = \prod_{t=1}^{s}\exp[-\frac{a_t\alpha_t + b_t\alpha_t^{-1}}{2}](\alpha_t)^{p-1}\times\frac{(a_t/b_t)^{p/2}}{2K_p(\sqrt{a_tb_t})} = \prod_{t=1}^{s}q(\alpha_t) \tag{3.29}$$

where $a_t = \mu_{\omega_{s,t}}^2 + \sigma_{\omega_{s,t}}^2$; $\mu_{\omega_{s,t}}$ and $\sigma_{\omega_{s,t}}^2$ represent the tth element of $\boldsymbol{\mu}_{\omega_s}$ and the tth diagonal element of $\mathbf{COV}_{\omega_s}$, respectively; $b_t = E(\gamma)$ is the posterior mean of γ, i.e., $E(\gamma) = \int \gamma q(\gamma)d\gamma$, as shown later in this section; $q(\alpha_t)$ =$\exp[-\frac{a_t\alpha_t + b_t\alpha_t^{-1})}{2}])(\alpha_t)^{p-1}\times\frac{(a_t/b_t)^{p/2}}{2K_p(\sqrt{a_tb_t})}$ is a generalized inverse Gaussian (GIG) distribution (e.g., Zhao et al. 2015; Dumitru 2017). $K_{p+1}(\cdot)$ represents modified Bessel function of the second kind with parameter p = -1/2 in this chapter. With the PDF of α_t, i.e., $q(\alpha_t)$, the expectation of α_t is computed as

$$E(\alpha_t) = \frac{\sqrt{b_t}K_{p+1}(\sqrt{a_tb_t})}{\sqrt{a_t}K_p(\sqrt{a_tb_t})} \tag{3.30}$$

which is used to compute $\mathbf{COV}_{\omega_s}$ in Eq. (3.28).

Following the same procedure, $q(\tau)$ is derived as a Gamma distribution, which is expressed as

$$q(\tau) = \frac{(d_n)^{c_n}}{\Gamma(c_n)}\tau^{c_n-1}\exp(-\tau d_n) \tag{3.31}$$

where c_n = $M/2+c_0$ and $d_n = d_0 + 1/2[(\boldsymbol{y})^{\mathrm{T}}(\boldsymbol{y}) - 2\boldsymbol{\mu}_{\omega_s}^{\mathrm{T}}\mathbf{A}_s^{\mathrm{T}}\boldsymbol{y} + E[(\boldsymbol{\omega}_s)^{\mathrm{T}}\mathbf{A}_s^{\mathrm{T}}\mathbf{A}_s(\boldsymbol{\omega}_s)]$. Note that $E[(\boldsymbol{\omega}_s^{1D})^{\mathrm{T}}\mathbf{A}_s^{\mathrm{T}}\mathbf{A}_s(\boldsymbol{\omega}_s^{1D})]$ may be further derived as $E[(\boldsymbol{\omega}_s)^{\mathrm{T}}\mathbf{A}_s^{\mathrm{T}}\mathbf{A}_s(\boldsymbol{\omega}_s)] = \boldsymbol{\mu}_{\omega_s}^{\mathrm{T}}\mathbf{A}_s^{\mathrm{T}}\mathbf{A}_s\boldsymbol{\mu}_{\omega_s} + T_r(\mathbf{A}_s\mathbf{COV}_{\omega_s}\mathbf{A}_s^{\mathrm{T}})$ based on Petersen and Pedersen (2012), where "$T_r(\cdot)$" is the trace operation, and it represents a sum of all diagonal elements of a matrix. In accordance with the definition of expectation and Eq. (3.31), $E(\tau)$ is derived as

$$E(\tau) = \frac{c_n}{d_n} \tag{3.32}$$

which is used in computation of both the $\boldsymbol{\mu}_{\omega_s}$ and $\mathbf{COV}_{\omega_s}$ in Eq. (3.28).

In addition, $E(\gamma)$ is needed when computing the statistics of α_t, as mentioned previously in Eq. (3.29). Computation of $E(\gamma)$ requires the PDF of γ, i.e., $q(\gamma)$, which is derived to follow a Gamma distribution as follows:

$$q(\gamma) = \frac{(\gamma_b)^{\gamma_a}}{\Gamma(\gamma_a)} \gamma^{(\gamma_a - 1)} \exp(-\gamma\gamma_b) \tag{3.33}$$

where $\gamma_a = s+a_0$ and $\gamma_b = b_0 + \sum_{t=1}^{s} E(\alpha_t^{-1})$. Based on the $q(\gamma)$ in Eq. (3.33), $E(\gamma)$ is derived as

$$E(\gamma) = \gamma_a / \gamma_b \tag{3.34}$$

It is worth noting that $E(\alpha_t^{-1})$ represents the posterior mean of α_t^{-1}, and it cannot be evaluated directly even when $q(\alpha_t)$ is available in Eq. (3.29). This is because $1/\alpha_t$ is a nonlinear function of α_t. To address this problem, the PDF of $1/\alpha_t$, i.e., $q(\alpha_t^{-1})$, is derived as

$$\begin{aligned} q(\alpha_t^{-1}) &= \frac{(a_t / b_t)^{p/2}}{2K_p(\sqrt{a_t b_t})} \exp[-\frac{a_t\alpha_t^{-1} + b_t\alpha_t}{2}])(\alpha_t^{-1})^{p-1} \times \left| \frac{d(\alpha_t^{-1})}{\alpha_t} \right| \\ &= \frac{(a_t / b_t)^{p/2}}{2K_p(\sqrt{a_t b_t})} \exp[-\frac{a_t\alpha_t^{-1} + b_t\alpha_t}{2}])(\alpha_t^{-1})^{-p-1} \end{aligned} \tag{3.35}$$

It is seen from Eq. (3.35) that $(1/\alpha_t)$ follows a GIG with parameter b_t, a_t, and $-p$. Therefore, the mean of $(1/\alpha_t)$, i.e., $E(\alpha_t^{-1})$, is derived as:

$$E(\alpha_t^{-1}) = \frac{\sqrt{a_t} K_{-p+1}(\sqrt{a_t b_t})}{\sqrt{b_t} K_{-p}(\sqrt{a_t b_t})} \tag{3.36}$$

3.4.2 Implementation Procedure of VBI

The derivations and discussions in Section 3.4.1 above show that $\boldsymbol{\mu}_{\omega_s}$ and $\mathbf{COV}_{\omega_s}$ are dependent on $E(\tau)$ and $E(\alpha_t)$ (see Eq. 3.28), which in turn depend on $\boldsymbol{\mu}_{\omega_s}$, $\mathbf{COV}_{\omega_s}$, and $E(\gamma)$ (see Eqs. 3.30, 3.32, and 3.34). As a result, an iteration process is needed among $\boldsymbol{\mu}_{\omega_s}$, $\mathbf{COV}_{\omega_s}$, $E(\tau)$, $E(\alpha_t)$, and $E(\gamma)$ to obtain final results of interest, i.e., $\boldsymbol{\mu}_{\omega_s}$ and $\mathbf{COV}_{\omega_s}$.

The implementation procedure with six steps is summarized next after nontrivial elements in $\boldsymbol{\omega}$ (i.e., $\boldsymbol{\omega}_s$) have been identified by the modified OMP in Section 3.2:

Step 1: Initialize all parameters of interest. For example, set parameters a_0, b_0, and d_0 as $a_0 = b_0 = d_0 = 10^{-4}$. c_0 is taken as $c_0 = 1$. Set initial values for $E(\alpha_t)$, $E(\gamma)$, and $E(\tau)$ as $E(\alpha_t) = 10^{-4}$, $E(\gamma) = 10^{-4}$, and $E(\tau) = 100/\text{var}(\boldsymbol{y})$, respectively. var($\boldsymbol{y}$) represents the variance of sparse measurements. Note that the setting $E(\tau) = 100/\text{var}(\boldsymbol{y})$ is suggested by Tipping (2001) and Ji et al. (2008) to ensure a good starting point during the iteration.

Step 2: Update $\boldsymbol{\mu}_{\boldsymbol{\omega}_s}$ and $\mathbf{COV}_{\boldsymbol{\omega}_s}$ in Eq. (3.28) using $E(\alpha_t)$ and $E(\tau)$. In the first iteration, both $E(\alpha_t)$ and $E(\tau)$ can be taken as 10^{-4}, as defined in Step 1; for subsequent iterations, updated $E(\alpha_t)$ and $E(\tau)$ should be used.

Step 3: Update $E(\alpha_t)$ using Eq. (3.30). Note that $E(\alpha_t)$ is a function of parameters a_t and b_t, which further depend on $\boldsymbol{\mu}_{\boldsymbol{\omega}_s}$ and $\mathbf{COV}_{\boldsymbol{\omega}_s}$ in Step 2. a_t and b_t are computed, respectively, as $a_t = \mu^2_{\omega_{s,t}} + \sigma^2_{\omega_{s,t}}$ and $b_t = E(\gamma)$. In the first iteration, $E(\gamma)$ is taken as 10^{-4}, as initialized in Step 1, and updated value should be used in subsequent iterations.

Step 4: Update $E(\tau)$ using Eq. (3.32), which is a function of parameters c_n and d_n. c_n is computed as a constant, i.e., $c_n = M/2+c_0$, whereas d_n depends on sparse measurements $\boldsymbol{y}$, $\boldsymbol{\mu}_{\boldsymbol{\omega}_s}$ and $\mathbf{COV}_{\boldsymbol{\omega}_s}$ in Step 2. Specifically, d_n is updated as $d_n = d_0 + 1/2[\boldsymbol{y}^{\mathrm{T}}\boldsymbol{y} - 2\boldsymbol{\mu}^{\mathrm{T}}_{\boldsymbol{\omega}_s}\mathbf{A}^{\mathrm{T}}_s\boldsymbol{y} + E(\boldsymbol{\omega}^{\mathrm{T}}_s\mathbf{A}^{\mathrm{T}}_s\mathbf{A}_s\boldsymbol{\omega}_s)]$, where $E[(\boldsymbol{\omega}_s)^{\mathrm{T}}\mathbf{A}^{\mathrm{T}}_s\mathbf{A}_s(\boldsymbol{\omega}_s)] = \boldsymbol{\mu}^{\mathrm{T}}_{\boldsymbol{\omega}_s}\mathbf{A}^{\mathrm{T}}_s\mathbf{A}_s\boldsymbol{\mu}_{\boldsymbol{\omega}_s} + T_r(\mathbf{A}_s\mathbf{COV}_{\boldsymbol{\omega}_s}\mathbf{A}^{\mathrm{T}}_s)$.

Step 5: Update $E(\gamma)$ using Eq. (3. 34), which is a function of γ_a and γ_b. γ_a is a constant, i.e., $\gamma_a = s+a_0$, whereas γ_b is computed as $\gamma_b = b_0 + \sum_{t=1}^{s} E(\alpha_t^{-1})$, where $E(\alpha_t^{-1})$ is expressed in Eq. (3.36). Note that the same a_t and b_t as those in Step 3 should be used in this step.

Step 6: Go back to Step 2 and continue the iteration until a prescribed criterion is satisfied. For example, the relative difference between measurements $\boldsymbol{y}$ and reconstructed measurements $\hat{\boldsymbol{y}} = \mathbf{A}_s\boldsymbol{\mu}_{\boldsymbol{\omega}_s}$, e.g., $r_E = \sum_{k=1}^{M}(y_k - \hat{y}_k)^2 / \sum_{k=1}^{M}(y_k)^2$, does not decrease with further iteration. y_k and $\hat{y}_k$ represent the kth element of $\boldsymbol{y}$ and $\hat{\boldsymbol{y}}$, respectively.

Consider, for example, the sparse measurements in Figure 3.2b. Once the nontrivial elements of $\boldsymbol{\omega}_s$ are identified from the modified OMP (i.e., $\mathbf{A}_s$ is obtained from matrix $\mathbf{A}$), Bayesian estimation of $\boldsymbol{\omega}_s$ can be carried out following the procedure described above. Figures 3.6a to 3.6d, respectively, show the posterior PDFs of four nontrivial elements (i.e., $\omega_{s,1}$, $\omega_{s,2}$, $\omega_{s,3}$, and $\omega_{s,4}$) by solid lines. For comparison, each subplot in Figure 3.6 also includes the underlying true values by a dashed line. It is seen that the posterior PDFs of the four elements all peak strongly at the dashed line, indicating that they are well learned from sparse measurements by the VBI method discussed in this section. Besides, Figure 3.6 shows that $\omega_{s,1}$, $\omega_{s,2}$, $\omega_{s,3}$, and $\omega_{s,4}$ vary

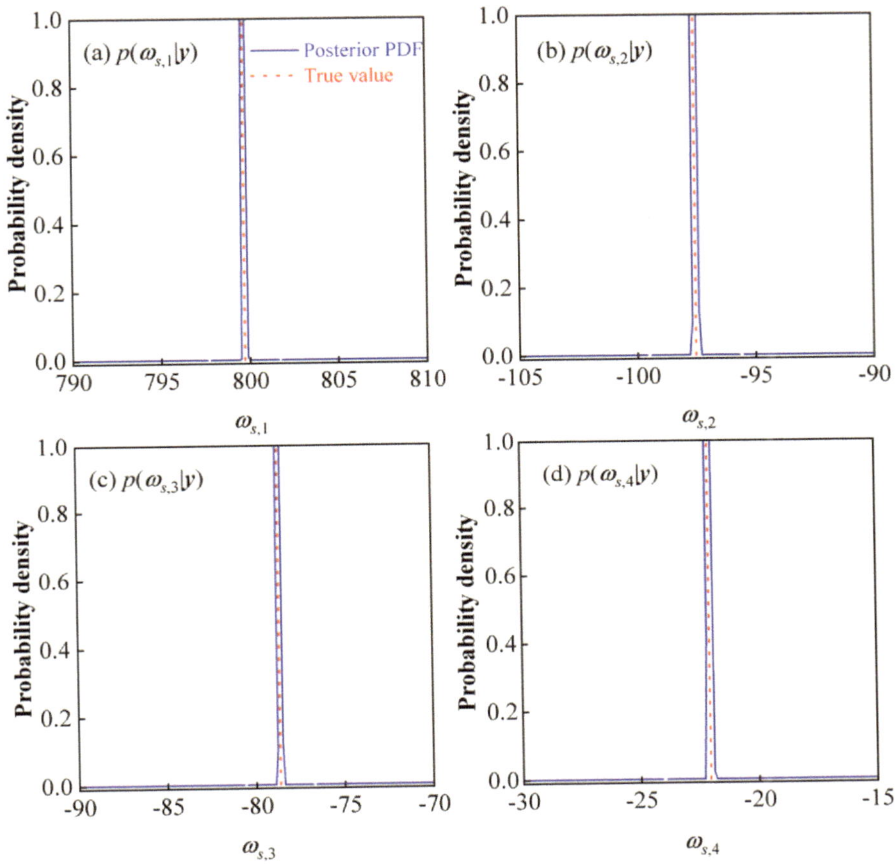

FIGURE 3.6 Posterior PDFs of the first four elements in $\boldsymbol{\omega}_s$: (a) to (d) for $\omega_{s,1}$ to $\omega_{s,4}$, respectively.

with a standard deviation (SD) of around 0.05, implying high reliability associated with these coefficients estimated by sparse measurements.

3.5 ANALYTICS OF 1D SPARSE MEASUREMENTS

3.5.1 Mathematical Derivations

In this section, 1D soil property profile $\boldsymbol{f}$ is predicted from sparse measurements using the OMP-VBI approach, as discussed in Sections 3.3 and 3.4. To differentiate 2D and 3D cases later in this chapter, nontrivial coefficients corresponding to 1D profile are denoted as $\boldsymbol{\omega}_s^{1D}$, with a superscript "1D" to highlight 1D cases. Similarly, sparse measurements in 1D case are denoted as $\boldsymbol{y}^{1D}$. Given $\boldsymbol{y}^{1D}$, nontrivial elements in $\boldsymbol{\omega}_s^{1D}$ can be identified by OMP with a modified matrix $\mathbf{A}_s$, followed by Bayesian estimation of these elements, with mean and covariance derived by Eq. (3.28). Because $\boldsymbol{f}$ is a linear function of $\boldsymbol{\omega}_s^{1D}$, i.e., $\boldsymbol{f} = \mathbf{B}_s\boldsymbol{\omega}_s + \varepsilon\boldsymbol{I}$, uncertainty associated with $\boldsymbol{f}$ can be derived analytically from $\boldsymbol{\omega}_s^{1D}$. $\mathbf{B}_s$ represents the first s columns of $\mathbf{B}$, and only nontrivial elements in $\boldsymbol{\omega}_s^{1D}$ are identified and estimated by the OMP-VBI method discussed in Sections 3.3 and 3.4. The trivial elements in $\boldsymbol{\omega}_s^{1D}$ are taken as zero because of their minimal contribution to the prediction of $\boldsymbol{f}$. Note that $\boldsymbol{\omega}_s^{1D}$ follows a multivariate normal distribution with a mean $\boldsymbol{\mu}_{\omega_s}$ and covariance matrix $\mathbf{COV}_{\omega_s}$ given by Eq. (3.28), and hence the statistics of predicted profile $\boldsymbol{f}$, denoted as $\hat{\boldsymbol{f}}$, is derived analytically as (e.g., Johnson and Wichern 2002):

$$\begin{cases} \boldsymbol{\mu}_{\hat{f}} = E(\mathbf{B}_s\boldsymbol{\omega}_s^{1D} + \varepsilon\boldsymbol{I}) = \mathbf{B}_s\boldsymbol{\mu}_{\omega_s^{1D}} \\ \mathbf{COV}_{\hat{f}} = \mathbf{B}_s\mathbf{COV}_{\omega_s^{1D}}\mathbf{B}_s^{\mathrm{T}} + \sigma_\varepsilon^2\mathbf{I}_N \end{cases} \tag{3.37}$$

where $\mathbf{I}_N$ is the identity matrix with a dimension of $N\times N$; σ_ε^2 is the variance of lumped uncertainty ε, and mean value of σ_ε^2 will be derived later in this section. $\boldsymbol{\mu}_{\hat{f}}$ represents the mean value of spatially varying geotechnical property of interest, whereas diagonal elements of $\mathbf{COV}_{\hat{f}}$ represent the propagated uncertainty from $\boldsymbol{\omega}_s^{1D}$ and ε. Let $\boldsymbol{\sigma}_{\hat{f}}^2 = diag(\mathbf{COV}_{\hat{f}})$ denote a column vector with the same length as $\boldsymbol{\mu}_{\hat{f}}$, and it represents the variances of the reconstructed spatially varying geotechnical property. Because magnitudes of $\boldsymbol{\sigma}_{\hat{f}}^2$ indicate uncertainty associated with the predicted $\boldsymbol{f}$, $\boldsymbol{\mu}_{\hat{f}}$ and $\boldsymbol{\sigma}_{\hat{f}}^2$ may be used to compute a credible interval (CI) to quantify corresponding confidence associated with the predicted $\boldsymbol{f}$. For example, a 95% CI associated with the predicted $\boldsymbol{f}$ can be obtained as by $\boldsymbol{\mu}_{\hat{f}} \pm 1.96\boldsymbol{\sigma}_{\hat{f}}$. It implies that there is around 95% posterior probability that the true spatially varying profile lies within this region defined by its upper and lower bounds (Zhao et al. 2018). Note that the statistics of $\boldsymbol{\omega}_s^{1D}$ in Eq. (3.28) are data-driven, and that no autocorrelation structure is predetermined during the derivation and computation. As a result, the statistics of $\hat{\boldsymbol{f}}$ in Eq. (3.37) are data-driven too, and the selection of autocorrelation structures and parameters estimation are bypassed.

As discussed previously, $\mathbf{COV}_{\hat{f}}$ is a function of σ_ε^2 (see Eq. 3.37), which is considered as a random variable and updated together with $\boldsymbol{\omega}_s^{1D}$. Specifically, the reciprocal of σ_ε^2, i.e., $\sigma_\varepsilon^2 = \tau^{-1}$, is taken as a random variable, and is updated together with $\boldsymbol{\omega}_s^{1D}$, which is derived to follow a Gamma distribution (see Eq. 3.31). As a result, σ_ε^2 follows an inverse Gamma distribution (see Chapter 4 in Ang and Tang 2007).

$$q(\sigma_\varepsilon^2) = \frac{(d_n)^{c_n}}{\Gamma(c_n)}(\sigma_\varepsilon^2)^{-c_n-1}\exp(-d_n / \sigma_\varepsilon^2) \tag{3.38}$$

Based on Eq. (3.38), the mean value of σ_ε^2 is derived as

$$\mu_{\sigma_\varepsilon^2} = \frac{d_n}{c_n - 1} \tag{3.39}$$

During implementation of the method, the mean value of σ_ε^2 may be used to replace σ_ε^2 in Eq. (3.27) when computing the covariance matrix $\mathbf{COV}_{\hat{f}}$.

3.5.2 Interpolation Using 1D Cone Penetration Test (CPT) Dataset

In this section, the OMP-VBI approach discussed in this chapter is illustrated using real-life CPT data, i.e., the tip resistance q_c data. Note that CPT data is usually recorded with a high resolution along depth. As a result, q_c profile is therefore almost continuous, it can be used to validate the profile reconstructed/predicted by the OMP-VBI method when only a limited number of q_c data points are used as input. Note that OMP-VBI is equally applicable to interpret other soil property profiles (e.g., a standard penetration test, SPT, and profile) and quantify their statistical uncertainty, as illustrated later in Section 3.5.3.

In this example, a set of CPT tip resistance q_c data obtained within a sand layer from Christchurch, New Zealand (NZGD 2016) is adopted to illustrate the 1D OMP-VBI approach, as shown in Figure 3.7 by a solid line. This set of q_c data is recorded with an interval of 0.02 m. For illustration, 512 data points of this q_c profile (extending from 4.02 m to 14.24 m, i.e., h = 10.22) are used in this example. Furthermore, to illustrate the OMP-VBI approach, M = 10 measurement data on q_c is taken as $\mathbf{y}^{1D}$, as shown by open circles in Figure 3.7. Note that these ten measured q_c data points versus depth are used as input to interpolate the q_c profile shown by the solid line. The original q_c profile with 512 points is used for comparison and validation only.

Suppose that the resolution of interest is $\eta = 0.02$ m. The number N is first calculated as $N = h/\eta+1= 512$, with which 1D basis function matrix $\mathbf{B}$ is constructed using, e.g., the "*dctmtx*" function in MATLAB™. Then, construct the sensing matrix $\mathbf{\Psi}$ using the depths of measurement data $\mathbf{y}^{1D}$. Determine the maximum number of columns s as s = 1M~1.4M, e.g., s = 1.2M = 12 in this example. $\mathbf{B}_s$ is obtained as the first s=12

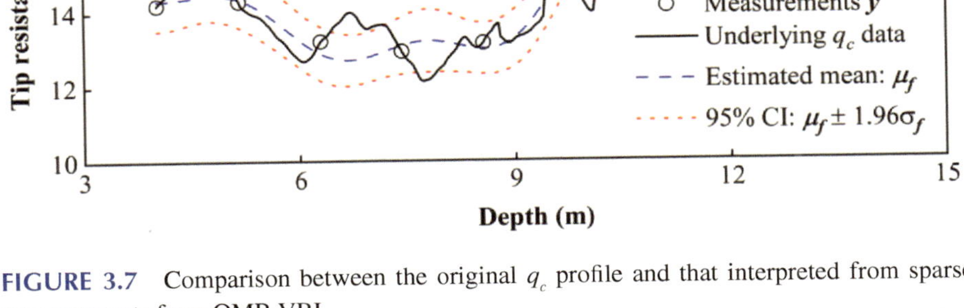

FIGURE 3.7 Comparison between the original q_c profile and that interpreted from sparse measurements from OMP-VBI.

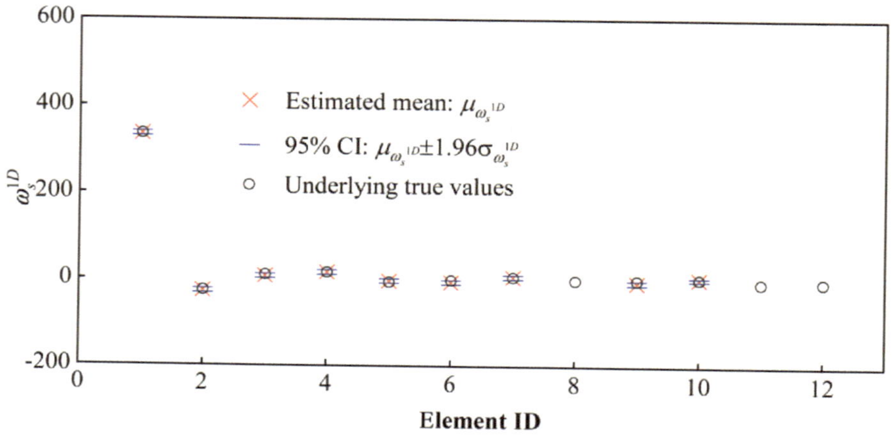

FIGURE 3.8 Statistical results of nontrivial elements in $\boldsymbol{\omega}_s^{1D}$ obtained from OMP-VBI.

columns of **B**, using which $\mathbf{A}_s$ is obtained as $\mathbf{A}_s = \boldsymbol{\Psi} \cdot \mathbf{B}_s$. Using $\mathbf{A}_s$ and $\boldsymbol{y}^{1D}$, nontrivial elements in $\boldsymbol{\omega}_s$ can be identified using the modified OMP method and procedure in Section 3.2. During the identification process, the relative error threshold $\varsigma = 0.01$ is adopted. After these nontrivial elements in $\boldsymbol{\omega}_s$ are identified, VBI described in Section 3.4 is used to estimate the identified $\boldsymbol{\omega}_s$ values probabilistically. Following the procedure in Section 3.4.2, mean and SD of $\boldsymbol{\omega}_s^{1D}$ are obtained (see Eq. 3.28). Figure 3.8 presents the mean and 95% CI of the identified nontrivial elements in $\boldsymbol{\omega}_s^{1D}$ by crosses and bars, respectively. For comparison, Figure 3.8 also shows that the underlying coefficients by open circles. It is seen from Figure 3.8 that (1) all crosses are very consistent with the open circles and (2) open circles fall within the computed 95% CI. These observations indicate that nontrivial elements in $\boldsymbol{\omega}_s^{1D}$ can be well learned statistically from sparse measurements shown in Figure 3.7.

With the identification and probabilistic estimation of nontrivial elements in $\boldsymbol{\omega}_s^{1D}$, the q_c profile with a length of N can be obtained from $\boldsymbol{\mu}_{\hat{f}}$ by Eq. (3.37). The interpolated q_c is shown in Figure 3.7 by a dashed line, whereas the uncertainty is represented by two dotted lines in the same figure. For comparison, Figure 3.7 also includes the true q_c profile from Christchurch by a solid line. It is seen from Figure 3.7 that the dashed line varies in a trend quite similar to the open circles and the solid line, indicating that the q_c profile interpolated from the OMP-VBI method is consistent with the real variation of q_c with depths. Table 3.1 summarizes the statistics of the original and interpreted q_c profiles. The mean and SD of the original q_c profile are calculated as 14.81 MPa and 1.56 MPa, respectively. On the other hand, the mean and SD of the interpreted q_c profile are 14.79 MPa and 1.52 MPa, respectively. The absolute difference between the mean is 0.02 MPa (with a relative difference of about 0.14%), while that of SD is 0.04 MPa (with a relative difference of about 2.56%). The differences are relatively small. This further indicates that the interpolated q_c profile

TABLE 3.1
Comparison of Statistics between the Original and Predicted of q_c Profile in Christchurch

Statistics	Mean (MPa)	SD (MPa)
Original profile	14.81	1.56
Predicted profile from ten data points	14.79	1.52
Absolute difference	0.02	0.04
Relative difference	0.14%	2.56%

(with 512 data points) from a limited number of measurement data points (i.e., ten measurement data points) is reasonable.

Figure 3.7 shows that some local variations of q_c are not reconstructed from $\boldsymbol{y}^{1D}$. This is because the number of measurement data $\boldsymbol{y}^{1D}$ is too limited (e.g., $M = 10 << N = 512$). The interpreted q_c profile contains statistical uncertainty, which is explicitly and objectively quantified by VBI (see $\mathbf{COV}_{\hat{f}}$ in Eq. 3.27). The diagonal components of $\mathbf{COV}_{\hat{f}}$ represent the variances of the corresponding elements of $\hat{\boldsymbol{f}}$. Figure 3.7 also includes the quantified uncertainty associated with $\hat{\boldsymbol{f}}$ by two dotted lines showing the 95% CI. It is seen from Figure 3.7 that although some local variations of q_c are not perfectly captured by OMP-VBI using only ten measurement data points, most local variations of the original q_c profile fall within the 95% CI.

3.5.3 Effect of Number of Measurements

In this section, different numbers of measurement data points, i.e., M = 10, 20, 50, and 100, are used to predict q_c profile and quantify the corresponding uncertainty. Measurement data for these four scenarios are represented by open circles in Figures 3.9a to 3.9d, respectively. For each M scenario (i.e., $M = 10, 20, 50, 512$), the q_c profile is predicted and the uncertainty is quantified by OMP-VBI, following the same procedure as shown in Section 3.5.2. Results are shown in Figures 3.9a to 3.9d, corresponding to $M = 10, 20, 50$, and 100, respectively. In Figure 3.9, the dashed lines represent the interpolated q_c profile, the solid lines represent the original q_c profile, and the regions defined by the dotted lines represent 95% CI. Figure 3.9 shows that as the number of measurement data points increases, the interpolated profiles become increasingly similar to the true q_c profile. When $M \geq 50$, the reconstructed q_c profiles are almost identical to the original one (with N = 512 points). More importantly, Figure 3.9 shows that as the number of measurement data M increases, the width of the 95% CI for the interpolated profile decreases quickly. This suggests that as the number of measurement data M increases, the statistical uncertainty is effectively reduced. To more clearly illustrate this point, SD, i.e., $\sigma_{\hat{f},i}$ (i=1, 2, …, 512), for each data point of the interpolated q_c profile is plotted in Figure 3.10. M = 10, 20, 50, and 100 scenarios are represented by a solid line with open squares, open triangles, plus symbols, and open circles, respectively. The vertical axis in Figure 3.10 represents $\sigma_{\hat{f}}$, varying from 0.0 MPa to around 1.0 MPa, while the horizontal axis represents

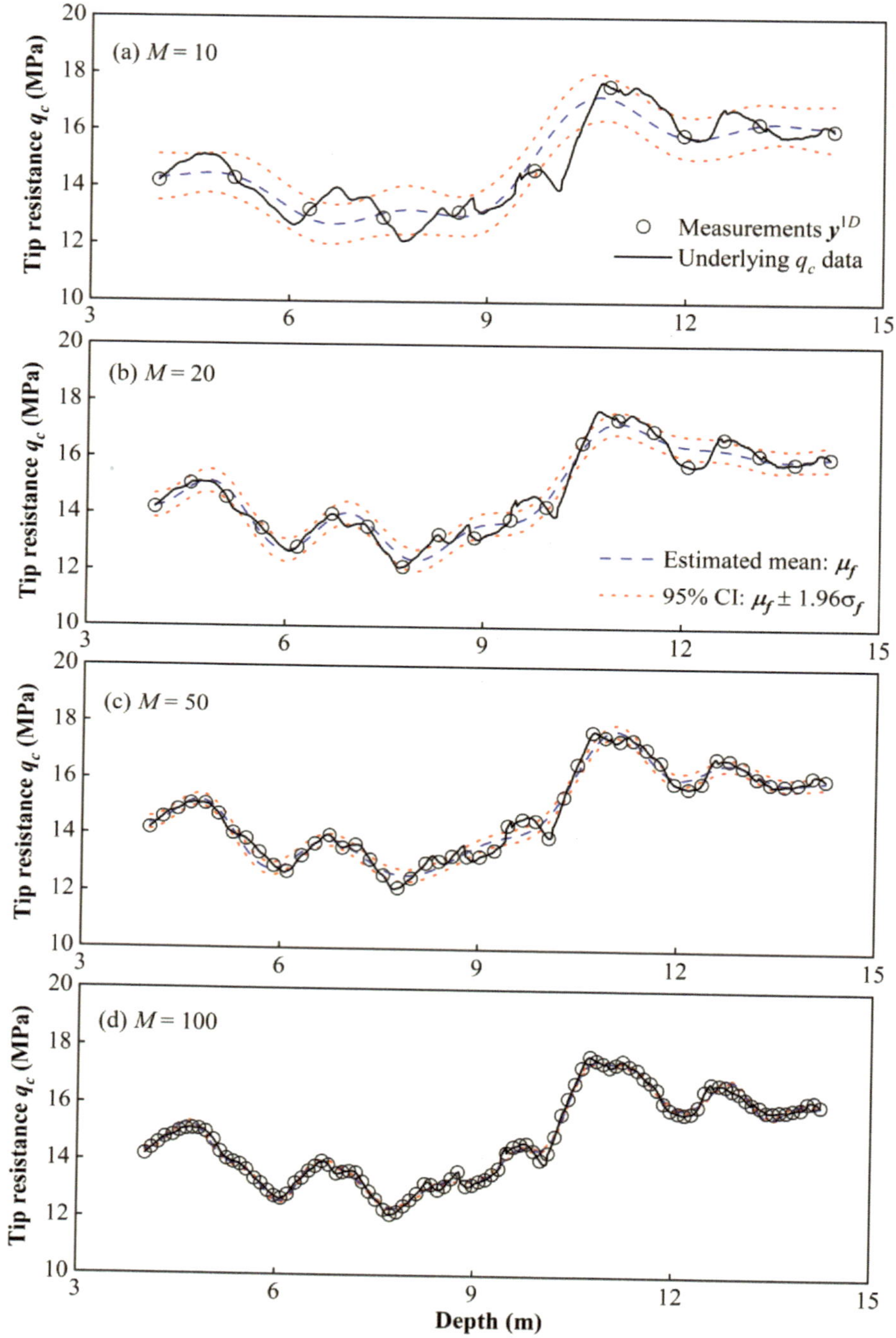

FIGURE 3.9 Effect of number of measurements on the interpreted profile by OMP-VBI: (a) to (d), respectively, for $M = 10$, 20, 50, and 100.

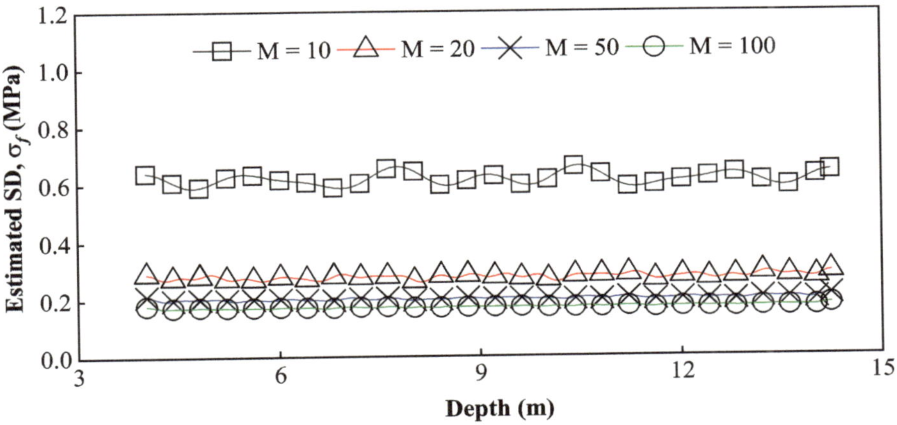

FIGURE 3.10 Reduction of statistical uncertainty with an increase of measurements.

the depths. Figure 3.10 shows that, as the number of measurement data M increases, the statistical uncertainty contained in the interpreted profiles decreases quickly. For the scenario of M = 50, the statistical uncertainty in the interpolated profile is reduced to small values, which reflects a small magnitude of lumped uncertainty of residual and measurement errors associated with q_c. As a result, further increases of M do not significantly reduce the uncertainty (see the M = 100 scenario in Figure 3.10). The quantified statistical uncertainty provides an explicit and objective measure on the accuracy and reliability of the interpreted soil property profile. Moreover, the method allows geotechnical engineers to observe how the statistical uncertainty in the interpreted profile evolves as the number of measurement data increases.

3.5.4 Application Example: Interpretation of Soil Undrained Shear Strength Profile

In this section, the OMP-VBI method described above is applied to interpret a profile of soil undrained shear strength S_u along depth from sparse measurements. In this example, the undrained shear strength S_u data were measured by vane shear tests performed in the Louiseville site, Singapore. This site contains a 60-m-thick deposit of high-plasticity clay, and the upper 14 m of this deposit was characterized in details (Leroueil et al. 2002). Since the upper 1.7 m of the deposit is a weathered crust, S_u data within this thin layer is not considered here. Based on the study by Leroueil et al. (2002), a total number of 26 S_u data with a resolution of around 0.5 m are obtained and shown in Figure 3.11 by open circles. In this example, M = 12 S_u data are taken as input to OMP-VBI with an approximately equal sampling interval, as shown in Figure 3.11 by solid circles. The remaining 26–12 = 14 S_u data are used for testing and validation. Given the M = 12 S_u measurements mentioned previously, the mean S_u profile with a high resolution, say η = 0.1 m, is obtained from OMP-VBI. Furthermore, the uncertainty associated with the predicted S_u profile, i.e., 95% CI in terms $\boldsymbol{\mu}_{\hat{f}} \pm 1.96\boldsymbol{\sigma}_{\hat{f}}$, are obtained too. Figure 3.11 shows the mean S_u profile and corresponding 95% CI profile by a dashed and two dotted lines, respectively. It is

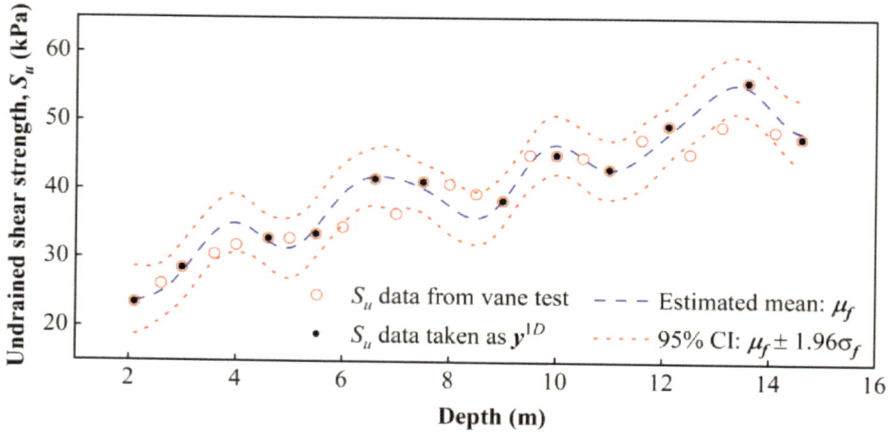

FIGURE 3.11 S_u profile interpreted from sparse measurements by OMP-VBI.

seen from Figure 3.11 that the dashed line follows a trend very consistent with the open circles (i.e., all S_u data from vane test), and that most of the open circles fall within the region defined by the two dotted lines. Besides, it is seen from Figure 3.11 that the dashed line plots very close to the solid circles, implying well-trained OMP-VBI model. For the 14 testing S_u data points, the difference between the predicted and original S_u data from vane tests are computed in terms of absolute relative error, i.e., $|y-\hat{y}|/y\times 100\%$. A small relative error of around 7.8% is obtained. All these observations indicate that OMP-VBI described in this chapter works reasonably well for statistical interpretation of sparse measurements.

3.6 ANALYTICS OF 2D SPARSE MEASUREMENTS

3.6.1 Sparse Representation of 2D Geo-data

In this section, the OMP-VBI method for 1D cases is extended to *2D* sparse measurements in geotechnical site characterization. Similar to the 1D CS, a 2D spatially/temporally varying signal can also be expressed concisely as a weighted summation of a limited number of 2D basis functions. Let an $N_1 \times N_2$ matrix **F** represent a 2D signal, i.e., a 2D profile, with a high resolution. In a context of signal decomposition, **F** is expressed as (e.g., Fang et al. 2012):

$$\mathbf{F} = \mathbf{B}_1^{1D}\boldsymbol{\Omega}(\mathbf{B}_2^{1D})^{\mathrm{T}} = \sum_{i=1}^{N_1}\sum_{j=1}^{N_2}\mathbf{B}_{i,j}^{2D}\omega_{i,j}^{2D} = \sum_{t=1}^{N_1\times N_2}\mathbf{B}_t^{2D}\omega_t^{2D} = \sum_{t=1}^{N}\mathbf{B}_t^{2D}\omega_t^{2D} \tag{3.40}$$

where $\mathbf{B}_t^{2\mathrm{D}}$ is the *t*th 2D basis function and $\omega_t^{2\mathrm{D}}$ is the weight corresponding to $\mathbf{B}_t^{2\mathrm{D}}$. The superscript "2*D*" here is used to highlight the 2D cases and differentiate from the "1D" cases in Section 3.5. Note that $\mathbf{B}_t^{2\mathrm{D}}$ is also a matrix, with the same dimension as **F**.

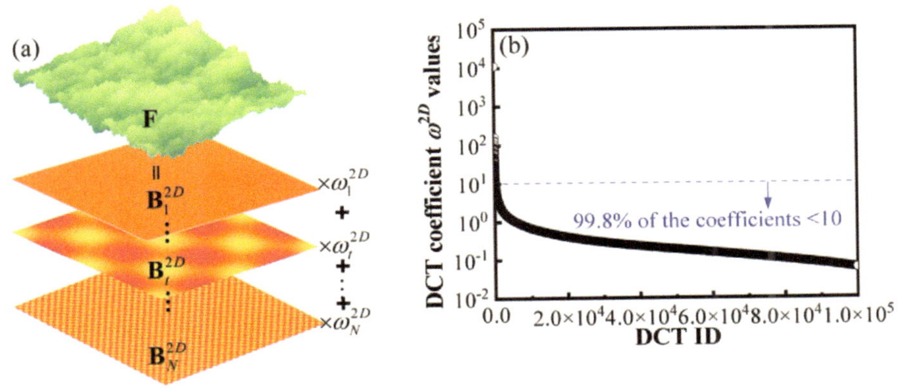

FIGURE 3.12 Decomposition of 2D signal in the context of compressive sampling (CS).

Similar to the 1D spatially or temporally varying signal, most elements of ω_t^{2D} are almost zero under a properly selected 2D basis function system $\mathbf{B}_t^{2D}$ ($t = 1, 2, \ldots, N$), where $N = N_1 \times N_2$ in the 2D case. Figure 3.12a illustrates this idea of signal decomposition, while Figure 3.12b shows the absolute values of ω_t^{2D} for t varying from $t = 1$ to 10^5. It is seen from Figure 3.12b that most elements of ω_t^{2D} are very close to zero and contribute negligibly to the reconstruction of **F**, indicating again that **F** can also be reconstructed reasonably from sparse measurements.

In this study, $\mathbf{B}_t^{2D}$ is constructed from two 1D basis function matrices, i.e., $\mathbf{B}_1$ and $(\mathbf{B}_2)^T$. The matrices $\mathbf{B}_1$ and $\mathbf{B}_2$ have a dimension of $N_1 \times N_1$ and $N_2 \times N_2$, respectively, which can be constructed via 1D DCT with N_1 and N_2, respectively, and the aid of built-in functions in some commercial software, such as "*dctmtx*" in MATLAB (MathWorks 2024) or "*scipy.fft.dct*" in Python. $\mathbf{\Omega}$ is a weight matrix that includes all the ω_t^{2D} ($t = 1, 2, \ldots, N_1 \times N_2$) as its elements. In linear algebra, $\mathbf{B}_1\mathbf{\Omega}(\mathbf{B}_2)^T$ in Eq. (3.40) can be expanded as a weighted summation of matrices with a dimension of $N_1 \times N_2$, i.e., $\sum_{i=1}^{N_1}\sum_{j=1}^{N_2} \mathbf{B}_{i,j}^{2D}\omega_{i,j}^{2D}$. $\mathbf{B}_{i,j}^{2D}$ is the 2D basis function matrix constructed from the product of the ith column of $\mathbf{B}_1$ and the jth row of $(\mathbf{B}_2)^T$. $\omega_{i,j}^{2D}$ represents the element in the ith row and jth column of $\mathbf{\Omega}$, indicating the weight corresponding to $\mathbf{B}_{i,j}^{2D}$. For example, $\mathbf{B}_{1,2}^{2D}$ represents a 2D basis function constructed from the product of the first column of $\mathbf{B}_1$ and the second row of $(\mathbf{B}_2)^T$, and $\omega_{1,2}^{2D}$ is the element at the first row and second column of matrix $\mathbf{\Omega}$, representing the weight corresponding to $\mathbf{B}_{1,2}^{2D}$. For mathematical convenience in derivation, the subscripts i and j of $\mathbf{B}_{i,j}^{2D}$ and $\omega_{i,j}^{2D}$ are replaced by t ($t = 1, 2, \ldots, N_1 \times N_2$), and $\mathbf{B}_t^{2D}$ and ω_t^{2D} ($t = 1, 2, \ldots, N_1 \times N_2$) are used hereafter. Note that the subscript "t" is defined by numbering all elements in $\mathbf{\Omega}$ column by column in an ascending order. Within the same column, "t" increases as the row number increases.

For example, $\omega_{1,2}^{2D}$ is defined as $\omega_{1+N_1}^{2D}$. When using such notations, the framework of the 2D CS is equivalent to that of 1D CS, where $\mathbf{B}_t^{2D}$ and ω_t^{2D} play the same roles as $\mathbf{B}_i^{1D}$ and $\boldsymbol{\omega}_i^{1D}$, respectively.

As most components in $\boldsymbol{\Omega}$ (i.e., ω_t^{2D}, $t = 1, 2, \ldots, N_1 \times N_2$) are almost zero or trivial, the 2D signal **F** of interest can be reconstructed approximately if the nontrivial coefficients in $\boldsymbol{\Omega}$ can be determined from sparse measurement data **Y**. For example, Figure 3.13 shows locations of a set of sparse measurement data **Y** by open circles. Similar to the relationship between $\boldsymbol{y}^{1D}$ and $\boldsymbol{f}$, where $\boldsymbol{y}^{1D}$ is a sub-vector of $\boldsymbol{f}$, **Y** is a sub-matrix of **F**. The relation between **Y** and **F** can be used to estimate the non-trivial coefficients in $\boldsymbol{\Omega}$ ($t = 1, 2, \ldots, N_1 \times N_2$) (e.g., Fang et al. 2012):

$$\mathbf{Y} = \boldsymbol{\Psi}^{2D} \circ \mathbf{F} = \boldsymbol{\Psi}^{2D} \circ \left(\sum_{t=1}^{N} \mathbf{B}_t^{2D} \omega_t^{2D}\right) = \sum_{t=1}^{N} \mathbf{A}_t^{2D} \omega_t^{2D} \tag{3.41}$$

where $\boldsymbol{\Psi}^{2D}$ represents an $N_1 \times N_2$ sensing matrix with "1" at locations having measurements and "0" otherwise. Note that "o" in Eq. (3.41) represents a Hadamard product, or the element-wise product (e.g., Petersen and Pedersen 2012). It takes two matrices of the same dimensions as input and returns a matrix of the multiplied corresponding elements, which is different from the conventional matrix product.

For derivation convenience, the measurements in **Y**, corresponding to locations of 1 in $\boldsymbol{\Psi}^{2D}$, are rewritten as a column vector $\boldsymbol{y}^{2D}$ by stacking measurements from left to right columns. Similarly, elements of $\mathbf{A}_t^{2D}$ at the same locations are rewritten as a column vector $\boldsymbol{A}_j$, forming a new matrix $\mathbf{A}^{2D}$. Note that $\mathbf{A}^{2D}$ has a dimension of $M \times N$, which is different from the dimension of $\mathbf{A}_t^{2D}$. M represents the total number of measurements, while $N = N_1 \times N_2$. In this case, the 2D problem reduces to a 1D problem with $\boldsymbol{y}^{2D} = \mathbf{A}^{2D}\boldsymbol{\omega}^{2D}$. $\boldsymbol{\omega}^{2D}$ represents ω_t^{2D} ($t = 1, 2, \ldots, N$) in a

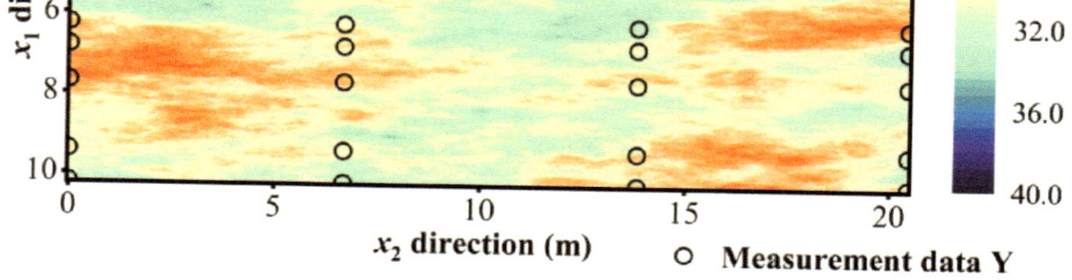

FIGURE 3.13 Spatially varying 2D geo-data and the sparse measurement data **Y**.

column vector format. Because there are only a limited number of measurements, e.g., M = 30~40 in Figure 3.13 for the 2D scenario, it is difficult to identify the nontrivial elements from hundreds and thousands of elements ($N = N_1 \times N_2$). In this case, a modification should be made to $\mathbf{A}^{2D}$ and $\boldsymbol{\omega}^{2D}$ before a reasonable estimation of $\boldsymbol{\omega}^{2D}$ from sparse measurements, similar to the modification made in Section 3.2.

Since the trends are of primary interest in geotechnical site investigation, the high-frequency components may be removed before estimation of $\boldsymbol{\omega}^{2D}$ as follows: Project the available measurements onto both the vertical and horizontal directions, and compute the relative distance between any two most adjacent measurements along these directions. Next, calculate the mean sampling intervals, Δ_v and Δ_h, by averaging the relative distances obtained previously along the x_1 and x_2 directions, respectively. Subsequently, s_1 and s_2 may be approximated as $s_1 \approx l_1/\Delta_v + 1$ and $s_2 \approx l_2/\Delta_h + 1$, where l_1 and l_2 represent the horizontal and vertical dimensions of the site of interest, respectively. This approach ensures that only the relatively low-frequency components of the 2D basis functions are retained for the reconstruction of the spatially varying geotechnical properties. After that, select the first s_1 and s_2 columns of $\mathbf{B}_1$ and $\mathbf{B}_2$, respectively, which are denoted as $\mathbf{B}_{1,s}$ and $\mathbf{B}_{2,s}$. Note that $\mathbf{B}_1$ and $\mathbf{B}_2$ characterize the spatial variation along x_1 and x_2 directions, respectively. $\mathbf{B}_{1,s}$ and $\mathbf{B}_{2,s}$, respectively, correspond to spatially varying trends along x_1 and x_2 directions with relatively low frequencies after the high-frequency components along x_1 and x_2 directions are removed. In this case, matrix $\mathbf{A}_t^{2\mathrm{D}}$ in Eq. (3.41) is constructed with t varying from $t = 1$ to $s_1 s_2$, based on which $\mathbf{A}^{2\mathrm{D}}$ is rewritten as $\mathbf{A}_s^{2D}$. Correspondingly, $\boldsymbol{\omega}^{2D}$ reduces to $\boldsymbol{\omega}_s^{2D}$, with "$s$" as a subscript to highlight that high-frequency components have been removed. $\mathbf{Y} = \sum_{t=1}^{N} \mathbf{A}_t^{2\mathrm{D}} \omega_t^{2\mathrm{D}}$ in Eq. (3.41) hence may be rewritten as $\mathbf{Y} \approx \sum_{t=1}^{s} \mathbf{A}_t^{2D} \omega_t^{2D}$, and $\mathbf{y}^{2D} = \mathbf{A}^{2D}\boldsymbol{\omega}^{2D}$ may be rewritten as $\mathbf{y}^{2D} \approx \mathbf{A}_s^{2D} \boldsymbol{\omega}_s^{2D} + \varepsilon^{2D} \boldsymbol{I}_M$. $\boldsymbol{I}_M$ is a column vector with a length of M; ε^{2D} represents the residuals between vectorized measurements and re-estimated measurements, and it is taken as a random variable as that in Sections 3.3 and 3.4.

3.6.2 Estimation and Reconstruction of 2D Spatial Data

After the manipulation described in Section 3.6.1, CS-based analytics of 2D spatial data are similar to that of 1D data. The nontrivial elements in $\boldsymbol{\omega}_s^{2\mathrm{D}}$ can be readily identified by OMP, followed by probabilistic estimation via VBI. The statistics of $\boldsymbol{\omega}_s^{2\mathrm{D}}$ in terms of mean and covariance matrix are given as follows:

$$\begin{cases} \boldsymbol{\mu}_{\boldsymbol{\omega}_s^{2D}} = \mathbf{COV}_{\boldsymbol{\omega}_s^{2D}} (\mathbf{A}_s^{2D})^{\mathrm{T}} \mathbf{y}^{2D} E(\tau) \\ \mathbf{COV}_{\boldsymbol{\omega}_s^{2D}} = [(\mathbf{A}_s^{2D})^{\mathrm{T}} \mathbf{A}_s^{2D} E(\tau) + E(\mathbf{D}^{\alpha})]^{-1} \end{cases} \tag{3.42}$$

Note that the statistics of $\boldsymbol{\omega}_s^{2D}$ in Eq. (3.42) have exactly the same expression as Eq. (3.28). "2D" in Eq. (3.42) is used to highlight the 2D cases here. It should be noted, however, that construction of matrix $\mathbf{A}_s^{2D}$ is not trivial, as described in detail in Section 3.6.1. With the statistics of $\boldsymbol{\omega}_s^{2D}$ in Eq. (3.42), a 2D geotechnical cross-section can be obtained with a high resolution by utilizing Eq. (3.40). As only a small number of coefficients and corresponding basis functions contribute significantly to the 2D profile, **F** in Eq. (3.40) may be rewritten as

$$\mathbf{F} = \sum_{t=1}^{s_1 s_2} \mathbf{B}_{s,t}^{2D} \omega_{s,t}^{2D} + \varepsilon^{2D} \mathbf{I}_{N_1 \times N_2} \tag{3.43}$$

where 2D basis function $\mathbf{B}_{s,t}^{2D}$ is constructed from products of columns of $\mathbf{B}_{1,s}$ and rows of $\mathbf{B}_{2,s}$, as described in the last paragraph of Section 3.6.1; $\omega_{s,t}^{2D}$ represents the *t*th element of $\boldsymbol{\omega}_s^{2D}$, corresponding to $\mathbf{B}_{s,t}^{2D}$. ε^{2D} represents the residuals between the true 2D profile and the reconstructed one, statistics of which have been derived from sparse measurements $\boldsymbol{y}^{2D}$ as those in 1D cases (see Eq. 3.39). $\mathbf{I}_{N_1 \times N_2}$ is an identity matrix, which shares the same dimension as **F**. Following the definition of mean value, mean value of reconstructed 2D profile is derived as

$$\boldsymbol{\mu}_{\hat{\mathbf{F}}} = E(\hat{\mathbf{F}}) = \sum_{t=1}^{s_1 s_2} \mathbf{B}_{s,t}^{2D} \omega_{s,t}^{2D} = \sum_{t=1}^{s_1 s_2} \mathbf{B}_{s,t}^{2D} \mu_{\omega_{s,t}^{2D}} \tag{3.44}$$

where $\boldsymbol{\mu}_{\omega_{s,t}^{2D}}$ represents the mean of the *t*th element in $\boldsymbol{\omega}_s^{2D}$; $\boldsymbol{\mu}_{\hat{\mathbf{F}}}$ represents the best estimate of the 2D signal **F** from **Y**, and it has the same dimension as **F**, i.e., $N_1 \times N_2$. Eq. (3.44) shows that each element of $\boldsymbol{\mu}_{\hat{\mathbf{F}}}$, i.e., $\mu_{\hat{F}_{i,j}}$ ($i = 1, 2, \ldots, N_1$ and $j = 1, 2, \ldots, N_2$), can be readily derived as $\boldsymbol{\mu}_{\hat{F}_{i,j}} = \sum_{t=1}^{s} \mathrm{B}_{s,t_{i,j}}^{2D} \mu_{\hat{\omega}_{s,t}^{2D}}$, where $\mathrm{B}_{s,t_{i,j}}^{2D}$ represents the element at the *i*th row and the *j*th column in the *t*th basis matrix function $\mathbf{B}_{s,t}^{2D}$, corresponding to the *t*th nontrivial element $\omega_{s,t}^{2D}$. As $\boldsymbol{\mu}_{\omega_s^{2D}}$ is a function of all measurements, $\boldsymbol{\mu}_{\hat{F}_{i,j}} = \sum_{t=1}^{s} \mathrm{B}_{t_{i,j}}^{2D} \boldsymbol{\mu}_{\omega_{s,t}^{2D}}$ means that all sparse measurement data $\boldsymbol{y}^{2D}$ contribute to $\mu_{\hat{F}_{i,j}}$ (e.g., Wang and Zhao 2017; Zhao, Hu, et al. 2018).

The uncertainty associated with $\hat{\mathbf{F}}$ can also be derived from the statistics of $\boldsymbol{\omega}_s^{2D}$, i.e., $\boldsymbol{\mu}_{\omega_s}^{2D}$ and $\mathbf{COV}_{\omega_s}^{2D}$ obtained in a Bayesian framework, in accordance with the definition of variance as below (e.g., Johnson and Wichern 2002; Zhao, Hu, et al. 2018):

$$Var(\hat{\mathbf{F}}) = \sum_{t=1}^{s_1 s_2} (\mathbf{B}_{s,t}^{2D} \circ \mathbf{B}_{s,t}^{2D}) \sigma_{\omega_{s,t}^{2D}}^2 + \tau^{-1} \mathbf{I}_{N_1 \times N_2} \tag{3.45}$$

where "o" in Eq. (3.45) represents the Hadamard product, as introduced in Section 3.6.1; $\sigma_{\omega_{s,t}^{2D}}^2$ is the *t*th diagonal element of $\mathbf{COV}_{\omega_s}^{2D}$, representing the estimated variance for the tth element of $\boldsymbol{\omega}_s^{2D}$; τ^{1} represents the variance of ε^{2D}. In practice, expectation of τ^{1}, i.e., the variance of $\sigma^2_{\varepsilon^{2D}}$ would be used, which has been derived in Eq. (3.39). Note that $Var(\hat{\mathbf{F}})$ in Eq. (3.45) is a matrix that has the same dimension as $\hat{\mathbf{F}}$, and each element of $Var(\hat{\mathbf{F}})$ in Eq. (3.45) represents the estimated variance at the corresponding location. For example, the element in the ith row and jth column of $Var(\hat{\mathbf{F}})$, i.e., $Var(\hat{F}_{i,j})$, represents the variance of the element of $\hat{\mathbf{F}}$ indexed by the ith row and jth column.

A close examination of Eq. (3.45) shows that $Var(\hat{F}_{i,j})$ is a weighted summation of all diagonal elements of $\mathbf{COV}_{\hat{\boldsymbol{\omega}}_s^{2D}}$, which is a function of measurements $\mathbf{y}^{2D}$. In other words, Eq. (3.45) also shows that the variance of each element of the interpreted spatially varying geo-data comes from all sparse measurement data $\mathbf{y}^{2D}$. Given available $\boldsymbol{\mu}_{\hat{\mathbf{F}}}$ and $Var(\hat{\mathbf{F}})$, a 95% CI region can be defined as $\boldsymbol{\mu}_{\hat{\mathbf{F}}} \pm 1.96\sqrt{Var(\hat{\mathbf{F}})}$. The 95% CI region implies that there is around a 95% posterior probability that the true spatially varying 2D profile lies within the region defined by its upper and lower bounds. This is similar to that in 1D cases.

Consider, for example, the sparse measurements shown by open circles in Figure 3.13. Following the method discussed at the last paragraph of Section 3.6.1, Δ_v and Δ_h are approximated as $\Delta_v = 0.9$ m and $\Delta_h = 6.6$ m. Then, s_1 and s_2 are approximated as $s_1 \approx 10.2/\Delta_v + 1 = 12$, and $s_2 \approx 20.4/\Delta_h + 1 = 4$, respectively. After that, the first $s_1 = 4$ columns of $\mathbf{B}_1$ are written as $\mathbf{B}_{1,s}$; whereas the first $s_2 = 10$ columns of $\mathbf{B}_2$ are denoted as $\mathbf{B}_{2,s}$. With $\mathbf{B}_{1,s}$ and $\mathbf{B}_{2,s}$, and measurement matrix $\boldsymbol{\Psi}^{2D}$, the matrix $\mathbf{A}_s^{2D}$ is obtained, and measurement data shown by open circles in Figure 3.13 is rewritten as a column vector $\mathbf{y}^{2D}$. Following the descriptions above in this section, statistics of $\boldsymbol{\omega}_s^{2D}$ are obtained using Eq. (3.42). With $\boldsymbol{\mu}_{\boldsymbol{\omega}_s}^{2D}$ and $\mathbf{COV}_{\boldsymbol{\omega}_s}^{2D}$, spatially varying 2D geo-data can be obtained, with mean and spatial distribution of SD shown in Figure 3.14b and 3.14c, respectively. For comparison, Figure 3.14a also shows the true $2D$ profile $\mathbf{F}$. It is seen that the spatially varying 2D profile in Figure 3.14b is consistent with that shown in Figure 3.14a, although some local variations of the true 2D profile in Figure 3.14a are not captured (see the absolute residuals shown in Figure 3.14d). This is because the number of measurement data $\mathbf{Y}$ used for interpretation is limited, and the interpreted results contain significant statistical uncertainty. The statistical uncertainty associated with $\mathbf{F}$ resulted from the sparse measurement data is explicitly quantified by Eq. (3.45) and expressed as the estimated variance of $\mathbf{F}$ at location (i, j). Then, taking square root of Eq. (3.45) for each data point leads to an SD surface, as shown in Figure 3.14c. Figure 3.14c shows that estimated SDs are relatively small at locations with measurements (e.g., see a comparison between Figures 3.14a and 3.14c). This is reasonable because information at these locations is well learned by OMP-VBI. In contrast, the estimated SDs at location without measurements are relatively large, indicating the data-driven characteristics of the OMP-VBI method.

3.6.3 Application Example: Interpretation of 2D Profile from Sparse Measurements

In this section, the OMP-VBI method is applied to a set of real-life 2D data, which characterizes electrical conductivity of topsoil from an experimental field in Germany.

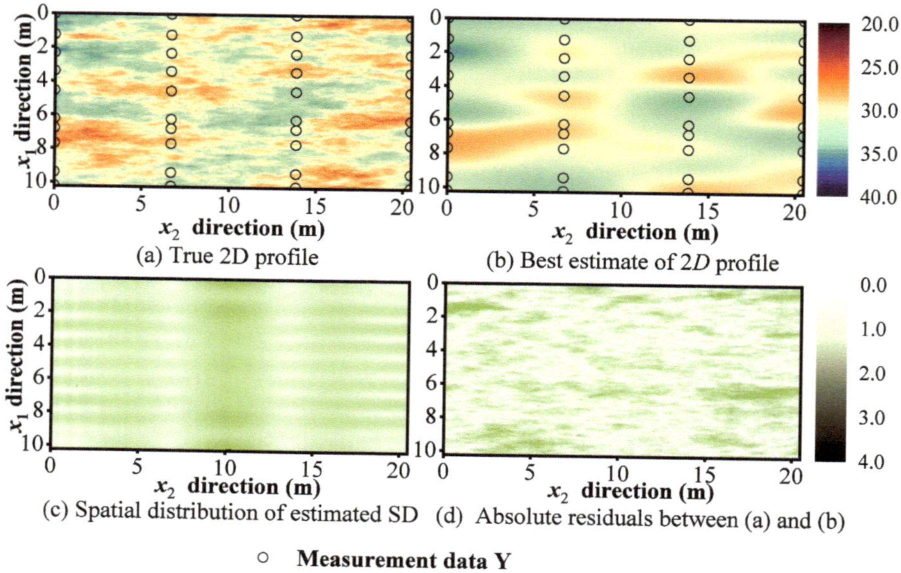

FIGURE 3.14 Comparison between the true 2D profile and that estimated from sparse measurements: (a) true 2D profile; (b) best estimate of spatially varying 2D profile; (c) spatial distribution of estimated SD; and (d) absolute residuals between (a) and (b).

The electrical conductivity data with a unit of milli Siemens per meter (i.e., $mS \cdot m^{-1}$) are derived from EM-38 electrical conductivity meters, and they were downloaded from European Soil Data Centre (ESDAC 2017). In this example, a horizontal plane with an area of 12.6 m × 15.8 m is selected for illustration and validation. The easting coordinate ranges from 700,004.8 m to 700,020.6 m, while the northing coordinate ranges from 569,7244.2 m to 569,7256.8 m. Note that the original data set used in this example contains 64×80 = 5,120 data points with a resolution of 0.2 m over each direction, as shown in Figure 3.15a by a color map. For clear presentation, the coordinates in Figure 3.15 are reset to start with "0". In Figure 3.15, "0" in easting direction means 700,004.8 m, while "0" in the northing direction means 569,7244.2 m. Only $M = 30$ scattered data points are extracted from the original data, as shown in Figure 3.15 by open circles, and used as input for the 2D OMP-VBI approach. The remaining 5,120–30 = 5,090 data points are used for comparison and validation.

Following the implementation procedure illustrated in Sections 3.6.1 and 3.6.2, the spatially varying 2D profile of the electricity conductivity of soil over easting and northing coordinates is interpreted, using the $M = 30$ data points and their coordinates as input. The spatial resolutions of interest over both easting and northing directions are taken as 0.2 m (i.e., $\eta_1 = \eta_2 = 0.2$ m). The best estimate of the spatially varying 2D profile of the electricity conductivity of soil is shown in Figure 3.15b. Figure 3.15b also includes the measurement data **Y** by circles. It is seen from this figure that the best estimate of the 2D profile of the electrical conductivity of soil from OMP-VBI method is in good agreement with the sparse measurement data **Y** and the original

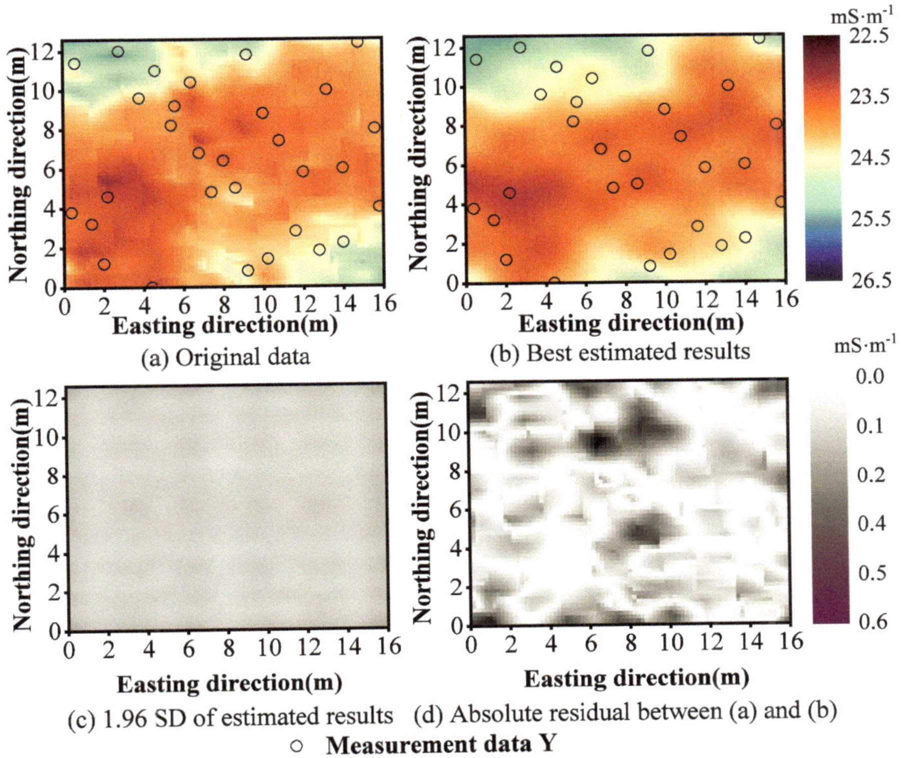

FIGURE 3.15 Comparison between the original data of soil electrical conductivity and results estimated from 30 measurement data points using 2D BCS: (a) original data; (b) best estimate results; (c) 1.96 SD of estimated results; and (d) absolute residual between (a) and (b).

electrical conductivity data shown in Figure 3.15a. The electrical conductivity of soil in both Figures 3.15a and 3.15b has a tendency to decrease first, from the upper left corner to the lower right corner of the figures, followed by an increase. For a quantitative evaluation and comparison, the statistics (e.g., the mean and SD) of the original and best estimate of *2D* electrical conductivity profile interpreted from **Y** are summarized in Table 3.2. The mean and SD of the original *2D* profile of electrical conductivity are 24.02 mS·m^{-1} and 0.43 (mS·m^{-1}), respectively. In contrast, the corresponding mean and SD of the best estimate of 2D profile of electrical conductivity data are 24.05 mS·m^{-1} and 0.47 mS·m^{-1}. Note that there is only 0.03 mS·m^{-1} absolute difference between the mean (with a relative difference of 0.12%) and 0.04 mS·m^{-1} absolute difference between the SD (with a relative difference of 6.98%). The small differences between the original and best estimate of 2D profile of soil electrical conductivity once again demonstrate the effectiveness of the OMP-VBI method. In addition, the OMP-VBI method provides explicit quantification of statistical uncertainty in the results interpreted from sparse measurement data. The quantified uncertainty in terms of 1.96 SD surface is shown in Figure 3.15c, whereas Figure 3.15d

TABLE 3.2
Summary of Mean and Standard Deviation (SD) of 2D Profile of Soil Electrical Conductivity

Statistics	Mean ($mS \cdot m^{-1}$)	SD ($mS \cdot m^{-1}$)
Original 2D profile with 64×80 points	24.02	0.43
Best estimate of 2D profile from 2D BCS	24.05	0.47
Absolute difference	0.03	0.03
Relative difference	0.12%	6.98%

presents the absolute residuals between original 2D profile and the predicted result from OMP-VBI. It is seen from Figure 3.15c that the uncertainty is relatively small at almost all locations, indicating that the predicted 2D profile in Figure 3.15b is reasonable. This is further demonstrated in Figure 3.15d, where absolute residuals between the original 2D profile and that predicted from sparse data by the OMP-VBI method are relatively small at almost all locations. These observations once again demonstrate that the OMP-VBI approach performs reasonably well in interpreting 2D geotechnical property profile from sparse measurements and properly quantifies the statistical uncertainty associated with the interpretation.

3.7 ANALYTICS OF 3D SPARSE MEASUREMENTS

3.7.1 Sparse Representation of 3D Geo-data

In this section, the OMP-VBI approach is extended to deal with sparse measurements on 3D geo-data. Similar to the 1D and 2D cases, a spatially varying 3D profile for a quantity $\underline{\mathbf{F}}$ of interest is expressed as weighted summation of 3D basis functions, i.e.,

$$\underline{\mathbf{F}} = \sum_{t=1}^{N} \omega_t^{3D} \underline{\mathbf{B}}_t^{3D} \tag{3.46}$$

where $\underline{\mathbf{F}}$ has a dimension of $N_1 \times N_2 \times N_3$. $N = N_1 \times N_2 \times N_3$. $\underline{\mathbf{B}}_t^{3D}$ is the tth 3D basis function, while ω_t^{3D} is the weight corresponding to $\underline{\mathbf{B}}_t^{3D}$. $\underline{\mathbf{B}}_t^{3D}$ has the same dimension as $\underline{\mathbf{F}}$. Similar to the 1D/2D cases in the previous sections, most elements of ω_t^{3D} are almost zero under a properly selected 3D basis function system $\underline{\mathbf{B}}_t^{3D}$ ($t = 1, 2, \ldots, N$), e.g., 3D DCT basis functions.

Consider, for example, a 3D dataset of Q shown in Figure 3.16a. The 3D dataset may be decomposed using Eq. (3.46), where $\underline{\mathbf{B}}_t^{3D}$ is constructed from discrete cosine functions. Then, the absolute values of all elements of $\boldsymbol{\omega}^{3D}$ are shown in Figure 3.16b in a descending order of their absolute values. Figure 3.16b shows that almost all elements are close to zero except the first few dozens. Figure 3.16 indicates again that spatially varying 3D profile of Q can be well reconstructed if the nontrivial elements of $\boldsymbol{\omega}^{3D}$ can be properly determined directly from sparse measurements.

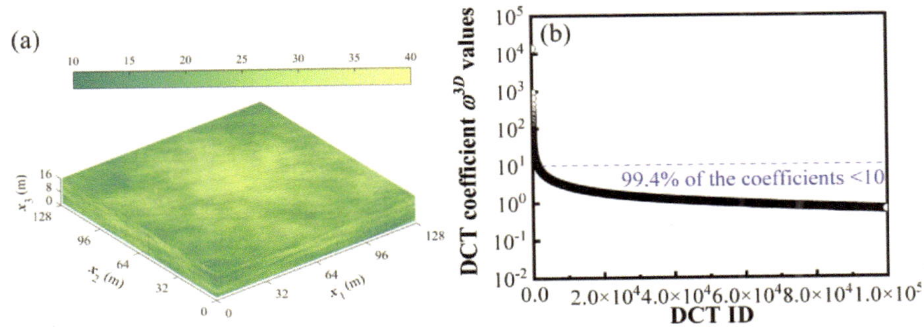

FIGURE 3.16 An example of 3D dataset and DCT coefficients in a descending order.

As shown in Eq. (3.46) and the observation above, the 3D basis function $\underline{\mathbf{B}}_t^{3D}$ is a key element of the method, which is constructed from three 1D orthonormal matrices, e.g., discrete cosine basis function matrix, $\mathbf{B}^1$, $\mathbf{B}^2$, and $\mathbf{B}^3$ along direction x_1, x_2, and x_3, respectively. $\mathbf{B}_1$, $\mathbf{B}_2$, and $\mathbf{B}_3$ have a dimension of $N_1 \times N_1$, $N_2 \times N_2$, and $N_3 \times N_3$, respectively. In this study, columns of $\mathbf{B}_1$, $\mathbf{B}_2$, and $\mathbf{B}_3$ represent orthonormal discrete cosine functions, and they can be computed column by column from the formula for discrete cosine functions (e.g., Caiafa and Cichocki 2013). Alternatively, they may be constructed readily using built-in function in some commercial software, such as "dctmtx" in MATLAB (MathWorks 2024) or "*scipy.fft.dct*" in Python. Note that only N_1, N_2, and N_3 are used as input to generate transpose of $\mathbf{B}_1$, $\mathbf{B}_2$, and $\mathbf{B}_3$ as output. Then, a 3D basis function can be constructed using columns of $\mathbf{B}_1$, $\mathbf{B}_2$, and $\mathbf{B}_3$. Specifically, a 3D basis function is constructed as $\underline{\mathbf{B}}_{i,j,k}^{3D} = \mathrm{b}_i^1 \otimes \mathrm{b}_j^2 \otimes \mathrm{b}_k^3$ ($i = 1, 2, \ldots, N_1$; $j = 1, 2, \ldots, N_2$; $k = 1, 2, \ldots, N_3$). $\boldsymbol{b}_i^1$, $\boldsymbol{b}_j^2$, and $\boldsymbol{b}_k^3$ are the ith, jth, and kth columns of $\mathbf{B}^1$, $\mathbf{B}^2$, and $\mathbf{B}^3$, respectively; "⊗" represents an outer product, and each element of $\underline{\mathbf{B}}_{i,j,k}^{3D}$ is defined as $\underline{\mathbf{B}}_{i,j,k}^{3D}(m,n,l) = b_i^1(m) b_j^2(n) b_k^3(l)$ (e.g., Zhao et al. 2018). Note that, from hereon, the subscript "(i, j, k)" of the 3D basis function is changed to t for derivation convenience and conciseness. "t" is numbered in an increasing order of i first, followed by j and k, respectively. This is similar to that in 2s cases.

As most components of ω_t^{3D} ($t = 1, 2, \ldots, N$) are almost zero or trivial, the 3D profile $\underline{\mathbf{F}}$ of interest can be reconstructed approximately if the nontrivial coefficients of ω_t^{3D} can be determined from sparse measurement data. Similar to the relationship between $\boldsymbol{y}$ and $\boldsymbol{f}$ for 1D or 2D cases, where $\boldsymbol{y}$ or $\mathbf{Y}$ is a subvector of $\boldsymbol{f}$ or $\mathbf{F}$; $\underline{\mathbf{Y}}$ is a sub-matrix of $\underline{\mathbf{F}}$. The relation between $\underline{\mathbf{Y}}$ and $\underline{\mathbf{F}}$ can be used to estimate the nontrivial coefficients in ω_t^{2D} ($t = 1, 2, \ldots, N_1 \times N_2$) (e.g., Caiafa and Cichocki 2013):

$$\underline{\mathbf{Y}}^{3D} = \underline{\mathbf{\Psi}}^{3D} \circ \underline{\mathbf{F}} = \underline{\mathbf{\Psi}}^{3D} \circ \left(\sum_{t=1}^{N} \underline{\mathbf{B}}_t^{3D} \cdot \omega_t^{3D}\right) = \sum_{t=1}^{N} \underline{\mathbf{A}}_t^{3D} \omega_t^{3D} \tag{3.47}$$

where $\underline{\mathbf{\Psi}}^{3D}$ represents an $N_1 \times N_2 \times N_3$ sensing matrix with the value of "1" at locations having measurements and the value of "0" otherwise. $\underline{\mathbf{A}}_t^{3D} = \underline{\mathbf{\Psi}}^{3D} \circ \underline{\mathbf{B}}_t^{3D}$.

For derivation convenience, the measurements in $\underline{\mathbf{Y}}$, corresponding to entries with the value of "1" in $\underline{\mathbf{\Psi}}^{3D}$, is rewritten as an M-length column vector $\boldsymbol{y}^{3D}$ by stacking all available measurements. M represents the number of measurements. Similarly, elements of $\underline{\mathbf{A}}_t^{3D}$ at the same locations are re-written as a column vector $\boldsymbol{A}_j$, forming a new matrix $\mathbf{A}^{3D}$. In this case, the 3D problem reduces to a 1D problem with $\boldsymbol{y}^{3D} = \mathbf{A}^{3D}\boldsymbol{\omega}^{3D}$. Note that only the elements of $\boldsymbol{\omega}^{3D}$ corresponding to relatively low frequencies, denoted as $\boldsymbol{\omega}_s^{3D}$, are considered and estimated by the OMP-VBI approach. The potential basis functions corresponding to $\boldsymbol{\omega}_s^{3D}$ can be determined in a manner similar to that discussed in detail in the last paragraph of Section 3.6.1. Specifically, project the measurements onto the horizontal plane (i.e., the x_1-x_2 plane) and the vertical direction (i.e., the x_3 axis). Compute the distances between each measurement location and its two nearest adjacent measurement locations within the horizontal plane. Additionally, compute the distances between each measurement depth and its nearest adjacent measurement depth along the vertical axis. Next, determine a representative sampling interval within each direction, denoted as Δ_h for the horizontal plane and Δ_v for the vertical axis. These intervals may be obtained by calculating the mean value of the previously computed distances within both the horizontal plane and along the vertical direction. Then, s_1 and s_2 may approximated as $s_1 \approx l_1 / \Delta_h + 1$ and $s_2 \approx l_2 / \Delta_h + 1$, whereas s_3 is approximated as $s_3 \approx l_3 / \Delta_v + 1$. l_1 and l_2 are two horizontal dimensions of the site of interest along the x_1 and x_2 directions, whereas l_3 represents the depth of site to be investigated. After that, the first s_1, s_2, and s_3 columns of $\mathbf{B}_1$, $\mathbf{B}_2$, and $\mathbf{B}_3$, are selected for computation, which are denoted as $\mathbf{B}_{1,s}$, $\mathbf{B}_{2,s}$, and $\mathbf{B}_{3,s}$, respectively. Note that $\mathbf{B}_{1,s}$, $\mathbf{B}_{2,s}$, and $\mathbf{B}_{3,s}$, respectively, correspond to spatially varying trends along x_1, x_2, and x_3 directions with relatively low frequencies, whereas the high-frequency varying components are removed. In this case, matrix $\underline{\mathbf{A}}_t^{3D}$ in Eq. (3.47) is constructed for t varying from $t = 1$ to $s_1s_2s_3$ and $\mathbf{A}^{3D}$ is written as $\mathbf{A}_s^{3D}$. The 3D basis function $\underline{\mathbf{B}}_t^{3D}$ in Eq. (3.46) is rewritten as $\underline{\mathbf{B}}_{s,t}^{3D}$ for t varying from 1 to $s_1s_2s_3$. Furthermore, Eq. (3.47) may be rewritten as $\underline{\mathbf{Y}} \approx \sum_{t=1}^{s_1s_2s_3} \underline{\mathbf{A}}_t^{3D}\omega_{s,t}^{3D}$; and $\boldsymbol{y}^{3D} = \mathbf{A}^{3D}\boldsymbol{\omega}^{3D}$ mentioned previously may be rewritten as $\boldsymbol{y}^{3D} = \mathbf{A}_s^{3D}\boldsymbol{\omega}_s^{3D} + \varepsilon^{3D}\boldsymbol{I}_M$. ε^{3D} reflects a lumped uncertainty of measurement errors and residuals between measurements $\boldsymbol{y}^{3D}$ and reestimated one via $\mathbf{A}_s^{3D}\boldsymbol{\omega}_s^{3D}$. ε^{3D} is taken as a Gaussian random variable with a zero mean and unknown variance. $\boldsymbol{I}_M$ has the identical number of elements to $\boldsymbol{y}^{3D}$, and its elements all have a value of "1".

3.7.2 Estimation and Reconstruction of 3D Spatial Data

After the manipulation in the previous section, the 3D data may be analyzed in a way similar to that of 1D data. In this case, nontrivial elements in $\boldsymbol{\omega}_s^{3D}$ can be identified and estimated probabilistically by the OMP-VBI method as discussed in Sections 3.2–3.4. The mean and covariance matrix of $\boldsymbol{\omega}_s^{3D}$ are obtained as follows:

$$\begin{cases} \boldsymbol{\mu}_{\boldsymbol{\omega}_s^{3D}} = \mathbf{COV}_{\boldsymbol{\omega}_s^{3D}} (\mathbf{A}_s^{3D})^{\mathrm{T}} \boldsymbol{y}^{3D} E(\tau) \\ \mathbf{COV}_{\boldsymbol{\omega}_s^{3D}} = [(\mathbf{A}_s^{3D})^{\mathrm{T}} \mathbf{A}_s^{3D} E(\tau) + E(\mathbf{D}^{\alpha})]^{-1} \end{cases} \tag{3.48}$$

With the statistics of $\boldsymbol{\omega}_s^{3\mathrm{D}}$ from Eq. (3.48), the 3D geotechnical profile can be obtained with a high resolution. Specifically, the 3D profile $\underline{\mathbf{F}}$ of interest can be approximately using $\boldsymbol{\omega}_s^{3D}$ and corresponding 3D basis functions, as follows:

$$\underline{\mathbf{F}} = \sum_{t=1}^{s_1 s_2 s_3} \underline{\mathbf{B}}_{s,t}^{3D} \omega_{s,t}^{3D} + \varepsilon^{3D} \underline{\mathbf{I}} \tag{3.49}$$

where the weight $\omega_{s,t}^{3D}$ corresponds to $\underline{\mathbf{B}}_{s,t}^{3D}$; $\underline{\mathbf{I}}$ is a 3D identity matrix with the same dimension as $\underline{\mathbf{F}}$. Subsequently, the 3D profile with a high resolution can be predicted in a statistical manner, with its mean expressed as follows:

$$\underline{\boldsymbol{\mu}}_{\hat{\underline{\mathbf{F}}}} = E(\hat{\underline{\mathbf{F}}}) = \sum_{t=1}^{s_1 s_2 s_3} \underline{\mathbf{B}}_{s,t}^{3D} E(\omega_{s,t}^{3D}) = \sum_{t=1}^{s_1 s_2 s_3} \underline{\mathbf{B}}_{s,t}^{3D} \mu_{\omega_{s,t}^{3D}} \tag{3.50}$$

where $\mu_{\omega_{s,t}^{3D}}$ represents the mean of the tth element in $\boldsymbol{\omega}_s^{3D}$; $\underline{\boldsymbol{\mu}}_{\hat{\underline{\mathbf{F}}}}$ represents the best estimate of the 3D profile estimated from $\boldsymbol{y}^{3D}$, and it has the same dimension as $\underline{\mathbf{F}}$, i.e., $N_1 \times N_2 \times N_3$. A combination of Eqs. (3.50) and (3.48) shows that each element of $\underline{\boldsymbol{\mu}}_{\hat{\underline{\mathbf{F}}}}$, i.e., $\underline{\mu}_{\hat{\underline{F}}_{i,j,k}}$ ($i = 1, 2, \ldots, N_1, j = 1, 2, \ldots, N_2$, and $k = 1, 2, \ldots, N_3$), can be readily derived as $\underline{\mu}_{\hat{\underline{F}}_{i,j,k}} = \sum_{t=1}^{s_1 s_2 s_3} \underline{\mathrm{B}}_{s,t_{i,j,k}}^{3D} \mu_{\omega_{s,t}^{3D}}$, where $\underline{\mathrm{B}}_{s,t_{i,j,k}}^{3D}$ represents the element indexed by (i, j, k) in the tth basis function $\underline{\mathbf{B}}_{s,t}^{3\mathrm{D}}$. In other words, each element in reconstructed 3D profile is a weighted average of all nontrivial elements $\boldsymbol{\mu}_{\boldsymbol{\omega}_s^{3D}}$. Furthermore, note that $\boldsymbol{\mu}_{\boldsymbol{\omega}_s^{3D}}$ (i.e., $\mu_{\omega_{s,t}^{2\mathrm{D}}}$, $t = 1, 2, \ldots, s$) is a function of all measurements. As a result, Eq. (3.50) suggests that all sparse measurement data $\boldsymbol{y}^{3D}$ contribute to $\underline{\boldsymbol{\mu}}_{\hat{\underline{\mathbf{F}}}}$, reflecting the data-driven characteristics of the OMP-VBI method in interpreting 3D geo-data.

In addition, the uncertainty associated with $\hat{\underline{\mathbf{F}}}$ can be further derived from the statistics of $\boldsymbol{\omega}_s^{3\mathrm{D}}$, i.e., $\boldsymbol{\mu}_{\boldsymbol{\omega}_s^{3D}}$ and $\mathbf{COV}_{\boldsymbol{\omega}_s^{3D}}$ obtained in the Bayesian framework. In accordance with the definition of variance and the computation rule of Hadamard product, variance of $\hat{\underline{\mathbf{F}}}$ is computed as (e.g., Johnson and Wichern 2002; Petersen and Pedersen 2012):

$$Var(\hat{\underline{\mathbf{F}}}) = \sum_{t=1}^{s_1 s_2 s_3} (\mathbf{B}_{s,t}^{3D} \circ \mathbf{B}_{s,t}^{3D}) \sigma_{\omega_{s,t}^{3D}}^2 + (\tau^{3D})^{-1} \mathbf{I}_{N_1 \times N_2} \tag{3.51}$$

where $Var(\hat{\underline{\mathbf{F}}})$ represents the estimated variance associated with $\hat{\underline{\mathbf{F}}}$, and it has the same dimension as $\hat{\underline{\mathbf{F}}}$; $\sigma_{\omega_{s,t}^{3D}}^2$ is the tth diagonal element of $\mathbf{COV}_{\boldsymbol{\omega}_s}^{3D}$, representing the estimated variance for the tth element of $\boldsymbol{\omega}_s^{3D}$; and $(\tau^{3D})^{-1}$ represents the variance of ε^{3D}. In practice, expectation of $(\tau^{3D})^{-1}$, i.e., the variance of $\sigma^2_{\varepsilon^{3D}}$ may be used, which is derived as $\mu_{\sigma^2_{\varepsilon^{3D}}} = d_n / (c_n - 1)$ (see Eq. 3.39). It is noted again that both the mean and variance of $\hat{\underline{\mathbf{F}}}$, i.e., $\boldsymbol{\mu}_{\hat{\underline{\mathbf{F}}}}$ and $Var(\hat{\underline{\mathbf{F}}})$, are dependent on measurements collected in geotechnical site characterization, and, hence, they are data-driven, without assumption on the autocorrelation structures of the 3D profile $\underline{\mathbf{F}}$ (e.g., stationary assumption).

3.7.3 Illustrative Example: Interpretation of 3D Profile from Sparse Measurements

In this section, a simulated 3D profile of a quantity Q with a dimension of 128×128×16, as shown in Figure 3.16a, is used to illustrate and validate the 3D OMP-VBI method discussed above. In this example, only sparse measurements are extracted from Figure 3.16a and used as input to the 3D BCS method, whereas the remaining data are used for comparison. For example, ten boreholes are drilled during site characterization, and ten measurements are obtained from each borehole. Figure 3.17 shows the locations of sparse measurements $\underline{\mathbf{Y}}$ by open circles. Following the method discussed in detail in the last paragraph of Section 3.6.1, the representative sampling interval within x_1-x_2 plane is computed as $\Delta_h \approx 37.3$ m, and that along the x_3 axis is $\Delta_v \approx 1$ m. Then, s_1, s_2, and s_3 are computed as $s_1 \approx l_1/\Delta_h + 1 = 4$, $s_2 = l_2/\Delta_h + 1 = 4$, and $s_3 = 14$. Therefore, the first four columns of $\mathbf{B}_1$ and $\mathbf{B}_2$ are, respectively, written as $\mathbf{B}_{1,s}$ and $\mathbf{B}_{2,s}$, whereas the first 14 columns of $\mathbf{B}_3$ are written as $\mathbf{B}_{3,s}$. With $\mathbf{B}_{1,s}$, $\mathbf{B}_{2,s}$, and $\mathbf{B}_{3,s}$, and measurement matrix $\mathbf{\Psi}^{3D}$, the matrix $\mathbf{A}_s^{3D}$ is obtained, and measurements shown

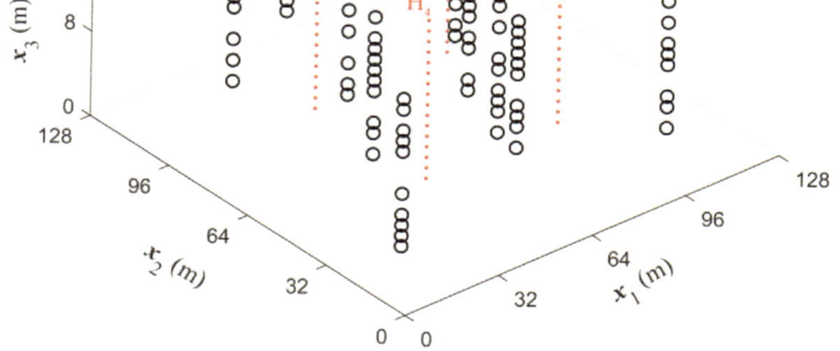

FIGURE 3.17 Locations of measurements and four boreholes H_1 to H_4 for testing in 3D space.

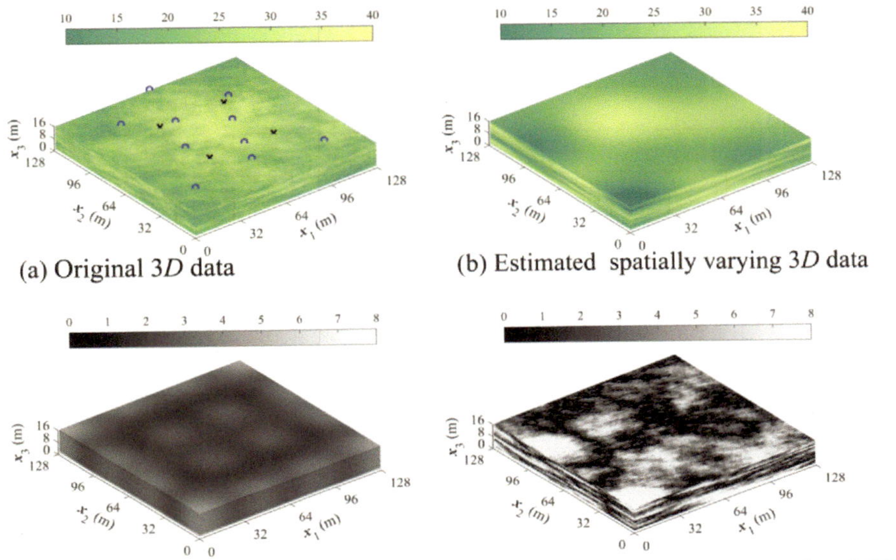

(a) Original 3D data (b) Estimated spatially varying 3D data

(c) Spatial distribution of estimated SD (d) Absolute residuals between (a) and (b)

FIGURE 3.18 Comparison between the original 3D data and that estimated from sparse measurements: (a) Original 3D data; (b) estimated spatially varying 3D data; (c) spatial distribution of estimated SD; and (d) absolute residuals between (a) and (b).

in Figure 3.17 by open circles are rewritten as a column vector $\boldsymbol{y}^{3D}$. Following the procedure described in this section, the statistics of $\boldsymbol{\omega}_s^{3D}$ are obtained, as shown in Eq. (3.48).

With $\boldsymbol{\mu}_{\boldsymbol{\omega}_s}^{3D}$ and $\mathbf{COV}_{\boldsymbol{\omega}_s}^{3D}$, spatially varying 3D signal of interest can be re-constructed. Figure 3.18b and 3.18c show spatial distribution of the mean and SD, respectively. For comparison, Figure 3.18a also shows the original 3D $\underline{\mathbf{F}}$. It is seen from Figure 3.18 that the 3D profile predicted from sparse measurements in Figure 3.18b is consistent with the trend shown in Figure 3.18a, although some local variations of the original 3D profile in Figure 3.18a are not perfectly captured (see Figure 3.18d). This is because the number of measurements used as input to the OMP-VBI method is relatively small when compared to the total number of measurements in $\underline{\mathbf{F}}$. In this example, the true 3D profile has a total number of 262,144 data points, and the interpolation results from sparse measurements contain significant statistical uncertainty. Note that the statistical uncertainty is explicitly quantified, as shown in Figure 3.18c by the estimated SD. It is seen from Figure 3.18c that the magnitude of the SD at most locations is slightly larger than or comparable to that of the absolute differences shown in Figure 3.18d. This observation implies that there is a high probability that the true values of the original 3D dataset fall within the range defined by the interpolated 3D dataset ± 1.96 interpolation SD (i.e., $\underline{\boldsymbol{\mu}}_{\mathrm{F}}(i, j, k) \pm 1.96\, \sigma_{\underline{\mathrm{F}}}(i, j, k)$). For instance, around 80% of elements of the original 3D dataset $\underline{\mathbf{F}}$ fall within the range defined by $\underline{\boldsymbol{\mu}}_{\mathrm{F}}(i, j,$

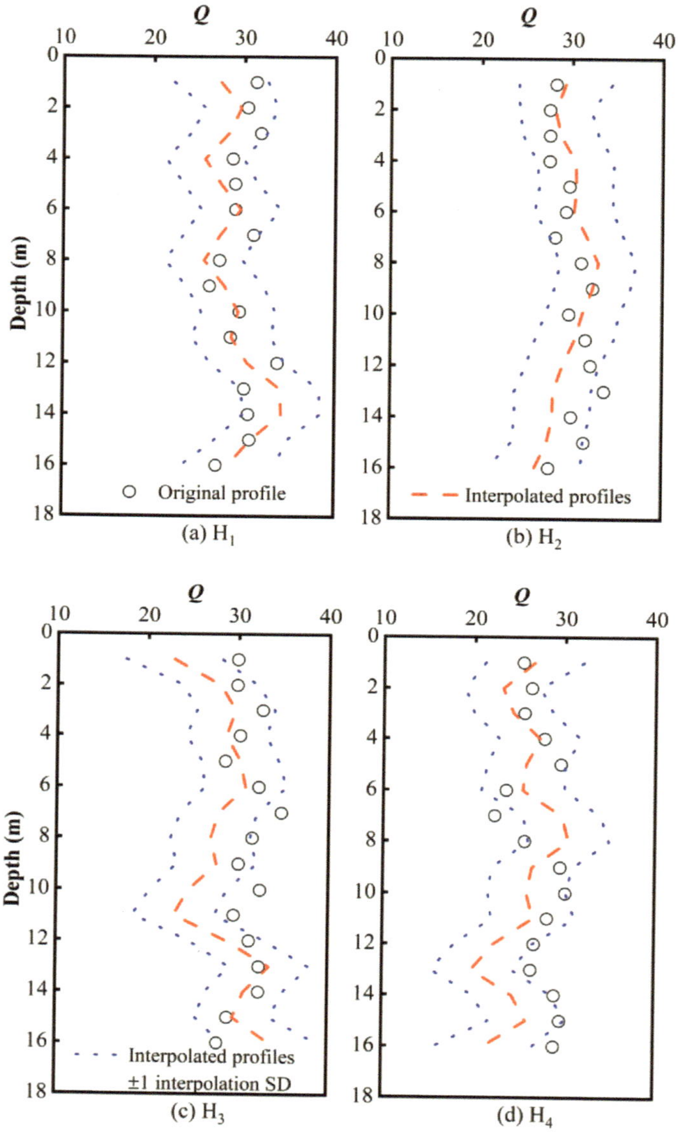

FIGURE 3.19 Comparison between the original and interpolated profiles at boreholes H_1, H_2, H_3, and H_4.

$k) \pm 1.96\ \sigma_E(i, j, k)$ in this example. Hence, the quantified interpolation SD can be used as an indicator to evaluate the reliability of the results obtained.

In addition to the above comparison, variations of Q along four boreholes H_1 to H_4 (see Figure 3.17) are extracted from the interpolated 3D dataset, and compared to those from the original 3D dataset, as shown in Figures 3.19a to 3.19d, respectively.

In Figure 3.19, the dashed line represents the interpolated profile from the OMP-VBI method, whereas the solid line represents the original profile of Q at each corresponding borehole. Figure 3.19 also includes the quantified interpolation uncertainty in terms of interpolated mean profile ± interpolation SD by a pair of dotted lines. Note that no measurement from boreholes H_1 to H_4 is used as input to OMP-VBI. It is seen from the figure that the dashed line is generally consistent with the solid line, and most local variations of the solid lines fall within the region defined by the two dotted lines. These agreements once again indicate that the OMP-VBI method provides a reasonable interpolation for Q at the locations where no measurements are taken as input, using only a small number of measurements. In addition, a close examination of each subplot in Figure 3.19 shows that the dashed lines in Figures 3.19a, b, and d (i.e., the H_1, H_2, and H_4 borehole cases) plot slightly closer to the solid line than those in the other subplot. This may be explained by noting that H_1, H_2, and H_4 are slightly closer than H_3 to the boreholes where measurements are available and used as input. The average distance among H_1 and the three nearest boreholes where measurements are available is around 48.2 m, and it is 47.5 m for H_2, and 43.0 for H_3. In contrast, the average distance between H_3 and the nearest boreholes where measurements are obtained is around 51.4 m, which are slightly larger than the distances for H_1, H_2, and H_4. This once again indicates the data-driven characteristics of the OMP-VBI method.

REFERENCES

Alan, V. Oppenheim, W. Schafer Ronald, and R. B. John. 1989. *Discrete-time signal processing*. Englewood Cliffs, NJ: Printice Hall Inc.

Ang, A., and Wilson H. Tang. 2007. *Probability concepts in engineering: Emphasis on applications to civil & environmental engineering*. 2nd ed. New York: Wiley.

Babacan, S. D., R. Molina, and A. K. Katsaggelos. 2010. Bayesian compressive sensing using laplace priors. *IEEE Transactions on Image Processing* 19 (1): 53–63. doi: 10.1109/TIP.2009.2032894

Beal, M. J. 2003. Variational algorithms for approximate Bayesian inference. PhD, The Gatsby Computational Neuroscience Unit, University of College London.

Bishop, C. M. 2006. *Pattern recognition and machine learning*. New York: Springer.

Bishop, C. M., and Michael E. Tipping. 2000. Variational relevance vector machines. *Proceedings of the sixteenth conference on uncertainty in artificial intelligence*, Stanford, CA.

Cai, T. T., and L. Wang. 2011. Orthogonal matching pursuit for sparse signal recovery with noise. *IEEE Transactions on Information Theory* 57 (7): 4680–4688. doi: 10.1109/TIT.2011.2146090

Caiafa, Cesar F., and Andrzej Cichocki. 2013. Multidimensional compressed sensing and their applications. *Wiley Interdisciplinary Reviews: Data Mining and Knowledge Discovery* 3 (6): 355–380. doi: doi:10.1002/widm.1108

Candès, E. J., and M. B. Wakin. 2008. An introduction to compressive sampling. *IEEE Signal Processing Magazine* 25 (2): 21–30. doi: 10.1109/MSP.2007.914731

Donoho, D. L. 2006. Compressed sensing. *IEEE Transactions on Information Theory* 52 (4): 1289–1306. doi: 10.1109/TIT.2006.871582

Donoho, D. L., Y. Tsaig, I. Drori, and J. L. Starck. 2012. Sparse solution of underdetermined systems of linear equations by stagewise orthogonal matching pursuit. *IEEE Transactions on Information Theory* 58 (2): 1094–1121. doi: 10.1109/TIT.2011.2173241

Dumitru, M. 2017. Sparsity enforcing priors in inverse problems via normal variance mixtures: Model selection, algorithms and applications. *arXiv preprint arXiv:1705.10354.*

ESDAC. 2017. European Soil Data Centre, European Commission, Joint Research Centre. http://esdac.jrc.ec.europa.eu/content/isoil-project-interactions-between-soil-related-sciences

Fang, Y., J. Wu, and B. Huang. 2012. 2D sparse signal recovery via 2D orthogonal matching pursuit. *Science China Information Sciences* 55 (4): 889–897.

Fenton, Gordon A., and D. Vaughan Griffiths. 2008. *Risk assessment in geotechnical engineering*. New York: Wiley.

Gelman, Andrew, John B. Carlin, Hal S. Stern, and Donald B. Rubin. 2004. *Bayesian data analysis*. Vol. 2. Boca Raton, FL: Chapman & Hall/CRC.

Ji, S., Y. Xue, and L. Carin. 2008. Bayesian compressive sensing. *IEEE Transactions on Signal Processing* 56 (6): 2346–2356. doi: 10.1109/TSP.2007.914345

Johnson, Richard Arnold, and Dean W. Wichern. 2002. *Applied multivariate statistical analysis*. 5th ed., Vol. 4. New Jersey, NJ: Prentice Hall.

Leroueil, S., K. Hamouche, F. Tavenas, M. Boudali, J. Locat, D. Virely, M. Roy, et al. 2002. Geotechnical characterization and properties of a sensitive clay from Quebec. *Proceedings of the Symposium on Characterisation and Engineering Properties of Natural Soils*, Singapore.

Mathworks (2024). MATLAB™: the language of technical computing. www.mathworks.com/products/matlab/ (accessed at 28/06/2024).

Mayne, Paul W., Barry R. Christopher, and J. DeJong. 2002. Subsurface investigations—geotechnical site characterization. In *FHWA NHI-01-031*. National Highway Institute, Federal Highway Administration, Washington, D.C.

Murphy, Kevin P. 2012. *Machine learning: A probabilistic perspective*. London, England: The MIT Press.

Needell, Deanna, and Roman Vershynin. 2010. Signal recovery from incomplete and inaccurate measurements via regularized orthogonal matching pursuit. *IEEE Journal of Selected Topics in Signal Processing* 4 (2): 310–316.

Nyquist, H. 1928. Certain topics in telegraph transmission theory. *Transactions of the American Institute of Electrical Engineers* 47 (2): 617–644. doi: 10.1109/T-AIEE.1928.5055024.

NZGD (New Zealand Geotechnical Database). 2016. Earthquake Commission (EQC), New Zealand, www.nzgd.org.nz (Last Modified 28/11/2016, accessed 28 May, 2024.)

Petersen, K. B., and M. S. Pedersen. 2012. *The matrix cookbook*. Technical University Denmark, Kongens Lyngby, Denmark, Technical Report.

Phoon, Kok-Kwang, and Fred H. Kulhawy. 1999. Characterization of geotechnical variability. *Canadian Geotechnical Journal* 36 (4): 612–624. doi: 10.1139/t99-038

Rao, K. Ramamohan, and Ping Yip. 1990. *Discrete cosine transform: Algorithms, advantages, applications*. New York: Academic Press.

Salomon, David. 2007. *Data compression: The complete reference*. 4th ed. New York: Springer Science & Business Media.

Shannon, Claude Elwood. 1948. A mathematical theory of communication. *The Bell System Technical Journal* 27 (3): 379–423.

Sivia, Devinderjit, and John Skilling. 2006. *Data analysis: A Bayesian tutorial*. New York: OUP Oxford.

Tipping, Michael E. 2001. Sparse bayesian learning and the relevance vector machine. *Journal of Machine Learning Research* 1: 211–244. doi: 10.1162/15324430152748236

Tropp, J. A., and A. C. Gilbert. 2007. Signal recovery from random measurements via orthogonal matching pursuit. *IEEE Transactions on Information Theory* 53 (12): 4655–4666. doi: 10.1109/TIT.2007.909108

Wang, Y., Z. Cao, and D. Q. Li. 2016. Bayesian perspective on geotechnical variability and site characterization. *Engineering Geology* 203:117–125. doi: org/10.1016/j.enggeo.2015.08.017

Wang, Y., and T. Zhao. 2016. Interpretation of soil property profile from limited measurement data: a compressive sampling perspective. *Canadian Geotechnical Journal* 53 (9): 1547–1559. doi: 10.1139/cgj-2015-0545

Wang, Y., and T. Zhao. 2017. Statistical interpretation of soil property profiles from sparse data using Bayesian compressive sampling. *Géotechnique* 67 (6): 523–536. doi: 10.1680/jgeot.16.P.143

Zhao, Q., L. Zhang, and A. Cichocki. 2015. Bayesian sparse Tucker models for dimension reduction and tensor completion. *arXiv preprint arXiv:1505.02343*.

Zhao, T., Y. Hu, and Y. Wang. 2018. Statistical interpretation of spatially varying 2D geo-data from sparse measurements using Bayesian compressive sampling. *Engineering Geology* 246: 162–175. doi: 10.1016/j.enggeo.2018.09.022

Zhao, T., S. Montoya-Noguera, K. K. Phoon, and Y. Wang. 2018. Interpolating spatially varying soil property values from sparse data for facilitating characteristic value selection. *Canadian Geotechnical Journal* 55 (2): 171–181. doi: org/10.1139/cgj-2017-0219.

Zhao, T., and Y. Wang. 2020. Non-parametric simulation of non-stationary non-gaussian 3D random field samples directly from sparse measurements using signal decomposition and Markov Chain Monte Carlo (MCMC) simulation. *Reliability Engineering & System Safety* 203: 107087. doi: org/10.1016/j.ress.2020.107087

Zhao, T., and Y. Wang. 2021. Statistical interpolation of spatially varying but sparsely measured 3D Geo-data using Compressive sensing and variational Bayesian inference. *Mathematical Geosciences* 53: 1171–1199. doi: org/10.1007/s11004-020-09913-x

4 Simulation of Spatial Data from Sparse Measurements

Tengyuan Zhao, Yu Wang, and Kok-Kwang Phoon

ABSTRACT

This chapter extends the Bayesian compressive sensing or sampling (BCS) methods introduced in Chapter 3 to random field simulation of spatial or temporal data directly from sparse measurements. Two random field generators are developed step by step: the BCS-KL generator, based on Karhunen-Loève (KL) expansion, and the BCS-MCMC generator, which integrates BCS with Markov Chain Monte Carlo (MCMC) simulation. Both generators can simulate random field samples (RFSs) directly from sparse measurements without a need for predetermined autocorrelation functions or parameter estimation for pre-selected autocorrelation models (e.g., autocorrelation length). Determining autocorrelation functions and estimating parameters typically demand extensive measurements, which are often unavailable in geotechnical practice. It is a unique capability of the BCS-based generators that RFSs can be directly simulated from sparse measurements for geotechnical site characterization. Due to the data-driven nature of BCS algorithms, the BCS-KL/BCS-MCMC generators can also simulate nonstationary, non-Gaussian RFSs directly from sparse measurements. Real-world data from South Parklands, Australia, are used to illustrate and validate their effectiveness. Results show that BCS-based generators successfully simulate 2D and 3D RFSs, ensuring consistency with site conditions, enhancing subsurface stratification, and improving the characterization of 3D soil variability."

4.1 INTRODUCTION

Geotechnical site characterization aims to provide a comprehensive understanding of the subsurface conditions, and it is indispensable in geotechnical and geological engineering. Characterization of the spatially varying geotechnical properties from sparse measurements remains a persistent and formidable challenge, because often only sparse measurements can be obtained from a specific project or site. As discussed in Chapter 3, Bayesian compressive sensing/sampling (BCS) offers an innovative and effective way to tackle this challenge. The BCS algorithms discussed in Chapter 3 not only provide mean or best estimate of the spatially varying geotechnical property

DOI: 10.1201/9781003381235-4

profile (i.e., spatial variability), but also quantify the statistical uncertainty associated with the interpretation (e.g., see Figures 3.11, 3.15, 3.18 in Chapter 3). When the interpreted profiles are adopted for subsequent geotechnical analysis and design, the spatial variability and associated statistical uncertainty propagate through the design and analysis models for the concerned geo-structures, such as slopes (e.g., Griffiths et al. 2009; Hicks and Spencer 2010), foundations (e.g., Fenton and Griffiths 2002, 2005; Griffiths and Fenton 2009), and retaining walls (e.g., Zevgolis and Bourdeau 2010; Beer et al. 2013). The effect of these uncertainties and variabilities on geo-structures may be quantified through an integration of analytical or numerical models for geotechnical design and analysis with random fields (e.g., Vanmarcke 1977; Vanmarcke et al. 1986) for modeling the uncertainties and spatial variability under a reliability-based design and analysis framework.

It is well recognized in the literature (e.g., Cao and Wang 2014; Stuedlein et al. 2012; Ching et al. 2016; Cami et al. 2020) that, for conventional random field modeling, extensive measurements are generally required to derive reliable random field parameters, especially the most appropriate type of autocorrelation function and its parameters (e.g., correlation length). However, the number of measurements in geotechnical site characterization is often sparse and limited from a specific site or project. The BCS algorithms described in Chapter 3 can be extended for direct simulation of random field samples (RFSs) from sparse measurements. Compared with the conventional random field simulations (e.g., Vanmarcke et al. 1986; Fenton and Griffiths 2008), BCS-based simulations are data-driven and do not require a determination of the type of autocorrelation function or its parameters, such as correlation length. Furthermore, the BCS-based simulation methods are also capable of simulating nonstationary and non-Gaussian RFSs directly from measurements.

The remainder of this chapter is organized as follows. Section 4.2 presents the two BCS-based methods for the simulation of one-dimensional (1D) spatially varying geotechnical properties from sparse measurements. Sections 4.3 and 4.4 demonstrate BCS-based simulation of nonstationary random fields without de-trending and simulation of non-Gaussian nonstationary random fields, respectively. Section 4.5 extends the BCS-based methods from 1D cases shown in Sections 4.2–4.4 to two-dimensional (2D) and three-dimensional (3D) cases. Then, the BCS-based simulation methods are applied to 2D and 3D real data for geotechnical site characterization in Section 4.6.

4.2 SIMULATION OF RANDOM FIELDS DIRECTLY FROM SPARSE MEASUREMENTS BY BCS

In this section, two BCS-based methods are presented for simulating RFSs of geotechnical properties from sparse measurements, including BCS-KL generator that is based on Karhunen-Loève (KL) expansion, and BCS-MCMC generator that integrates BCS with Markov Chain Monte Carlo (MCMC) simulation. Both generators can simulate spatially varying and autocorrelated RFSs directly from sparse measurements, and therefore the difficulties associated with random field modeling (i.e., a lack of extensive measurements) in site characterization are bypassed.

4.2.1 BCS-KL Random Field Generator

BCS-KL random field generator is developed based on the BCS derivation formulated in Chapter 3 and KL expansion. This section first introduces the basic idea of truncated KL expansion, followed by the development of BCS-KL generator for simulating RFSs directly from sparse measurements.

KL expansion is a widely used method for simulating RFSs with a given autocovariance matrix. During the implementation of the KL expansion, a continuous random process is approximated as a discrete random process. A discrete random process, e.g., a 1D discrete random field, may be characterized by a finite number, N, of correlated variables with coordinates $z_1, z_2, \dots z_N$, such as $\boldsymbol{X} = [X(z_1), X(z_2), \dots, X(z_N)]^{\mathrm{T}}$. $\boldsymbol{X}$ is a column vector with a length of N. The autocorrelation among elements of $\boldsymbol{X}$ may be characterized by a covariance matrix $\mathbf{C}$, and $\mathbf{C}$ is often defined as $\mathbf{C} = \sigma_X^2\mathbf{R}$ for a second-order stationary random field, where $X(z_1)$ to $X(z_N)$ follows an identical probability distribution with the same parameters, such as mean and variance. In conventional random field modeling, a stationary covariance function is frequently used to represent the covariance matrix. In KL expansion, the corresponding covariance function (or covariance matrix $\mathbf{C}$) is decomposed into a summation of a set of deterministic orthogonal functions and uncorrelated random variables, i.e., $\mathbf{C} = \sum_{i=1}^{N}(\boldsymbol{t}_i)\cdot(\boldsymbol{t}_i)^T\lambda_i$. $\boldsymbol{t}_i$ is the ith eigenfunction expressed in a column vector (see, e.g., Figure 4.1a); λ_i is the ith eigenvalue corresponding to $\boldsymbol{t}_i$, which indicates the weight of the corresponding eigenfunction (see, e.g., Figure 4.1b). Note that λ_i and eigenfunction $\boldsymbol{t}_i$ can be obtained by numerical or analytical eigen-decomposition of the covariance function or equivalently eigen-decomposition of the covariance matrix (e.g., Strang 2006).

Mathematically, the random vector $\boldsymbol{X}$ may be expressed as (e.g., Spanos and Ghanem 1989; Zhang and Ellingwood 1994):

$$\boldsymbol{X} = \boldsymbol{\mu}_X + \sum_{i=1}^{N}\sqrt{\lambda_i}\,\xi_i\boldsymbol{t}_i \tag{4.1}$$

where $\boldsymbol{\mu}_X$ is an N-length column vector, and it represents the mean of $\boldsymbol{X}$; N represents the number of KL terms and it approaches infinity if the random field of interest is a continuous one; ξ_i $(i = 1, 2, \dots, N)$ is a set of uncorrelated random variables, such as the standard Gaussian random variables (e.g., Spanos and Ghanem 1989; Van Trees et al. 2013). In the context of KL expansion, most eigenvalues λ_i $(i = 1, 2, \dots, N)$ are very small or even zero (see Figure 4.1b). Therefore, the first K $(K<<N)$ largest eigenvalues and corresponding eigenfunctions are able to accurately characterize the main feature of the covariance function of interest (e.g., Spanos and Ghanem 1989; Phoon et al. 2002b, a). In this case, a truncated KL expansion, expressed as

$$\boldsymbol{X} \approx \boldsymbol{\mu}_X + \sum_{i=1}^{K}\sqrt{\lambda_i}\,\xi_i\boldsymbol{t}_i \tag{4.2}$$

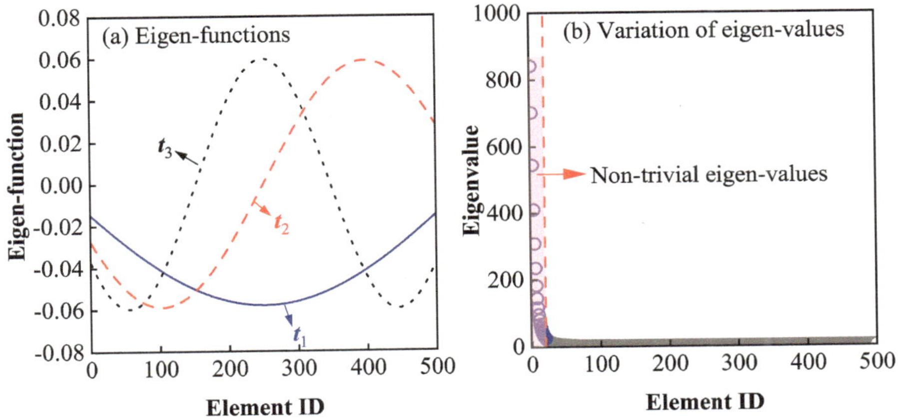

FIGURE 4.1 Examples of eigenfunctions and eigenvalues by eigen-decomposition of a covariance matrix **C** with an auto-covariance function $\rho(\Delta) = 3^2 \cdot \exp[-\Delta / 2]$.

is commonly adopted in RFS generation in practice. Note that λ_i in Eq. (4.2) is sorted in descending order, and the $\boldsymbol{t}_i$ corresponding to λ_i is also sorted accordingly. Once λ_i and $\boldsymbol{t}_i$ are obtained, RFSs of interest can be obtained by realizing K (K<<N) uncorrelated zero-mean and unit-variance random variables ξ_i, which is relatively trivial.

The KL expansion discussed above can be conveniently integrated with the BCS algorithms discussed in Chapter 3 to simulate spatially varying RFSs of $\boldsymbol{f}$ directly from sparse measurements. Specifically, both the mean $\boldsymbol{\mu}_{\hat{f}}$ and covariance matrix $\mathbf{COV}_{\hat{f}}$ associated with $\boldsymbol{f}$ have been derived in Chapter 3 (see Eq. 3.37). $\boldsymbol{\mu}_{\hat{f}}$ quantifies the spatial variation of a quantity of interest (e.g., a variation of soil properties along the depth). The off-diagonal elements in $\mathbf{COV}_{\hat{f}}$ reflect the statistical autocorrelation of the quantity (e.g., soil properties) at two different locations; whereas the diagonal elements in $\mathbf{COV}_{\hat{f}}$ represent the statistical uncertainty in the quantity (e.g., soil property) at a given location when interpreted from sparse data (e.g., Wang and Zhao 2017). Based on the derived mean $\boldsymbol{\mu}_{\hat{f}}$ and covariance matrix $\mathbf{COV}_{\hat{f}}$, a link with KL expansion can be established, and the statistical uncertainty due to the interpretation of sparse data can be explicitly considered in RFS generation using KL, as discussed below.

Specifically, Eq. (3.37) in Chapter 3 can be reformulated as a truncated KL expansion to generate RFSs of $\boldsymbol{f}$. The eigenvalues and eigenfunctions in Eq. (4.1) can be obtained conveniently by an eigen-decomposition of $\mathbf{COV}_{\hat{f}}$ from Eq. (3.37), which can be derived further from the eigen-decomposition of $\mathbf{COV}_{\omega_s^{1D}}$, i.e., as $\mathbf{COV}_{\omega_s^{1D}} = \mathbf{U}_s \Lambda \mathbf{U}_s^{\mathrm{T}} = \sum_{i=1}^{s} \boldsymbol{u}_i \cdot (\boldsymbol{u}_i)^{\mathrm{T}} \cdot \lambda_{\omega_{s,i}^{1D}}$. $\mathbf{U}_s$ is an orthonormal matrix, whose columns represent the eigenvectors of $\mathbf{COV}_{\omega_s^{1D}}$. Λ is a diagonal matrix, whose diagonal elements $\lambda_{\omega_s^{1D},1}, \lambda_{\omega_s^{1D},2}, \ldots$ are eigenvalues of $\mathbf{COV}_{\omega_s^{1D}}$. Substituting $\mathbf{COV}_{\omega_s^{1D}} = \mathbf{U}_s \Lambda \mathbf{U}_s^{\mathrm{T}}$ into $\mathbf{B}_s \mathbf{COV}_{\omega_s^{1D}} \mathbf{B}_s^{\mathrm{T}}$ leads to an eigen-decomposition of

$\mathbf{B}_s\mathbf{COV}_{\boldsymbol{\omega}_s^{1D}}\mathbf{B}_s^{\mathrm{T}} = \mathbf{V}\Lambda\mathbf{V}^{\mathrm{T}} = \sum_{i=1}^{s}\mathbf{V}_i \cdot (\mathbf{V}_i)^T \cdot \lambda_{\boldsymbol{\omega}_s^{1D},i}$, where $\mathbf{V} = \mathbf{B}_s\mathbf{U}$ is the eigenmatrix of $\mathbf{B}_s\mathbf{COV}_{\boldsymbol{\omega}_s^{1D}}\mathbf{B}_s^{\mathrm{T}}$; and Λ is a diagonal matrix, corresponding to the eigenvalues. $\boldsymbol{V}_i$ is the ith eigenfunction of $\mathbf{B}_s\mathbf{COV}_{\boldsymbol{\omega}_s^{1D}}\mathbf{B}_s^{\mathrm{T}}$. $\boldsymbol{V}_i$ and $\lambda_{\boldsymbol{\omega}_s^{1D},i}$ play the same role as $\boldsymbol{t}_i$ and λ_i in Eq. (4.1), respectively. In this case, the formula for generating RFSs of $\hat{\boldsymbol{f}}$ using KL may be derived as (e.g., Au and Wang 2014; Wang et al. 2018):

$$\hat{\boldsymbol{f}} = \boldsymbol{\mu}_{\hat{f}} + \sum_{i=1}^{s}\boldsymbol{V}_i \cdot \sqrt{\lambda_{\boldsymbol{\omega}_s^{1D},i}} \cdot \xi_i^f + \boldsymbol{\varepsilon} \tag{4.3}$$

where ξ_i^f has the same meaning as ξ_i in Eq. (4.2); $\boldsymbol{\varepsilon}$ is a column vector, representing N uncorrelated Gaussian variable with zero mean and variance shown in Eq. (3.39). Because Eq. (4.3) is based on the BCS results in Chapter 3, it is denoted as a BCS-KL random field generator. Similar to $\boldsymbol{X}$ in Eq. (4.2), $\hat{\boldsymbol{f}}$ can be simulated readily by generating a series of uncorrelated variables ξ_i^f. It is worth noting that eigen-decomposition of $\mathbf{COV}_{\hat{f}}$ reduces to eigen-decomposition of $\mathbf{COV}_{\boldsymbol{\omega}_s^{1D}}$, which only has a dimension of $s \times s$. This indicates that simulation of RFSs of $\hat{\boldsymbol{f}}$ is computationally efficient. Besides, an additional term $\boldsymbol{\varepsilon}$ exists in Eq. (4.3), and it models the lumped uncertainty discussed in Chapter 3. When simulating RFSs of $\hat{\boldsymbol{f}}$, $\boldsymbol{\varepsilon}$ is realized by simulation of N independent Gaussian random variables with a mean of zero and variance shown in Eq. (3.39).

As the simulated RFSs of $\hat{\boldsymbol{f}}$ may be sensitive to the statistical correlation that may exist between samples of ξ_i^f and $\xi_j^f (i \neq j$, and $i, j = 1, 2, \ldots, s)$, an orthogonalization technique, such as Cholesky decomposition, may be adopted to remove such correlation (e.g., Phoon et al. 2004). Suppose that N_{RFS} sets of theoretically uncorrelated ξ_i^f $(i = 1, 2, \ldots, S)$ samples are realized and stored in an $s \times N_{\mathrm{RFS}}$ matrix $\mathbf{G}$. The statistical covariance among columns of $\mathbf{G}$ (i.e., $\mathbf{COV}_{\mathbf{G}}$ with a dimension of $s \times s$) then can be calculated as $\mathbf{COV}_{\mathbf{G}} = \mathbf{G}\mathbf{G}^{\mathrm{T}} / N_{RFS} - \mathbf{G}\boldsymbol{I}_{N_{RFS}}(\boldsymbol{I}_{N_{RFS}})^{\mathrm{T}}\mathbf{G}^{\mathrm{T}} / (N_{RFS}^2)$ (e.g., Johnson and Wichern 2002). $\boldsymbol{I}_{N_{RFS}}$ is a unit column vector with a length of N_{RFS}. Subsequently, a Cholesky decomposition is conducted to $\mathbf{COV}_{\mathbf{G}}$, i.e., $\mathbf{COV}_{\mathbf{G}} = \mathbf{L}\mathbf{L}^{\mathrm{T}}$, where $\mathbf{L}$ is a lower triangular matrix with a dimension of $s \times s$. Then, using $\mathbf{G}$ and $\mathbf{L}$, realizations of N_{RFS} sets of truly uncorrelated $\xi_i^f (i = 1, 2, \ldots, s)$ can be obtained as $\hat{\mathbf{G}} = \mathbf{L}^{-1}\mathbf{G}$. Substituting the N_{RFS} sets of uncorrelated $\xi_i^f (i = 1, 2, \ldots, s)$ samples into Eq. (4.3) leads to N_{RFS} sets of $\hat{\boldsymbol{f}}$ samples.

Consider, for example, the CPT tip resistance q_c data shown in Section 3.5.2. With the sparse measurements shown in Figure 3.7, the mean $\boldsymbol{\mu}_{\hat{f}}$ and covariance matrix $\mathbf{COV}_{\hat{f}}$ may be derived as Eq. (3.37). Then, following the methods described above in this section, $\mathbf{COV}_{\hat{f}}$ may be eigen-decomposed through an eigen-decomposition of $\mathbf{COV}_{\boldsymbol{\omega}_s^{1D}}$. The first four eigenfunctions (or eigenvectors) are summarized in Figure 4.2a, whereas the $s = 7$ nontrivial eigenvalues are summarized in Figure 4.2b by open circles. With the available eigenfunctions and eigenvalues, RFSs of $\hat{\boldsymbol{f}}$ can be generated by simulating a series of truly uncorrelated ξ_i^f (i = 1, 2, …, s) and uncorrelated ε using Eq. (4.3). Figure 4.3 shows several examples of simulated RFSs of $\hat{f}$ by gray lines. Besides, with a large number of RFSs of $\hat{f}$, the mean and 95% confidence interval (CI) may be obtained readily at each location from the simulated RFSs, as shown in Figure 4.3 by a dashed line and two dotted lines, respectively. As expected, both the mean and 95% CI in Figure 4.3 are almost identical to that shown in Figure 3.7. This is because simulation of $\hat{f}$ in this section depends on the statistics $\boldsymbol{\mu}_{\hat{f}}$ and $\mathbf{COV}_{\hat{f}}$ shown in Eq. (3.37). Besides, a close examination of Figure 4.3 shows that the RFS reflects mainly the global varying trends of the underlying q_c profile but misses local details. This may be attributed to the small number of measurement data points used as input. As the number of measurements, M, increases (e.g., from 10 to 100), the RFSs simulated by the BCS-KL generator converge gradually to the

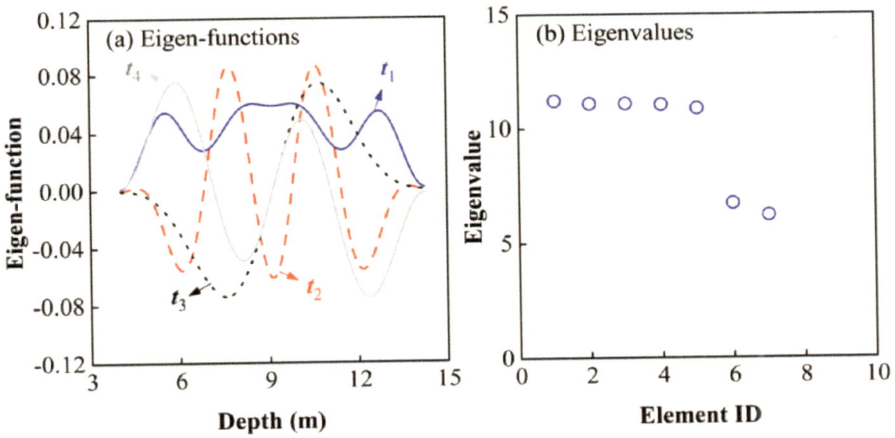

FIGURE 4.2 Eigenfunctions and eigenvalues of $\mathbf{COV}_{\hat{f}}$.

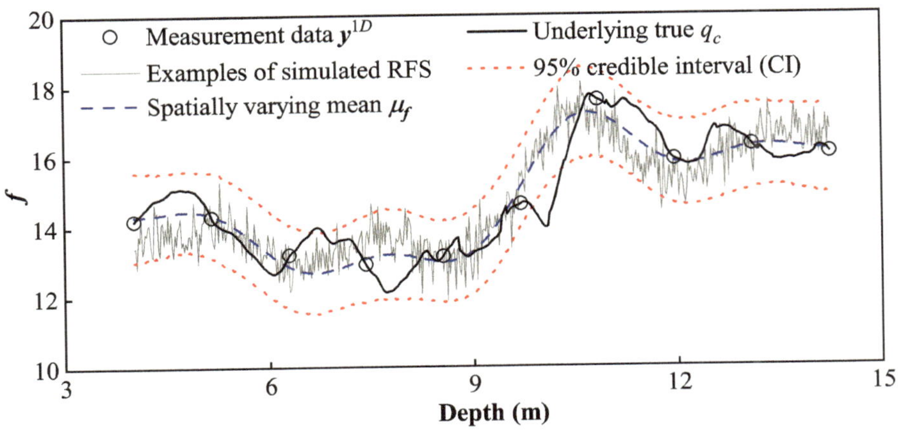

FIGURE 4.3 Examples and statistics of RFSs of $\hat{f}$ simulated by BCS-KL generator.

underlying q_c profile with local details, as shown in Figure 4.4a to Figure 4.4d. Such observations indicate the data-driven nature of the BCS-KL generator developed in this section. In addition, note that although only the equations for 1D data are derived and illustrated in this section, the derivations above are equally applicable to simulate 2D and 3D site-specific RFSs from 2D or 3D measurements. This can be achieved by eigen-decomposition of $\mathbf{COV}_{\boldsymbol{\omega}_s^{2D}}$ and $\mathbf{COV}_{\boldsymbol{\omega}_s^{3D}}$ in Eqs. (3.42) and (3.48), respectively, followed by simulation of uncorrelated standard Gaussian random variables.

4.2.2 BCS-MCMC Random Field Generator

The simulation of RFSs by BCS-KL generator in Section 4.2.1 depends on the BCS formulation results in Eq. (3.28), Eq. (3.42), and Eq. (3.48) in Chapter 3. These statistical results of nontrivial coefficients further depend on the mean of auxiliary variables, including the mean of $\boldsymbol{\alpha}$ and τ, which further depends on the mean of γ. The uncertainties associated with these variables, however, are not considered during estimation of $\boldsymbol{\omega}_s^{1D}$, $\boldsymbol{\omega}_s^{2D}$ or $\boldsymbol{\omega}_s^{3D}$. In this section, another data-driven method of simulating RFSs of $\hat{f}$ is discussed, which is based on the Markov Chain Monte Carlo (MCMC) simulation to quantify the posterior PDF in Eq. (3.13) in Chapter 3 with considerations of uncertainties of the auxiliary variables.

MCMC simulation is a numerical technique to simulate a (or many) sequence(s) of samples of a random variable(s) (e.g., $\boldsymbol{\omega}_s^{1D}, \boldsymbol{\alpha}, \gamma, \tau$) as Markov chain(s). When the Markov chains reach their stationary states, the samples simulated effectively depict PDF of the random variables of interest, including marginal posterior PDF of $\boldsymbol{\omega}_s^{1D}$, i.e., $p(\boldsymbol{\omega}_s^{1D} | \mathbf{y}^{1D})$. MCMC provides an effective and robust way to deal with PDFs that are too complicated or difficult to express analytically (e.g., Pérez et al. 2006). Because of its efficacy and robustness, MCMC has been increasingly used in many fields with a rapid development of computational power of personal computers in the past three decades (e.g., Pérez et al. 2006; Wang and Cao 2013; Mara et al. 2016; Zhao and Wang 2020). When random variables of interest are of high dimensions, such as $\boldsymbol{\omega}_s^{1D}, \boldsymbol{\alpha}, \gamma, \tau$ in this book with a dimension much larger than ten, Gibbs sampling may be adopted as it is efficient and straightforward to perform (e.g., Andrieu et al. 2003; Yuen and Ortiz 2018), as discussed in detail in the next section.

4.2.2.1 Gibbs Sampling Procedure

Gibbs sampling requires analytical expressions of conditional PDFs, such as $p(\boldsymbol{\omega}_s^{1D} | \boldsymbol{\alpha}, \tau, \gamma, \mathbf{y}^{1D})$, $p(\boldsymbol{\alpha} | \boldsymbol{\omega}_s^{1D}, \tau, \gamma, \mathbf{y}^{1D})$, $p(\gamma | \boldsymbol{\omega}_s^{1D}, \boldsymbol{\alpha}, \tau, \mathbf{y}^{1D})$, and $p(\tau | \boldsymbol{\omega}_s^{1D}, \boldsymbol{\alpha}, \gamma, \mathbf{y}^{1D})$ in Chapter 3, which will be derived in the next section. With those conditional PDFs, a standard procedure for Gibbs sampling is summarized in the following:

Step 1: Provide some initial values for $\boldsymbol{\alpha}$, τ, and γ, and update $p(\boldsymbol{\omega}_s^{1D} | \boldsymbol{\alpha}, \tau, \gamma, \mathbf{y}^{1D})$.

Step 2: Simulate one set of $\boldsymbol{\omega}_s^{1D}$ sample from $p(\boldsymbol{\omega}_s^{1D} | \boldsymbol{\alpha}, \tau, \gamma, \mathbf{y}^{1D})$ using a 1D random field generator. Then, substitute the simulated $\boldsymbol{\omega}_s^{1D}$ sample, τ, and γ into $p(\boldsymbol{\alpha} | \boldsymbol{\omega}_s^{1D}, \gamma, \tau, \mathbf{y}^{1D})$ from which a set of $\boldsymbol{\alpha}$ samples is generated. Similarly, substitute simulated $\boldsymbol{\omega}_s^{1D}$, $\boldsymbol{\alpha}$ sample and τ into $p(\tau | \boldsymbol{\omega}_s^{1D}, \boldsymbol{\alpha}, \gamma, \mathbf{y}^{1D})$, and a univariate τ sample is obtained. Then, using the newly simulated $\boldsymbol{\alpha}$, τ, and $\boldsymbol{\omega}_s^{1D}$ samples as input to $p(\gamma | \boldsymbol{\omega}_s^{1D}, \boldsymbol{\alpha}, \tau, \mathbf{y}^{1D})$, a new univariate γ sample is obtained.

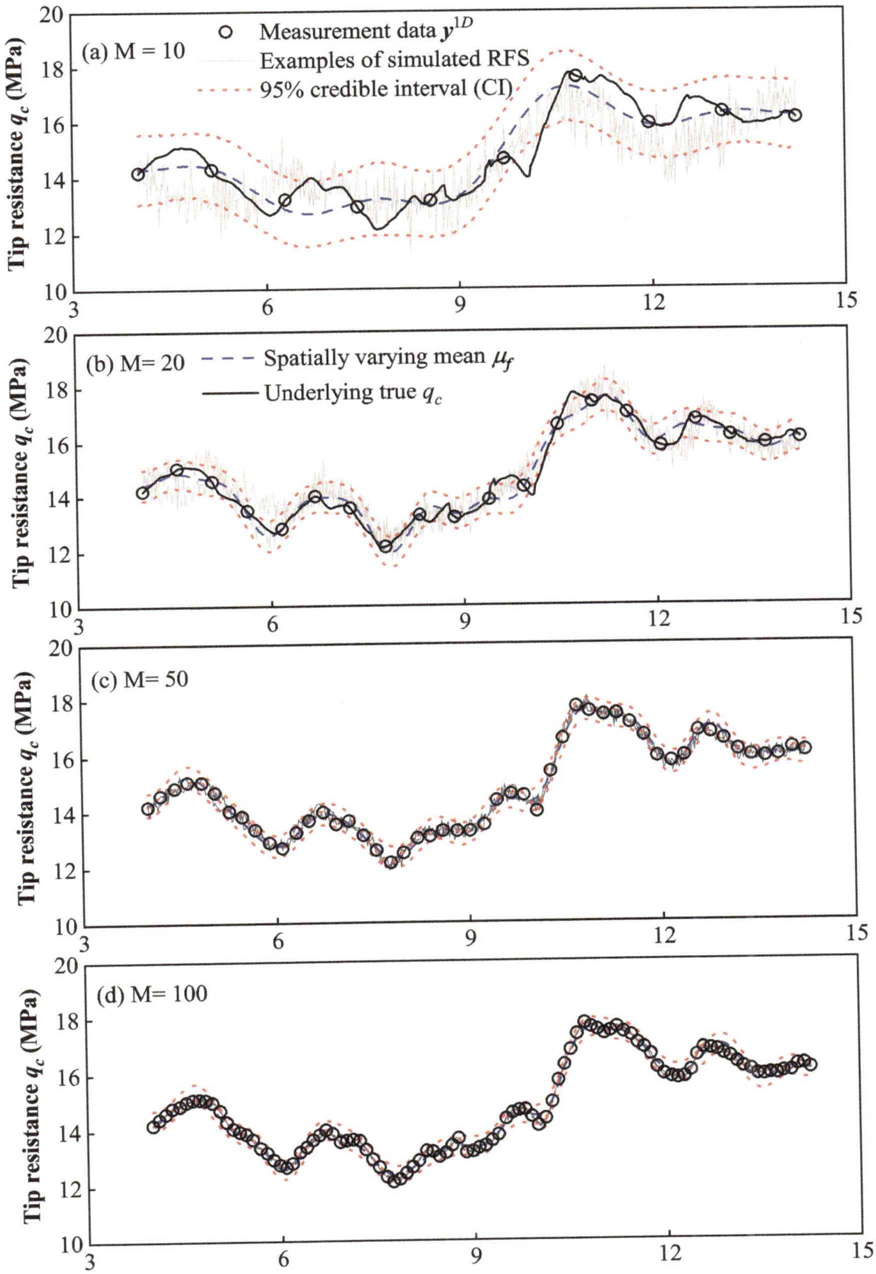

FIGURE 4.4 Statistical results of RFSs of q_c simulated with BCS-KL generator under different M scenarios: (a) to (d) for M = 10, 20, 50, and 100, respectively.

Step 3: Collect the newly obtained $\boldsymbol{\alpha}$, τ, and γ samples and substitute into $p(\boldsymbol{\omega}_s^{1D}|\boldsymbol{\alpha},\tau,\gamma,\boldsymbol{y}^{1D})$, and repeat Step 2 until the prescribed number of stationary MCMC samples is achieved.

Because MCMC samples of random variables of interest, e.g., $\boldsymbol{\omega}_s^{1D}$, at the beginning of iterations above are significantly affected by the initial values of $\boldsymbol{\alpha}$, τ, and γ, they may not accurately represent the desired distribution. Therefore, samples in such a burn-in period, e.g., the first few hundreds of $\boldsymbol{\omega}_s^{1D}$ samples, are often discarded in practice (e.g., Yuen et al. 2016). The MCMC sample number, n_{burn}, of the burn-in period may be determined by visualization plots of the MCMC samples or Geweke's Z statistics (e.g., Geweke 1992). In addition, the MCMC samples simulated are often found to exhibit statistical correlations, which in turn influence the characteristics of MCMC samples. Therefore, a thinning process is commonly used to pick a $\boldsymbol{\omega}_s^{1D}$ sample from every n_{ind} iteration. The number n_{ind} can be determined by calculating and observing the correlation plot along the MCMC sample chains.

As shown in the procedure of Steps 1 to 3 above, analytical expressions of the conditional PDFs are needed, including $p(\boldsymbol{\omega}_s^{1D}|\boldsymbol{\alpha},\tau,\gamma,\boldsymbol{y}^{1D})$, $p(\boldsymbol{\alpha}|\boldsymbol{\omega}_s^{1D},\gamma,\tau,\boldsymbol{y}^{1D})$, $p(\gamma|\boldsymbol{\omega}_s^{1D},\boldsymbol{\alpha},\tau,\boldsymbol{y}^{1D})$, and $p(\tau|\boldsymbol{\omega}_s^{1D},\boldsymbol{\alpha},\gamma,\boldsymbol{y}^{1D})$, which are derived step-by-step below.

4.2.2.2 Derivation of Conditional PDFs

The conditional PDFs, i.e., $p(\boldsymbol{\omega}_s^{1D}|\boldsymbol{\alpha},\tau,\gamma,\boldsymbol{y}^{1D})$, $p(\boldsymbol{\alpha}|\boldsymbol{\omega}_s^{1D},\gamma,\tau,\boldsymbol{y}^{1D})$, $p(\gamma|\boldsymbol{\omega}_s^{1D},\boldsymbol{\alpha},\tau,\boldsymbol{y}^{1D})$, and $p(\tau|\boldsymbol{\omega}_s^{1D},\boldsymbol{\alpha},\gamma,\boldsymbol{y}^{1D})$, are derived in this section. Consider, for example, $p(\boldsymbol{\omega}_s^{1D}|\boldsymbol{\alpha},\tau,\gamma,\boldsymbol{y}^{1D})$. In accordance with the rules of conditional probability (e.g., Ang and Tang 2007; Gelman et al. 2004), $p(\boldsymbol{\omega}_s^{1D}|\boldsymbol{\alpha},\tau,\gamma,\boldsymbol{y}^{1D})$ is expressed as:

$$p(\boldsymbol{\omega}_s^{1D}|\boldsymbol{\alpha},\tau,\gamma,\boldsymbol{y}^{1D})=\frac{p(\boldsymbol{\omega}_s^{1D},\boldsymbol{\alpha},\gamma,\tau|\boldsymbol{y}^{1D})}{p(\boldsymbol{\alpha},\tau,\gamma\,|\,\boldsymbol{y}^{1D})} \tag{4.4}$$

where $p(\boldsymbol{\omega}_s^{1D},\boldsymbol{\alpha},\gamma,\tau|\boldsymbol{y}^{1D})$ is the posterior joint PDF of $\boldsymbol{\omega}_s^{1D}$, $\boldsymbol{\alpha}$, γ, and τ updated by 1D measurements $\boldsymbol{y}^{1D}$, and it has been derived step by step in the Bayesian framework in Chapter 3 and reported as Eq. (3.13). Then, substituting Eq. (3.13) into Eq. (4.4), rearranging all the terms involving $\boldsymbol{\omega}_s^{1D}$, and grouping all the terms without $\boldsymbol{\omega}_s^{1D}$ as a constant "$const_1$" lead to:

$$\begin{aligned}p(\boldsymbol{\omega}_s^{1D}|\boldsymbol{\alpha},\tau,\gamma,\boldsymbol{y}^{1D})&=const_1\cdot\frac{1}{\sqrt{(2\pi)^s\det(\mathbf{COV}_{\boldsymbol{\omega}_s^{1D}})}}\\&\cdot\exp\left(-\frac{(\boldsymbol{\omega}_s^{1D}-\boldsymbol{\mu}_{\boldsymbol{\omega}_s^{1D}})^{\mathrm{T}}(\mathbf{COV}_{\boldsymbol{\omega}_s^{1D}})^{-1}(\boldsymbol{\omega}_s^{1D}-\boldsymbol{\mu}_{\boldsymbol{\omega}_s^{1D}})}{2}\right)\end{aligned} \tag{4.5}$$

Because of the property of conjugate pair, the conditional PDF of $\boldsymbol{\omega}_s^{1D}$, i.e., $p(\boldsymbol{\omega}_s^{1D}|\boldsymbol{\alpha},\tau,\gamma,\boldsymbol{y}^{1D})$, is a normal distribution with a mean of $\boldsymbol{\mu}_{\boldsymbol{\omega}_s^{1D}}$ and covariance matrix $\mathbf{COV}_{\boldsymbol{\omega}_s^{1D}}$. Note that the terms

$\frac{1}{\sqrt{(2\pi)^s \det(\mathbf{COV}_{\omega_s^{1D}})}} \exp\left(-\frac{(\boldsymbol{\omega}_s^{1D}-\boldsymbol{\mu}_{\omega_s^{1D}})^{\mathrm{T}}(\mathbf{COV}_{\omega_s^{1D}})^{-1}(\boldsymbol{\omega}_s^{1D}-\boldsymbol{\mu}_{s\omega_s^{1D}})}{2}\right)$ in Eq. (4.5) are a multivariate normal PDF. Therefore, "$const_1$" = 1 in Eq. (4.5). Note that $\boldsymbol{\mu}_{\omega_s^{1D}}$ and $\mathbf{COV}_{\omega_s^{1D}}$ fully characterize the PDF of $p(\boldsymbol{\omega}_s^{1D}|\boldsymbol{\alpha}, \tau, \gamma, \mathbf{y}^{1D})$, and $\boldsymbol{\mu}_{\omega_s^{1D}}$ and $\mathbf{COV}_{\omega_s^{1D}}$ are expressed as:

$$\begin{aligned}\boldsymbol{\mu}_{\omega_s^{1D}} &= \mathbf{COV}_{\omega_s^{1D}}\mathbf{A}_s\mathbf{y}^{1D}\tau \\ \mathbf{COV}_{\omega_s^{1D}} &= (\mathbf{A}_s^{\mathrm{T}}\mathbf{A}_s\tau+\mathbf{D}^{\alpha})^{-1}\end{aligned} \tag{4.6}$$

Since γ terms only contribute to "$const_1$" (=1), the PDF of $\boldsymbol{\omega}_s^{1D}$ therefore is not dependent on γ. In this case, $p(\boldsymbol{\omega}_s^{1D}|\boldsymbol{\alpha}, \tau, \gamma, \mathbf{y}^{1D})$ can be simplified as $p(\boldsymbol{\omega}_s^{1D}|\, \alpha, \tau, \mathbf{y}^{1D})$.

$p(\boldsymbol{\alpha}|\boldsymbol{\omega}_s^{1D}, \tau, \gamma, \mathbf{y}^{1D})$ can also be derived in a similar manner. First, it is expressed as $p(\tau|\boldsymbol{\omega}_s^{1D}, \tau, \gamma, \mathbf{y}^{1D}) = p(\boldsymbol{\omega}_s^{1D}\, \alpha, \gamma, \tau|\mathbf{y}^{1D}) / p(\boldsymbol{\omega}_s^{1D}, \tau, \gamma\,|\, \mathbf{y}^{1D})$ in accordance with the rules of conditional probability. Then, substituting Eq. (3.13) into the above equation, rearranging the terms involving $\boldsymbol{\alpha}$, and grouping all the terms without $\boldsymbol{\alpha}$ as a constant "$const_2$" lead to

$$p(\boldsymbol{\alpha}|\boldsymbol{\omega}_s^{1D}, \tau, \gamma, \mathbf{y}^{1D}) = const_2\prod_{t=1}^{s}\exp(-\frac{a_t\alpha_t+b_t\alpha_t^{-1}}{2})(\alpha_t)^{p-1}\times\frac{(a_t/b_t)^{p/2}}{2K_p(\sqrt{a_tb_t})} \tag{4.7}$$

A close examination of Eq. (4.7) shows that $\exp(-\frac{a_t\alpha_t+b_t\alpha_t^{-1}}{2})(\alpha_t)^{p-1}\times\frac{(a_t/b_t)^{p/2}}{2K_p(\sqrt{a_tb_t})}$ is a standard generalized inverse Gaussian distribution, where a_t = $\omega_{s,t}^{1D}$, $b_t = \gamma$, p = -1/2, and $K_p(\cdot)$ represents modified Bessel function of the second kind with parameter p. By definition of a PDF, $\int p(\boldsymbol{\alpha}|\boldsymbol{\omega}_s^{1D}, \tau, \gamma, \mathbf{y}^{1D})d\boldsymbol{\alpha}=1$. Since $\exp(-\frac{a_t\alpha_t+b_t\alpha_t^{-1}}{2})(\alpha_t)^{p-1}\times\frac{(a_t/b_t)^{p/2}}{2K_p(\sqrt{a_tb_t})}$ is also a PDF, its integration also equals 1. As a result, $const_2$ = 1. Since all the terms involving τ only contribute to "$const_2$" (=1), τ disappears in the Generalized inverse Gaussian PDF, and $p(\boldsymbol{\alpha}|\boldsymbol{\omega}_s^{1D}, \tau, \gamma, \mathbf{y}^{1D})$ can be simplified as $p(\boldsymbol{\alpha}|\boldsymbol{\omega}_s^{1D}, \gamma, \mathbf{y}^{1D})$.

Similarly, in accordance with the rules of conditional probability, $p(\tau|\boldsymbol{\omega}_s^{1D},\boldsymbol{\alpha}, \gamma, \mathbf{y}^{1D})$ and $p(\gamma|\boldsymbol{\omega}_s^{1D},\boldsymbol{\alpha}, \tau, \mathbf{y}^{1D})$ may be expressed as $p(\tau|\boldsymbol{\omega}_s^{1D}, \boldsymbol{\alpha}, \gamma, \mathbf{y}^{1D}) = p(\boldsymbol{\omega}_s^{1D},\boldsymbol{\alpha}, \gamma, \tau|\mathbf{y}^{1D}) / p(\boldsymbol{\omega}_s^{1D},\boldsymbol{\alpha}, \gamma\,|\, \mathbf{y}^{1D})$ and $p(\gamma|\,\boldsymbol{\omega}_s^{1D},\boldsymbol{\alpha}, \tau, \mathbf{y}^{1D}) = p(\boldsymbol{\omega}_s^{1D},\boldsymbol{\alpha}, \gamma, \tau|\mathbf{y}^{1D}) / p(\boldsymbol{\omega}_s^{1D},\boldsymbol{\alpha}, \tau\,|\, \mathbf{y}^{1D})$, respectively. Then, substitute Eq. (3.13) into each of the above equations, and rearrange the terms involving τ and γ, respectively, while the terms without τ and γ are derived to be a constant of 1. This leads to

$$p(\tau|\boldsymbol{\omega}_s^{1D},\boldsymbol{\alpha}, \gamma, \mathbf{y}^{1D}) = (d_n)^{c_n}\tau^{c_n-1}\exp(-d_n\tau)/\Gamma(c_n) \tag{4.8}$$

$$p(\gamma|\boldsymbol{\omega}_s^{1D},\boldsymbol{\alpha},\tau,\mathbf{y}^{1D}) = (\gamma_b)^{\gamma_a}\gamma^{\gamma_a-1}\exp(-\gamma_b\gamma)/\Gamma(\gamma_a) \tag{4.9}$$

where $c_n = M/2+c_0$, $d_n = d_0 + 1/2[(\mathbf{y}^{1D})^{\mathrm{T}}(\mathbf{y}^{1D}) - 2\boldsymbol{\mu}_{\boldsymbol{\omega}_s^{1D}}^{\mathrm{T}}\mathbf{A}_s^{\mathrm{T}}\mathbf{y}^{1D} + (\boldsymbol{\omega}_s^{1D})^{\mathrm{T}}\mathbf{A}_s^{\mathrm{T}}\mathbf{A}_s\boldsymbol{\omega}_s^{1D}]$, and $\gamma_a = s+a_0$ and $\gamma_b = b_0 + \sum_{t=1}^{s}\alpha_t^{-1}$. Eqs. (4.8) and (4.9) show that both conditional PDFs of τ and γ follow a Gamma distribution. In addition, note that $p(\tau|\boldsymbol{\omega}_s^{1D},\boldsymbol{\alpha},\gamma,\mathbf{y}^{1D})$ and $p(\gamma|\boldsymbol{\omega}_s^{1D},\boldsymbol{\alpha},\tau,\mathbf{y}^{1D})$ are equivalent to $p(\tau|\boldsymbol{\omega}_s^{1D},\mathbf{y}^{1D})$ and $p(\gamma|\boldsymbol{\alpha},\mathbf{y}^{1D})$, respectively, since the γ and $\boldsymbol{\alpha}$ terms, $\boldsymbol{\omega}_s^{1D}$ and τ terms, are not involved in $p(\tau|\boldsymbol{\omega}_s^{1D},\boldsymbol{\alpha},\gamma,\mathbf{y}^{1D})$ and $p(\gamma|\boldsymbol{\omega}_s^{1D},\boldsymbol{\alpha},\tau,\mathbf{y}^{1D})$, respectively. Given these conditional probability distributions, $\boldsymbol{\omega}_s^{1D}$ samples can be iteratively simulated using the procedure of Gibbs sampling described previously. After stationary MCMC samples of $\boldsymbol{\omega}_s^{1D}$ are obtained, RFSs of $\boldsymbol{f}$ may be generated by substituting each $\boldsymbol{\omega}_s^{1D}$ sample together with N realizations of ε into $\boldsymbol{f} = \mathbf{B}_s\boldsymbol{\omega}_s+\varepsilon\boldsymbol{I}$.

Consider, for example, the same case used in Section 4.2.1 and Section 3.5.2. With the sparse measurements in Figure 3.7 or Figure 4.3, spatially varying q_c profiles can be simulated without a need of determining autocorrelation function or its parameters, i.e., autocorrelation length, variance or mean. Following the discussion made in this section, MCMC samples of $\boldsymbol{\omega}_s^{1D}$, $\boldsymbol{\alpha}$, γ, and τ are iteratively simulated. For example, Figures 4.5a to 4.5d, respectively, show the MCMC chain of several elements of $\boldsymbol{\omega}_s^{1D}$, e.g., $\omega_{s,1}^{1D}$, $\omega_{s,2}^{1D}$, $\omega_{s,3}^{1D}$, and $\omega_{s,4}^{1D}$ by open circles. It is seen from these plots that MCMC samples quickly converge to almost stationary states after several iterations (see the open circles within the zoomed-in window in Figure 4.5a). In other words, there is a short burn-in period in this example.

To quantitatively determine the burn-in period after which representative MCMC samples can be collected, the Geweke's Z statistic is adopted in this example. As discussed by Geweke (1992), when the absolute value of Z statistics of MCMC samples is less than 1.96, i.e., $Z \le 1.96$, there is 95% confidence that MCMC samples reach a stationary state. Z statistics is defined as (e.g., Geweke 1992)

$$Z = \frac{|\boldsymbol{\theta}_1 - \boldsymbol{\theta}_2|}{\sqrt{1/n_1 \cdot Var(\boldsymbol{\theta}_1) - 1/n_2 \cdot Var(\boldsymbol{\theta}_2)}} \tag{4.10}$$

where $\boldsymbol{\theta}_1$ and $\boldsymbol{\theta}_2$, respectively, represent the burn-in and remaining parts of an MCMC chain of a variable of interest; n_1 and n_2 represent the number of samples in $\boldsymbol{\theta}_1$ and $\boldsymbol{\theta}_2$, respectively. $Var(\boldsymbol{\theta}_1)$ and $Var(\boldsymbol{\theta}_2)$ represent the variance associated with $\boldsymbol{\theta}_1$ and $\boldsymbol{\theta}_2$. Note that $n_1+n_2 = N_{\mathrm{MCMC}}$, where n_1 can vary from 1 to N_{MCMC}-1. Based on Eq. (4.10) and MCMC samples obtained, Z statistics for each variable can be computed. For example, Figures 4.6a to 4.6d summarize the Z statistics of MCMC samples of $\omega_{s,1}^{1D}$, $\omega_{s,2}^{1D}$, $\omega_{s,3}^{1D}$, and $\omega_{s,4}^{1D}$ corresponding to those shown in Figure 4.5 with n_1 varying from 1 to 1,000 for clear visualization. It is seen from Figure 4.6 that Z statistics dramatically drop to less than 1.96 after several MCMC iterations. In other words, $\omega_{s,1}^{1D}$, $\omega_{s,2}^{1D}$, $\omega_{s,3}^{1D}$, and $\omega_{s,4}^{1D}$ start to converge to stationary state after several MCMC iterations. To be conservative, the first 100 sets of MCMC samples may be taken as the burn-in period

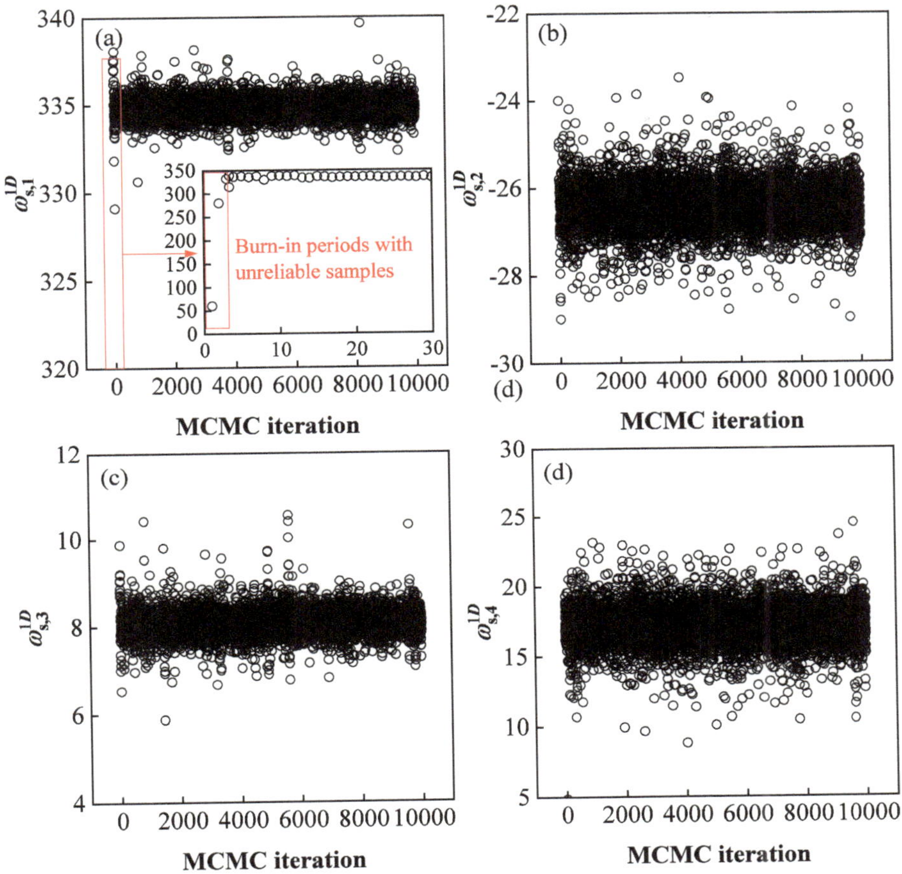

FIGURE 4.5 MCMC chains for (a) $\omega_{s,1}^{1D}$, (b) $\omega_{s,2}^{1D}$, (c) $\omega_{s,3}^{1D}$, and (d) $\omega_{s,4}^{1D}$.

and discarded before collection of stationary and representative MCMC samples for further analysis. A total of 10,100 sets of MCMC samples are simulated in this example. After discarding the first 100 sets of samples within the burn-in period, a total of 10,000 sets of MCMC samples are collected.

It is also worth noting that MCMC samples tend to be autocorrelated due to its simulation process. Therefore, it is preferable to reduce the autocorrelation among MCMC samples obtained by thinning the samples. One of the commonly used thinning methods is to pick up an MCMC sample after some iterations based on visual examination of experimental auto-covariance function (AF) of MCMC samples of a variable of interest, say, $\omega_{s,1}^{1D}$. Experimental AF may be computed from MCMC samples as (e.g., Christian et al. 1994; Wang et al. 2018):

$$\mathrm{AF}(\delta) = \frac{1}{N_{MCMC} - \delta} \sum_{j=1}^{N_{MCMC}-\delta} [(\omega_{s,1,j}^{1D} - \mu_{\omega,1})(\omega_{s,1,j+\delta}^{1D} - \mu_{\omega,1})] \tag{4.11}$$

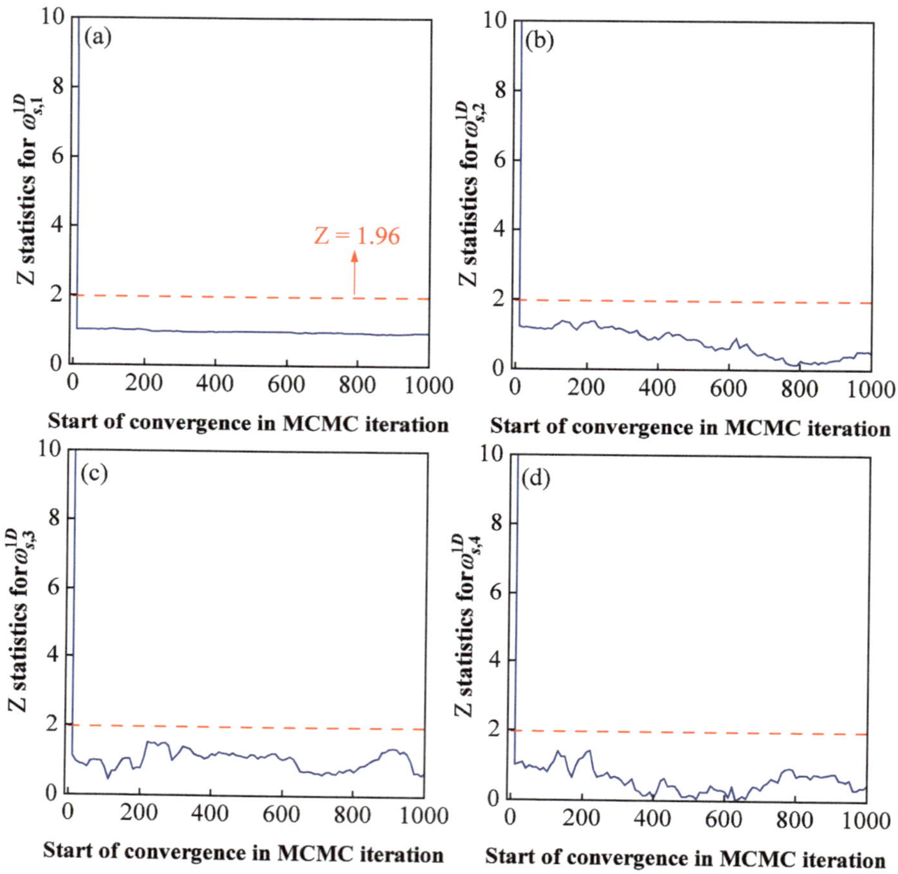

FIGURE 4.6 Determination of the start of convergence of MCMC by Z statistics (a) for $\omega_{s,1}^{1D}$, (b) for $\omega_{s,2}^{1D}$, (c) for $\omega_{s,3}^{1D}$, and (d) for $\omega_{s,4}^{1D}$.

where δ represents the lag distance in terms of integers; Figures 4.7a to 4.7d, respectively, show the autocorrelation function for $\omega_{s,1}^{1D}$, $\omega_{s,2}^{1D}$, $\omega_{s,3}^{1D}$, and $\omega_{s,4}^{1D}$ obtained by dividing AF obtained from Eq. (4.11) into the corresponding estimated variance, i.e.,$\sigma_{\omega,1}^2$, $\sigma_{\omega,2}^2$, $\sigma_{\omega,3}^2$, and $\sigma_{\omega,4}^2$. Figure 4.7 shows that the autocorrelation among MCMC samples of $\omega_{s,1}^{1D}$, $\omega_{s,2}^{1D}$, $\omega_{s,3}^{1D}$, and $\omega_{s,4}^{1D}$ reduces to less than 0.05 when the lag distance is larger than 2 or 3. In other words, all MCMC samples collected are almost independent and representative for a lag distance larger than 3.

Once these MCMC samples reach a stationary state, RFSs of interest can be obtained by a summation of $\mathbf{B}_s\boldsymbol{\omega}_s^{1D}$ and N realizations of ε. Samples of ε are realized from a Gaussian random variable with zero mean and variance of (τ^{-1}). Note that each set of $\boldsymbol{\omega}_s^{1D}$ sample corresponds to a τ sample when running the MCMC simulation. One set of $\boldsymbol{\omega}_s^{1D}$ and one set of N realizations of ε lead to one set of RFS, as shown in Figure 4.8a by a gray line. In this example, N_{MCMC} = 10,000 sets of MCMC samples are collected after discarding those within the burn-in period, leading to N_{MCMC} sets

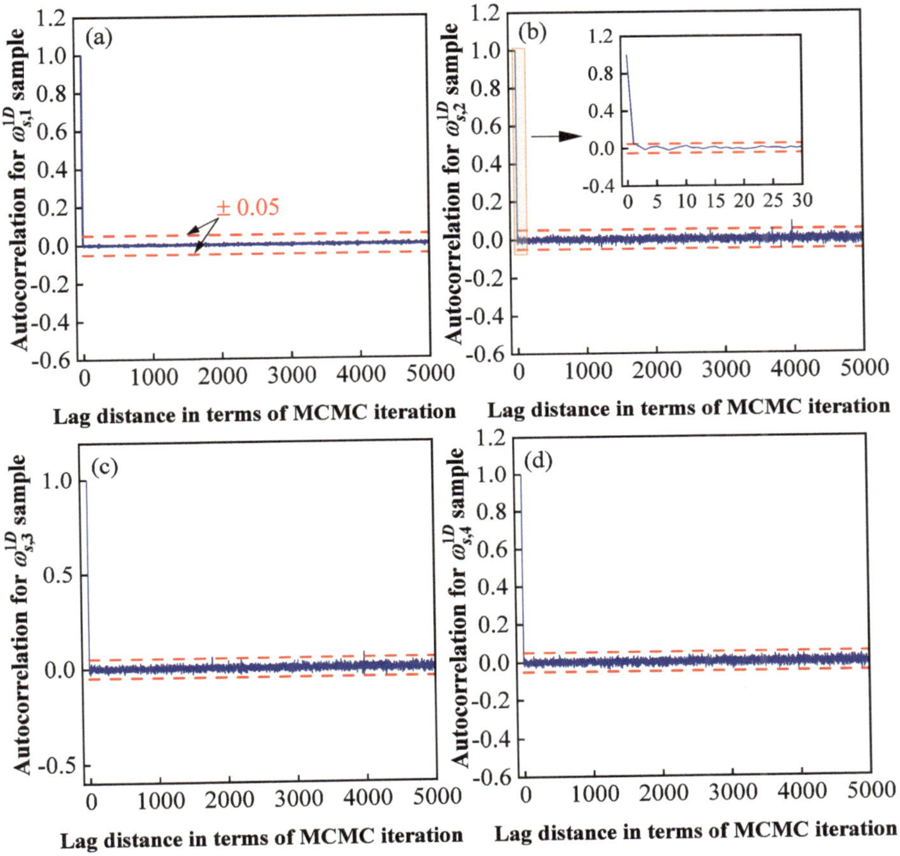

FIGURE 4.7 Autocorrelation for MCMC samples of (a) $\omega^{1D}_{s,1}$, (b) $\omega^{1D}_{s,2}$, (c) $\omega^{1D}_{s,3}$, and (d) $\omega^{1D}_{s,4}$.

of $\hat{f}$ samples. Then, the spatially varying mean and 95% CI can be obtained from the N_{MCMC} sets of $\hat{f}$ samples, as shown in Figure 4.8a by a dashed line and a pair of dotted lines. Similar to that shown in Figure 4.4a, both the gray line and the dashed lines are globally consistent with the measurements and underlying true q_c data. For further exploration, RFSs of q_c are simulated from the BCS-MCMC generator with different number, M, of measurement data points, i.e., M = 20, 50, and 100. Measurements corresponding to M = 20, 50, and 100 scenarios are shown in Figure 4.8b to Figure 4.8d by open circles, respectively. Then, following the same procedure, RFSs of q_c can be simulated, based on which results are shown in Figure 4.8b to Figure 4.8d with the same legends as those in Figure 4.8a. Figure 4.8 shows that as the number of measurements increases, the simulated RFSs of q_c become increasingly similar to the underlying q_c profile. Such observations indicate the data-driven nature of the BCS-MCMC generator. These observations from Figures 4.4 and 4.8 demonstrate that both the BCS-KL generator and BCS-MCMC generator work reasonably well and consistently.

4.2.3 Comparison between BCS-KL and BCS-MCMC Generators

A close examination of Figures 4.8 and 4.4 shows that the variability of RFSs from the BCS-MCMC generator is slightly larger than that from the BCS-KL generator, especially when M is relatively small (e.g., see Figure 4.8a and Figure 4.4a with $M = 10$). This may be attributed to the fact that more variabilities associated with the auxiliary variables (i.e., $\boldsymbol{\alpha}$, τ, γ) may propagate to $\boldsymbol{\omega}_s^{1D}$ and $\hat{\boldsymbol{f}}$ through the iterative procedure described in Section 4.2.2. In contrast, only the mean values of $\boldsymbol{\alpha}$, τ, and γ are used in estimation of $\boldsymbol{\omega}_s^{1D}$ in Section 4.2.1. As the number of measurements M increases, the uncertainties associated with $\boldsymbol{\alpha}$, τ, and γ quickly reduce. As a result, the varying 95% CI from BCS-MCMC generator quickly become almost identical to that from BCS-KL generator (see Figures 4.8b–4.8d and Figures 4.4b–4.4d for comparison). On the other hand, the RFSs simulated from the BCS-MCMC generator seem slightly more accurate than those from the BCS-KL generator, which may be justified by noting the consistency between the dashed line and solid line in Figures 4.8c–4.8d and Figures 4.4c–4.4d. This may be partly explained by an approximation of posterior PDFs performed for deriving Eq. (3.28), which is subsequently used as input to the BCS-KL generator. Although the BCS-MCMC generator is slightly more accurate, it is computationally more expensive than the BCS-KL generator. Consider, for example, the $M = 10$ case in Figure 4.4a and Figure 4.8a. It takes only 0.4 second for the BCS-KL generator to simulate 10,000 RFSs of $\hat{\boldsymbol{f}}$ in a commonly available desktop computer, while the BCS-MCMC generator takes around 11.4 seconds. In summary, both generators can be used to simulate realistic RFSs of geotechnical properties directly from sparse measurements with slight differences. The BCS-KL generator is more efficient in terms of computation, while the BCS-MCMC generator is more advantageous in simulating RFSs that are consistent with the underlying property profiles. Either generator may be used for simulating RFSs directly from sparse measurements in practice.

4.3 SIMULATION OF NONSTATIONARY RANDOM FIELDS WITHOUT DE-TRENDING

Two BCS-based random field generators, i.e., BCS-KL generator and BCS-MCMC generator, are presented in Section 4.2 for direct simulation of RFSs from sparse measurements. This section further illustrates their capability of dealing with geotechnical properties profiles, which often contain linear or nonlinear trends. Because of these trends, geotechnical site investigation data are inherently nonstationary. One commonly used method in conventional random field modeling is to transform data with a nonstationary trend of mean into stationary data via a removal of the trend (i.e., the so-called de-trending). De-trending is often performed using regression analysis, such as the least-squares method (e.g., Jaksa et al. 1997), on the available measurement data points, followed by stationary random field modeling of the residuals with a zero mean.

Although de-trending has been widely used to divide a random field with a trend in mean into a deterministic trend function and a stationary random field with a zero mean, it is a tricky process and deserves more rigorous attention (e.g., Fenton 1999b, a; Ching and Phoon 2017). Proper separation of the deterministic trend component

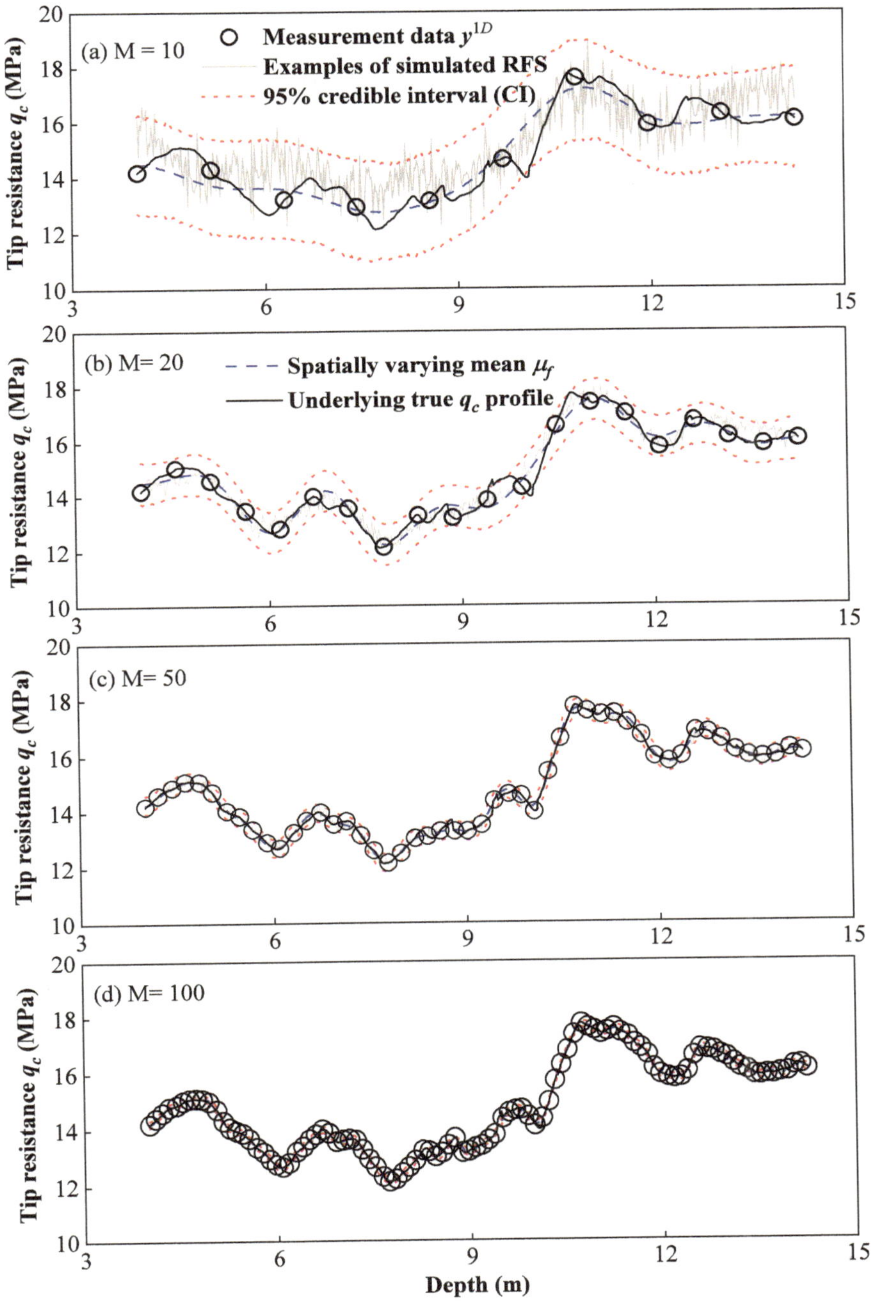

FIGURE 4.8 Statistical results of RFSs of q_c simulated by BCS-MCMC with different number, M, of measurements: (a) to (d) for M = 10, 20, 50, and 100, respectively.

and residuals (i.e., the stationary random field) might be difficult (e.g., Kulatilake 1991; Hu et al. 2001). The more variability of the data is modeled by the trend component, the less variability is left for the stationary random field modeling of the residuals (e.g., Fenton 1999b). Furthermore, de-trending is usually performed using a predetermined function form (e.g., linear or nonlinear) of trend function (e.g., Ching and Phoon 2017). However, the most appropriate form of trend function is often unknown in a real application. Using different forms of trend function may result in not only different deterministic trends, but also different parameters for the stationary random field (i.e., different random fields). Consider, for example, a polynomial function that has been widely used for de-trending (e.g., Jaksa et al. 1997; Stuedlein et al. 2012; Lloret-Cabot et al. 2014). Ching and Phoon (2017) showed that it is difficult to determine the most appropriate order for the polynomial function, particularly when there is only one realization (e.g., one set of measurement data). In addition, even when the form of trend function (e.g., the right order of polynomial function) can be correctly identified, the trend function and auto-covariance function estimated from the available measurements might not be reliable, particularly when the number of measurement data points is limited (e.g., Qi and Liu 2019).

The difficulties associated with de-trending in the simulation of RFSs with linear or nonlinear trend motivate the simulation of RFSs directly from sparse measurements without de-trending in this section. The BCS-KL or BCS-MCMC generators discussed and illustrated in Section 4.2 may be directly used to simulate nonstationary random fields without de-trending. Neither de-trending, selection of autocorrelation function, nor determination of correlation length is needed when using the BCS-KL or BCS-MCMC generators. Consequently, the difficulties associated with de-trending in the simulation of random fields with a linear or nonlinear trend are bypassed.

4.3.1 A Nonstationary Random Field Example

In this section, a numerical example is firstly used to demonstrate the issues associated with conventional random field simulation with de-trending. Consider, for example, a quantity Q that is modeled by a 1D lognormal random field with a quadratic trend in mean $\mu_Q = -4.8\times10^{-6}x^2 + 2.46\times10^{-3}x + 0.90$ and other random field parameters as follows: a constant variance $\sigma_Q^2 = 0.01$ and an exponential autocorrelation function ρ defined as

$$\rho_{i,j} = \exp(-2\Delta_{i,j} / \lambda_c) \tag{4.12}$$

where $\lambda_c = 51.2$ is the correlation length, and it quantifies the distance within which geo-properties are highly autocorrelated. Note that $\Delta_{i,j}$ in Eq. (4.12) is defined as the relative distance between location x_i and x_j (i, j = 1, 2, …, 512, for the 1D random field with a length of 512 in terms of integer). In other words, Q is nonstationary in the mean, but stationary in autocorrelation. "nonstationary in the mean" indicates that Q has a spatially varying mean; "stationary in autocorrelation" indicates that the autocorrelation between Q at different locations only depends on the relative distance between these locations. Given the parameters specified above, RFSs can be simulated using a random field generator, such as KL expansion as discussed in Eq.

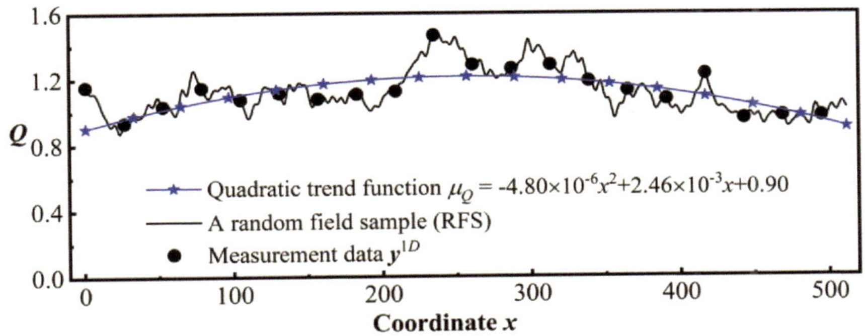

FIGURE 4.9 A random field sample (RFS) and limited measurement data y^{1D}.

(4.1) in Section 4.2.1. Figure 4.9 plots one example of the Q RFS by a solid line, together with the quadratic trend by a solid line with stars.

It is worth noting that, however, the random field parameters used for random field simulation in engineering practice are often unknown and need to be estimated from measurement data of Q before RFS generation (e.g., Stuedlein et al. 2012; Ching et al. 2016). The measured data are often sparse and limited (e.g., see the solid circles in Figure 4.9) due to equipment resolution, limited budget, time, or access, and sensor failure, etc. In addition, when the measurement data exhibits a significant trend, linear or nonlinear, de-trending is often performed first on the measured data to separate the measurement data into a trend component and residuals before estimating random field parameters from the residuals (i.e., a stationary random field).

4.3.2 Random Field Simulation Using Conventional Method with a Linear Trend

Conventional random field modeling is performed for comparison with the BCS-KL generator in Section 4.2.1. It starts with the de-trending by a commonly used linear function to remove the trend embedded in the measurement data $\boldsymbol{y}^{1D}$ with $M = 20$ data points, as shown in Figure 4.10 by solid circles. A linear regression is performed to estimate the linear trend function as $Q_{tr}(x) = -3.1\times10^{-5}x + 1.14$ and residuals $\boldsymbol{y}_r^{1D}$, as shown in Figures 4.10b and 4.10c, respectively. Then, the residuals $\boldsymbol{y}_r^{1D}$ are used to calculate the sample AF by Eq. (4.11), which is subsequently fitted to an assumed AF function type (e.g., exponential auto-covariance function) for random field parameter estimation. The sample AF obtained from $\boldsymbol{y}_r^{1D}$ is shown in Figure 4.11 by crosses.

Suppose that the auto-covariance function type is correctly known as an exponential auto-covariance function. Then, the sample AF is fitted to an exponential auto-covariance function using an ordinary least-squares method. The estimated variance and correlation length are obtained as $\sigma^2 = 0.0162$ and λ=79.87, respectively, in this example. When compared with the prescribed parameters (i.e., $\sigma_Q^2 = 0.01$ and $\lambda_c = 51.2$) of the original random field, the estimated variance (σ^2) and correlation length (λ_c) are significantly biased. It should also be noted that an exponential auto-covariance function has been used when fitting the sample AF. This is consistent with

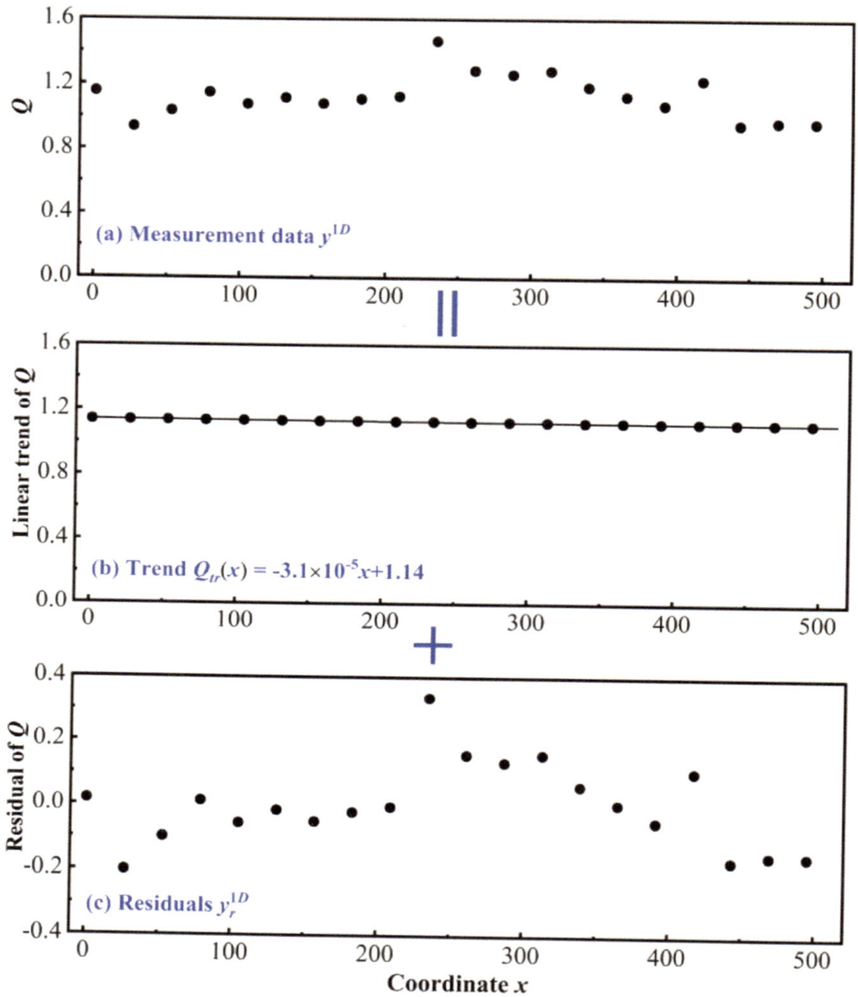

FIGURE 4.10 Illustration of de-trending procedure using measurement data y^{1D} (modified from Wang et al. 2019).

the prescribed auto-covariance function of the original random field. In practice, the type of auto-covariance function is generally unknown, and hence, selection of the proper type of auto-covariance function might be challenging and requires additional efforts (e.g., using Bayesian model selection method, see Cao and Wang 2014).

With the estimated linear trend function $Q_{tr}(x) = -3.1\times10^{-5}x + 1.14$ and random field parameters estimated for the residuals (i.e., $\sigma^2 = 0.0162$ and λ=79.87), samples of the nonstationary lognormal random field can be generated using conventional methods, such as KL expansion in Section 4.1. Figure 4.12 shows an example of an RFS generated by a gray solid line. Table 4.1 summarizes statistics (e.g., mean and SD) of this RFS and those of the original Q data. The relative difference between

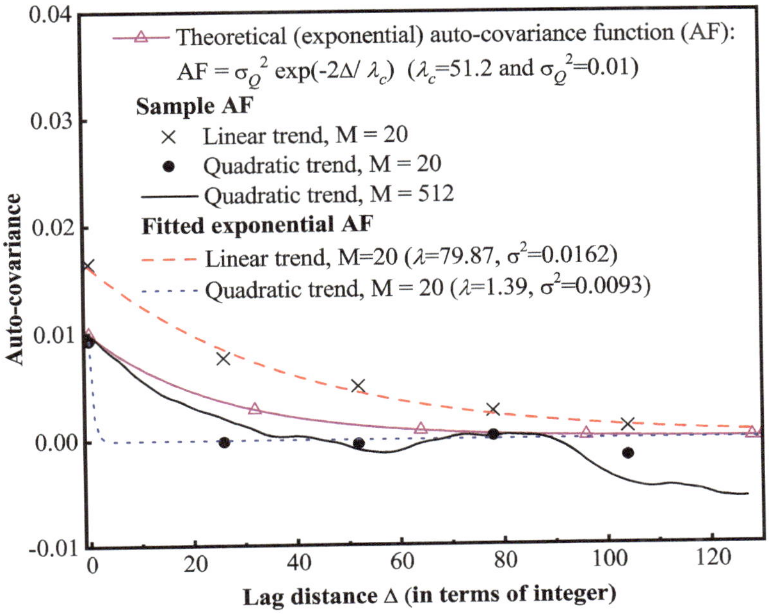

FIGURE 4.11 Sample and fitted auto-covariance functions after removal of a linear or quadratic trend (modified from Wang et al. 2019).

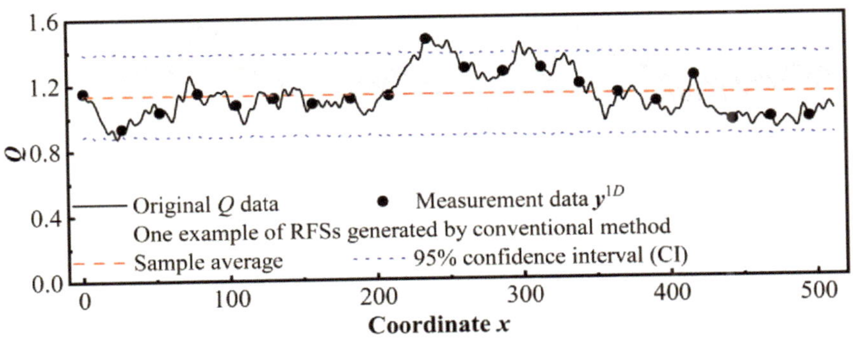

FIGURE 4.12 RFSs generated using conventional method with removal of a linear trend (modified from Wang et al. 2019).

these two sets of statistics is around 9.6% for mean and about 5.8% for SD, respectively. The difference is generally small, and the results are consistent among each other. A large number of RFSs can be easily simulated, and statistical analysis on the simulated RFSs can be performed. For example, $N_{RFS} = 500$ RFSs are generated, and the average of the simulated Q data at each x coordinate is calculated and shown by a dashed line in Figure 4.12. Note that this average is virtually identical to the linear trend function (i.e., $Q_{tr}(x) = -3.1\times10^{-5}x + 1.14$) estimated from measurement data

TABLE 4.1
Statistics of Random Field Samples and the Original Q Data

Statistics	Original Q Data	RFS from Conventional Method with Removal of		RFS from BCS-KL Generator
		Linear Trend	Quadratic Trend	
Mean	1.120	1.228 (9.64%)	1.135 (1.34%)	1.126 (0.54%)
Standard deviation, SD	0.137	0.129 (5.84%)	0.128 (6.57%)	0.151 (10.0%)

Note: $M = 20$. Numbers in bracket represent relative difference when compared with statistics of original Q data.

TABLE 4.2
Summary of Quadratic Regression Coefficients Estimated from y^{1D}

Regression Coefficients	Original Q Data	Measurement Data y^{1D}
a_2 ($\times 10^{-6}$)	−4.80	−4.22 (12.08%)
a_1 ($\times 10^{-3}$)	2.38	2.06 (13.44%)
a_0	0.93	0.97 (4.74%)

Note: $M = 20$. a_2, a_1, and a_0 are coefficients of quadratic function $Q_{tr}(x) = a_2x^2+a_1x+a_0$. The number in bracket represents relative difference when compared with coefficients of the original Q data.

y^{1D}. Using the N_{RFS} RFSs generated, a 95% confidence interval (CI) of the Q profile can also be obtained and shown by two dotted lines in Figure 4.12. The 95% CI of Q data at each x coordinate is defined as the range between the 2.5th and 97.5th percentile of the N_{RFS} Q data points at the x coordinate of interest. Figure 4.12 shows that the generated RFSs follow a slightly linear trend because of the de-trending process. This is different from the original Q data, which exhibit a quadratic trend with Q increasing before $x = 256$ and Q decreasing after $x = 256$ (see the solid line with stars in Figure 4.9). RFSs generated from the conventional method with removal of a linear trend might not be representative of the original data targeted in the simulation.

4.3.3 Random Field Simulation Using Conventional Method with a Quadratic Trend

In this section, a quadratic function is selected as the most appropriate form of trend function. This example illustrates the situation where the selected trend function fortuitously agrees with the actual trend function which is unknown in practice. A de-trending procedure similar to that shown in Figure 4.10 is carried out to obtain the best-fitted quadratic function from the measurement data (i.e., the solid circles in Figure 4.9). Table 4.2 summarizes the best-fitted coefficients obtained from regression analysis in the third column. For comparison, the regression coefficients of the original Q data (i.e., the solid line in Figure 4.9) are also included in the second column of Table 4.2. The difference between the coefficients estimated from $\mathbf{y}^{1D}$ and those of the original Q data varies from 4.74% to 13.44%. The coefficients of the quadratic trend estimated from measurement data $\mathbf{y}^{1D}$ are consistent with those of the original Q data, but not identical. Even when the form of trend function is correctly identified, the trend function coefficients estimated from the measurement data may not be identical to the actual coefficients because measurement data points used in the regression analysis are "noisy" and sparse.

The residuals $\mathbf{y}_r^{1D}$ obtained after removal of the quadratic trend are used together with Eq. (4.11) to calculate the sample AF, as shown by solid circles in Figure 4.11. Then the random field parameters are estimated by fitting the sample AF to the assumed auto-covariance function (i.e., exponential auto-covariance function) using the least-squares method. The parameters obtained in this example include λ=1.39 and σ^2=0.0093. When compared with λ_c = 51.2 in the prescribed random field, λ= 1.39 is grossly underestimated. This may be attributed partly to the limited number of measurement data points (i.e., M = 20 is relatively small when compared with N = 512) used in the analysis. In this case, the residuals of a limited number of measurement data points cannot effectively reflect autocorrelation of the original random field. It is curious that the selection of an incorrect linear trend produces significantly better estimates of σ^2 and λ. This illustrates the complications associated with de-trending. When more accurate estimation of random field parameters is pursued, other information shall be incorporated using, e.g., Bayesian methods (e.g., Qi and Liu 2019; Stuedlein et al. 2012).

Similar to Section 4.3.2, after the removal of quadratic trend and estimation of random field parameters from the residuals, samples of the lognormal random field can be generated using conventional methods, such as KL expansion. A large number of RFSs may be generated, and the sample average and 95% CI can be estimated accordingly. Similar to Figure 4.12, Figure 4.13 shows the sample average and 95% CI from N_{RFS} = 500 RFSs generated by a dashed line and a pair of dotted lines, respectively. It is obvious that the dashed line is consistent with the quadratic trend function of the original random field (see the solid line with stars in Figure 4.9).

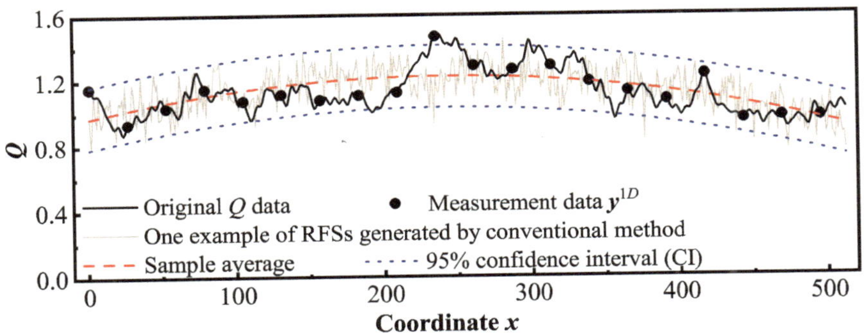

FIGURE 4.13 RFSs generated using conventional method with removal of a quadratic trend.

Figure 4.13 also shows an example RFS taken from the $N_{RFS} = 500$ RFSs by a gray solid line. Because λ=1.39 is used in the random field simulation, the gray solid line in Figure 4.13 varies rapidly and exhibits apparent randomness or very short correlation length. This is quite different from the pattern of the original Q data, which obviously is spatially varying and autocorrelated. Since most variability and autocorrelation of the limited measurement data points are considered by the quadratic trend function, less variability or autocorrelation is left for the residuals after de-trending (e.g., Fenton 1999a, b). It is also worth noting that, although the autocorrelated pattern of the original Q data is not reflected in the RFS, the mean and SD of RFS still agree well with those of the original Q data. Table 4.1 summarizes the mean and SD of this RFS in the fourth column. The relative difference is about 1.3% and 6.6% for the mean and SD, respectively. This is because the mean and SD of RFS are mainly affected by the trend function, rather than the residuals.

It is evident from Figure 4.13 that even when the correct form of trend function is used, the autocorrelated pattern of the random field might not be properly simulated by conventional methods with de-trending. To bypass the issues associated with de-trending, the BCS-KL generator discussed in Section 4.2.1 is used to simulate samples of random fields with a linear or nonlinear clear trend without performing de-trending.

4.3.4 Random Field Simulation Using BCS-KL Random Field Generator

In this section, RFSs are simulated directly from $\boldsymbol{y}^{1D}$ using the BCS-KL random field generator (See Section 4.2.1). No de-trending is needed when using the BCS-KL generators, as the method is data-driven and directly applicable to measurement data which might be either stationary or nonstationary. Therefore, the issues associated with de-trending shown previously for conventional methods are bypassed. Figure 4.14 shows an example of RFSs simulated by the BCS-KL generator by a gray line, together with the original Q data by a solid line. Table 4.1 also includes the mean and SD of this RFS sample, which are consistent with those estimated from the original Q data, with only about 0.5% and 10% relative differences for the mean and SD, respectively. With a large number of RFSs, the average and 95% CI of Q value along

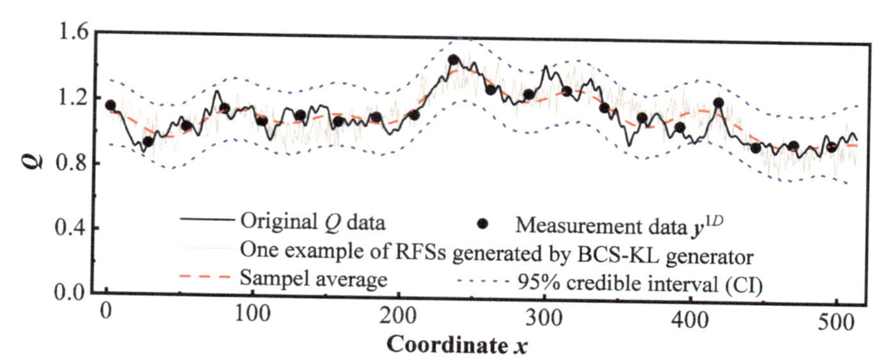

FIGURE 4.14 RFSs directly simulated from measurement data y^{1D} without de-trending using the BCS-KL generator.

the x-axis can also be obtained, as shown in Figure 4.14, using the same symbols as those in Figures 4.12 and 4.13. Figure 4.14 shows that the sample average follows a trend very similar to the original Q data, and almost all variations of the original Q data fall within the 95% CI obtained.

To further evaluate the trend in each RFS simulated by the BCS-KL generator, a quadratic regression is performed for each of the RFSs simulated here. In this example, $N_{RFS} = 500$ RFSs are generated directly from the BCS-KL generator, leading to N_{RFS} sets of quadratic regression coefficients. The statistics (i.e., mean and SD) of the N_{RFS} sets of regression coefficients are summarized in the third and fourth columns of Table 4.3, respectively. For comparison, Table 4.3 also includes the regression coefficients of the original data Q (i.e., the solid line with stars shown in Figure 4.9) in the second column. The relative errors vary from 0.75% to 6.46% between the mean of coefficients from the N_{RFS} RFSs generated by the BCS-KL generator and the coefficients of the original Q data. The differences are quite small, indicating that the coefficients of the quadratic trend in the RFSs generated by the BCS-KL generator are reasonable and accurate. In addition, the SD of the coefficients, listed in the fourth column in Table 4.3, may be used as an index to evaluate reliability of the trend reflected by the RFSs from the BCS-KL generator. When compared with the mean of the coefficients, the SD values vary from several percent to less than 20% of their respective mean values. The trend reflected by the RFSs from the BCS-KL generator is generally reliable. In summary, the comparison in Table 4.3 shows that, even with $M = 20$, the BCS-KL generator is able to properly generate RFSs that accurately and reliably reflect the trend of the original Q data.

To evaluate the autocorrelated pattern of each RFS simulated by the BCS-KL generator, sample AF is calculated for each RFS using Eq. (4.11), after removal of the mean quadratic trend (i.e., $Q_{tr}(x) = -4.49\times10^{-6}x^2+2.30x\times10^{-3}+0.94$ in Table 4.3) from each of the N_{RFS} RFSs. This leads to $N_{RFS} = 500$ AFs in total. Using the $N_{RFS} = 500$ AFs, the average and 95% CI for the 500 AFs can be obtained accordingly, and they are shown as a dashed line and a pair of dotted lines with solid circles, respectively, in Figure 4.15. It shows that the average AF (i.e., the dashed line with solid circles) follows a trend generally similar to the AF of the original Q data (i.e., the solid line) and almost all variations of the solid line fall within the 95% CI obtained. The comparisons above demonstrate that the RFSs generated by the BCS-KL generator are able to properly represent the autocorrelation pattern of the original Q data.

Although the RFSs generated from the BCS-KL generator are capable of characterizing the main characteristics of the underlying Q data, some of the local variations of the original Q data cannot be fully captured. For example, see the dashed line and the solid line in Figure 4.14 or the dashed line with solid circles and solid line in Figure 4.15. This may be attributed to a limited number (i.e., $M = 20$ out of 512) of measurement used as input. As M increases, the results shown above gradually improve, as shown in the next section.

4.3.5 Effect of Data Quantity on the RFSs Simulated from the BCS-KL Generator

In this section, the effect of measurement data quantity on the RFSs simulated from the BCS-KL generator is investigated, including the number, M, of measurement data

TABLE 4.3
Quadratic Regression Coefficients Estimated from RFSs Generated by the BCS-KL Generator under Different *M* Cases

Regression Coefficients	Original *Q* Data	RFSs from the BCS-KL Generator under Different *M* Cases							
		$M = 20$		$M = 64$		$M = 128$		$M = 512$	
		Mean	SD	Mean	SD	Mean	SD	Mean	SD
a_2 ($\times 10^{-6}$)	–4.80	–4.49 (6.46%)	0.90	–4.93 (2.71%)	0.21	–4.77 (0.63%)	0.13	–4.79 (0.21%)	0.003
a_1 ($\times 10^{-3}$)	2.38	2.30 (3.36%)	0.46	2.40 (0.84%)	0.11	2.37 (0.42%)	0.07	2.38 (0.00%)	0.001
a_0	0.93	0.94 (0.75%)	0.05	0.94 (0.75%)	0.012	0.93 (0.11%)	0.008	0.93 (0.00%)	0.001

Note: a_2, a_1, and a_0 are coefficients of quadratic function $Q_{tr}(x) = a_2x^2+a_1x+a_0$. The number in bracket represents relative error when compared with coefficients of the original *Q* data.

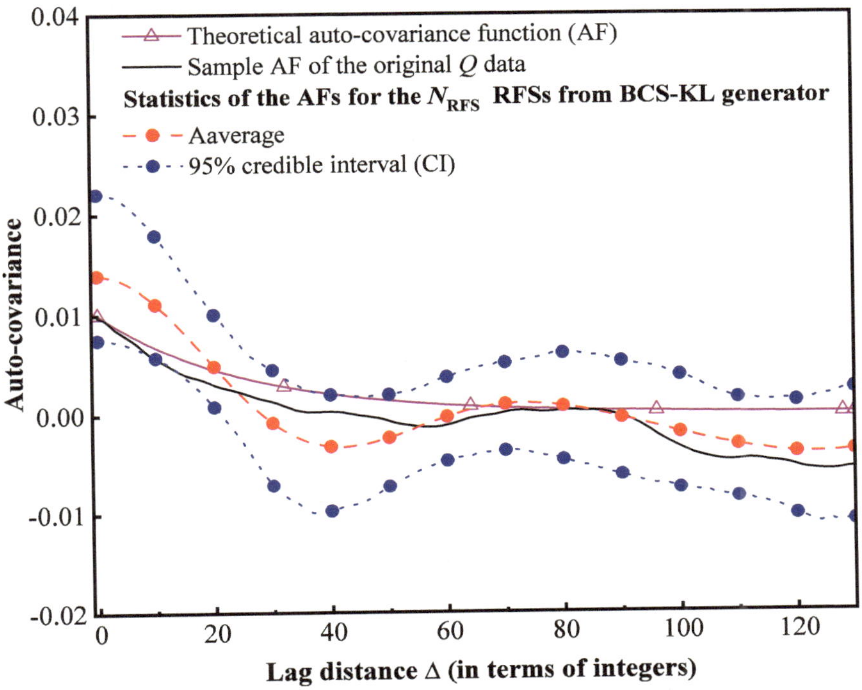

FIGURE 4.15 Sample auto-covariance functions for the RFSs generated from the BCS-KL random field generator.

points within one set of data, and the number, N_s, of sets of sparse measurements, which are discussed next.

4.3.5.1 Effect of Number, *M*, of Measurements

Four *M* scenarios with a constant $N_s = 1$ are adopted to explore the effect of *M* on the characteristics of RFSs simulated from the BCS-KL generator, including $M = 64$, 128, and 512, in addition to the $M = 20$ scenario described in Section 4.3.4. Following the same procedure for the $M = 20$ scenario, $N_B = 500$ RFSs are generated directly from y^{1D} using the BCS-KL generator for each *M* scenario. Figures 4.16a–d, respectively, show an example of RFSs generated by the BCS-KL generator by a gray line, together with the original *Q* data by a solid line for $M = 20$, 64, 128, and 512 scenarios. For each *M* scenario, the average and 95% CI of *Q* value along the *x*-axis are calculated using the $N_{RFS} = 500$ RFSs. The results are shown in Figures 4.16a–d for each *M* scenario, respectively, using the same symbols as those in Figure 4.14. It is shown that the sample averages become increasingly similar to the original *Q* data as *M* increases. Moreover, the region defined by the 95% CI becomes increasingly narrow as *M* increases. All these observations suggest that, as *M* increases, the RFSs simulated by the BCS-KL generator become increasingly realistic and representative of the original *Q* data.

The convergence of the sample average and 95% CI to the original *Q* data implies that the trend characterized by the RFSs generated using the BCS-KL generator

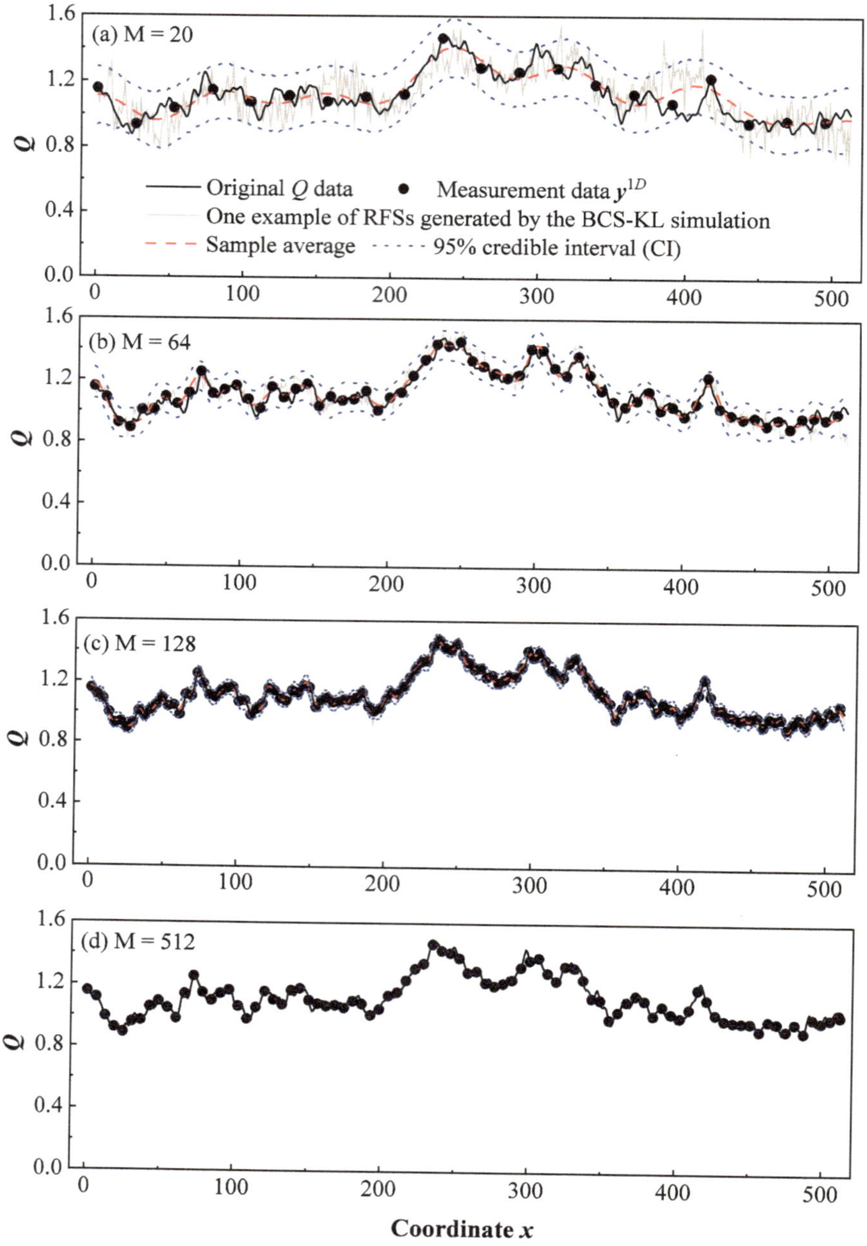

FIGURE 4.16 Direct generation of nonstationary RFSs using the BCS-KL simulation for different M scenarios: (a) $M = 20$, (b) $M = 64$, (c) $M = 128$, and (d) $M = 512$.

becomes increasingly similar to that of the original Q data with an increases of M. For each of the N_{RFS} = 500 RFSs in each M scenario, a quadratic trend function is estimated, leading to N_{RFS} sets of coefficients of quadratic function for each M scenario. Then, the mean and SD of each coefficient are calculated and summarized in Table 4.3. Table 4.3 shows that as M increases from M = 20 to 512, the mean of each coefficient becomes increasingly similar to the coefficient of the original Q data. Consider, for example, the coefficient a_2. As M increases from M = 20 to 512, the relative error between the mean of a_2 and the a_2 estimated directly from the original Q data decreases from 6.46% to 0.21%. In addition, the SD of a_2 also decreases from 0.90 to 0.003 as M increases. Similar observations can also be found in a_1 and a_0 (see Table 4.3). The decreases of the relative difference between the mean of coefficients and the coefficients of the original Q data, as well as the decreases in the SD of the coefficients, indicates that the accuracy and reliability of the trend function characterized by the RFSs from the BCS-KL generator improve as M increases.

Figure 4.17 shows the average and 95% CI of AFs estimated for each M scenario. The AFs and its statistics for M = 64, 128, and 512 scenarios are calculated using the same procedure as that for the M = 20 scenario described above. Figure 4.17 shows that as M increases, the average AF becomes increasingly similar to the AF of the original Q data. This suggests that the autocorrelation structure characterized by the RFSs from the BCS-KL generator becomes increasingly accurate as M increases. In addition, as M increases, the 95% CI (i.e., the dotted lines with solid circles in Figure 4.17) gradually converges to the sample AF of the original Q data (i.e., the solid line in Figure 4.17), indicating that the reliability of the estimated autocorrelation pattern from BCS-KL generator also improves. As M increases, the accuracy and reliability of the autocorrelation pattern characterized by the RFSs from the BCS-KL generator both improve.

It is worth pointing out that the trend and autocorrelation structure characterized by the RFSs generated by the BCS-KL generator approach to those of the original Q data, rather than those of the prescribed random field (see the solid line with open triangles in Figures 4.17). This is because one set of Q data, even with 512 data points, is not enough to represent the ensemble, i.e., an infinitely large population of realizations (none of which are identical) that is the theoretically equivalent to the prescribed lognormal random field with a quadratic trend. However, as the number, N_s, of sets of available measurement data increases, both the trend and the autocorrelation pattern converge to the prescribed random field, as discussed below.

4.3.5.2 Effect of Number, N_s, of Sets of Measurements

In this section, suppose that there are N_s = 10, 50, and 100 sets of sparsely measured data (e.g., M = 20), which can be taken as input to the BCS-KL generator discussed in Section 4.3.4. These measurements may be obtained by repeatedly measuring the quantity Q at the same locations, as shown by open circles in Figure 4.9. For each set of measurement data, N_B = 500 RFSs are generated using the BCS-KL generator. N_s sets of limited measurement data therefore lead to $N_s \times N_B$ = 5,000, 25,000, 50,000 RFSs in total, for N_s = 10, 50, and 100 scenarios, respectively. Then, the average and 95% CI of Q value along the x-axis are calculated, as shown in Figure 4.18b. Note that N_s = 1 scenario with M = 20 is included in Figure 4.18a for comparison. Each

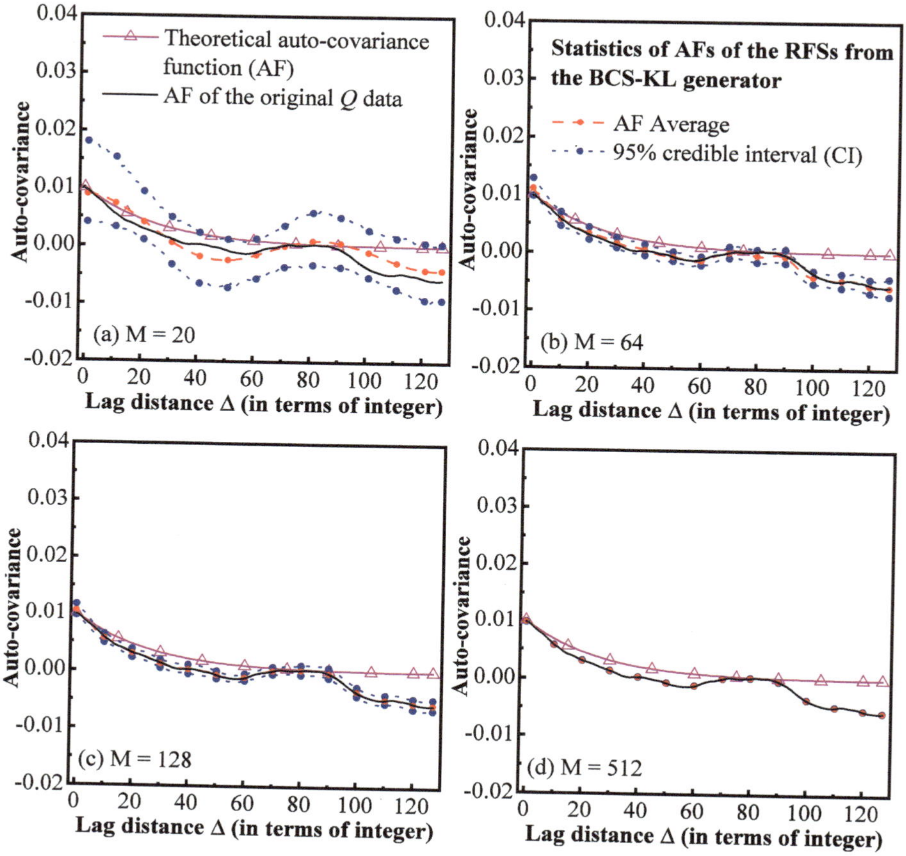

FIGURE 4.17 Auto-covariance function (AF) of the original Q data and those of the RFSs from BCS-KL generator under different M scenarios: (a) $M = 20$, (b) $M = 64$, (c) $M = 128$, and (d) M=512.

subplot in Figure 4.18 also includes the prescribed trend function (i.e., $\mu_Q = -4.8\times 10^{-6}x^2+2.46\times10^{-3}x+0.9$) shown as solid lines with stars.

Figure 4.18 shows that, as N_s increase, the dashed line (i.e., sample average) gradually converges to the solid line with stars (i.e., the prescribed trend function). This suggests that the trend characterized by the RFSs using the BCS-KL generator becomes increasingly consistent with the prescribed trend as N_s increases, even when only $M = 20$ measurement data points out of 512 data points in total are available in each data set. In addition, as N_s increases, the AF characterized by the RFSs also statistically converges to the prescribed AF (i.e., the theoretical exponential AF). After removal of the mean trend function for each N_s scenario (e.g., $Q_{tr} = -4.78\times10^{-6}x^2+ 2.41\times10^{-3}x+0.89$ for $N_s = 10$), the AFs are obtained for each of the $N_s \times N_B$ RFSs using Eq. (4.11). Then, the average and 95% CI for AF are calculated and shown in Figure 4.19 for $N_s = 1$, 10, 50, and 100 scenarios, using the same symbols as those in Figure 4.17. For comparison, each subplot in Figure 4.19 also includes the theoretical exponential AF by a solid line with open triangles. It is shown that as N_s increases,

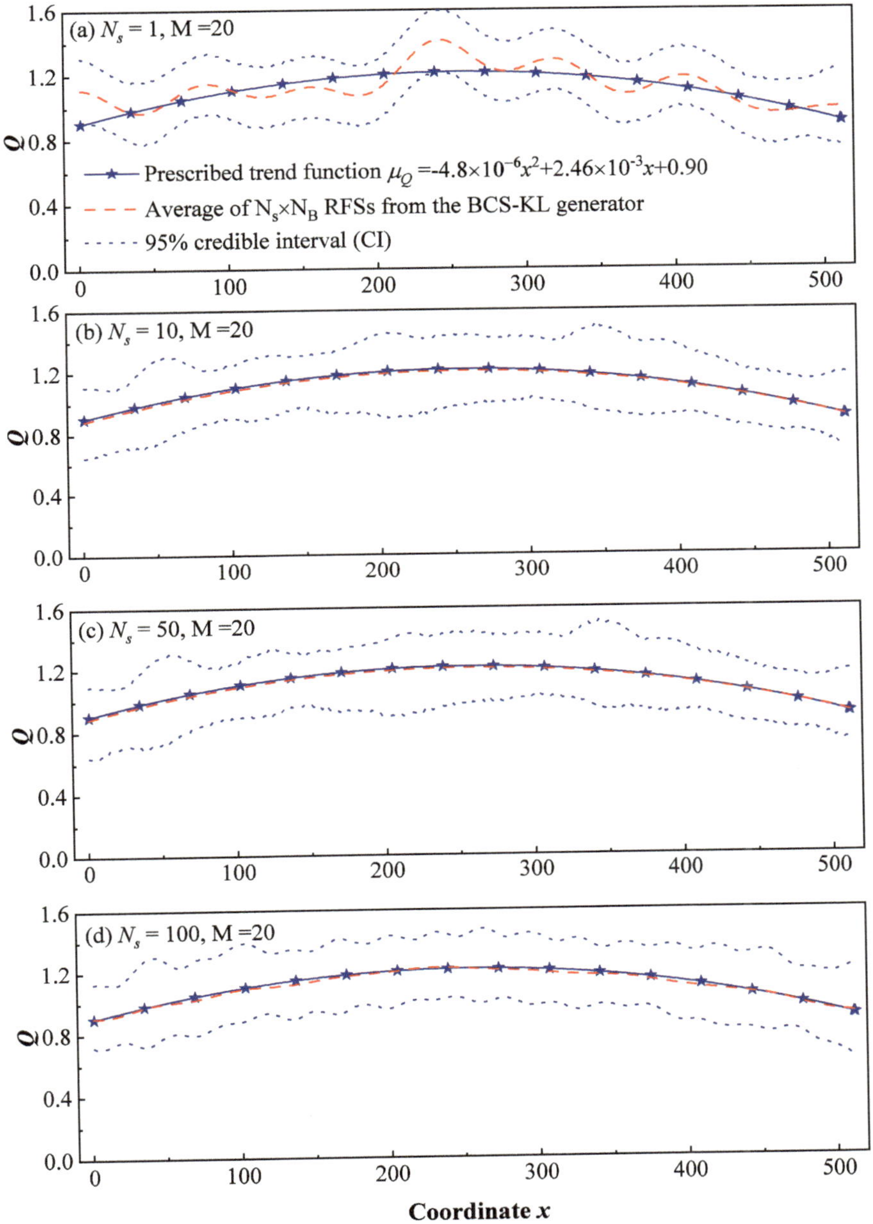

FIGURE 4.18 Statistics of nonstationary RFSs directly simulated from the BCS-KL generator for different N_s scenarios with M = 20: (a) N_s = 1, (b) N_s = 10, (c) N_s = 50, and (d) N_s = 100.

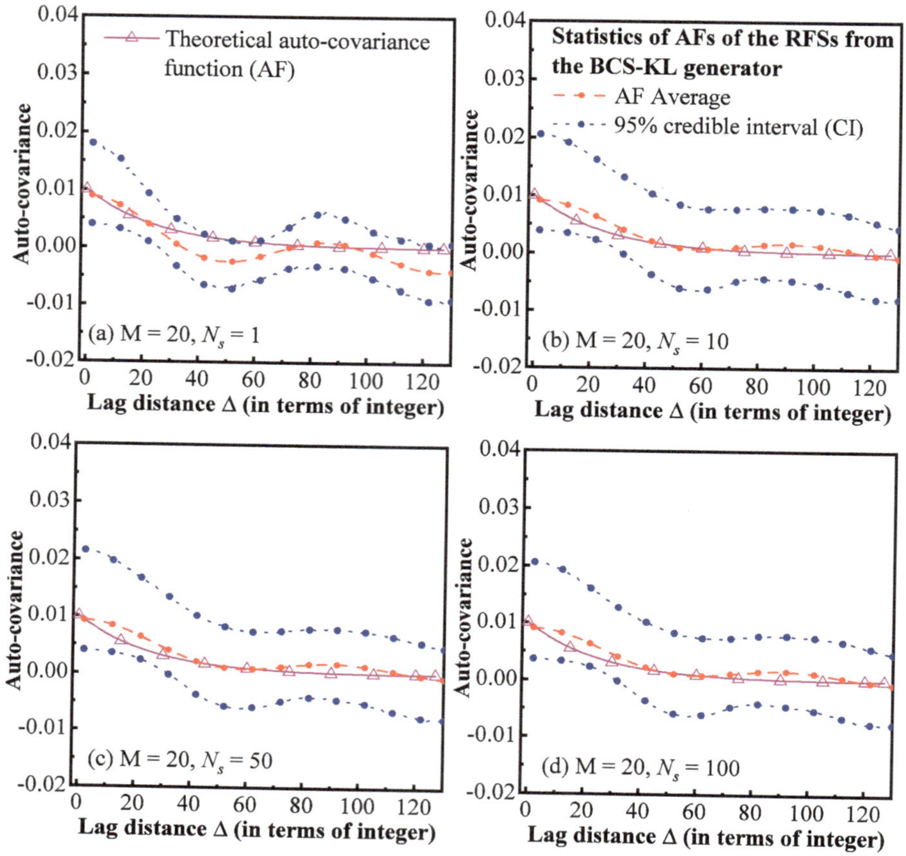

FIGURE 4.19 Auto-covariance function (AF) of the Q data and those of the BCS-KL-generated RFSs under different N_s scenarios with $M = 20$: (a) $N_s = 1$, (b) $N_s = 10$, (c) $N_s = 50$, and (d) $N_s = 100$.

the dashed line (i.e., the average AF) gradually approaches to the solid line with open triangles (i.e., the theoretical exponential AF).

It should be pointed out that the differences between the theoretical and estimated AFs are considerable, even when N_s is relatively large (see Figures 4.19c and d). This is due to the fact that only $M = 20$ data points are measured for each of the N_s datasets. If M increases to a relatively large value, the differences mentioned above are effectively reduced. To illustrate this point, three additional sets of simulations are performed with $N_s = 100$ sets of measurement data, each of which contains $M = 64$, 128, or 512 data points, respectively. The AF estimated for $N_s = 100$ with $M = 20$, 64, 128, and 512 are summarized in Figure 4.20, using the same symbols as those in Figure 4.19. Figure 4.20 shows that, as M increases, with N_s remained as 100, the dashed line (i.e., the average AF) gradually converges to the solid line with open triangles (i.e., the theoretical AF).

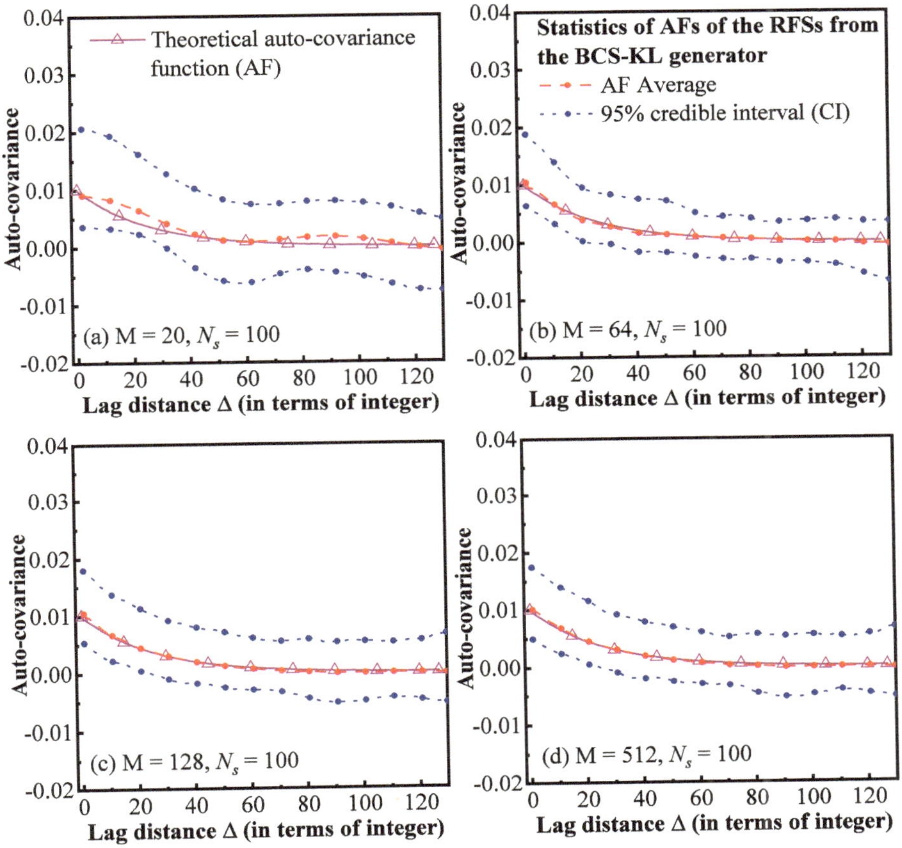

FIGURE 4.20 Auto-covariance (AF) of the Q data and those of the BCS-KL-generated RFSs under different M scenarios with $N_s = 100$: (a) $M = 20$, (b) $M = 64$, (c) $M = 128$, and (d) $M = 512$.

4.4 SIMULATION OF NON-GAUSSIAN NONSTATIONARY RANDOM FIELDS

The BCS-based random field generators, i.e., BCS-KL or BCS-MCMC generators, are applicable to not only Gaussian random fields with stationary correlation structure, but also those with nonstationary correlation structure and non-Gaussian probability distributions. The performance of BCS-based random field generators on non-Gaussian, nonstationary spatial or temporal data are illustrated in this section. A nonstationary random field with highly skewed non-Gaussian marginal cumulative distribution function (CDF) is first generated by conventional KL expansion. Then, sparse measurements are taken as input to the BCS-based generators, e.g., the BCS-KL generator. The resulting random field is subsequently compared with the original or theoretical one.

The random field of a quantity f simulated has a shifted exponential distribution, with the PDF expressed as follows:

$$p(f) = m\,\mathrm{e}^{-m(f-\theta)} \tag{4.13}$$

where m represents the decay factor and determines the rate at which the probability of an event decreases; θ is a location parameter, and it plays a crucial role in shifting the distribution along the axis. Unlike the standard exponential distribution, which starts at zero, the shifted exponential distribution starts at θ. The mean and SD of the shifted exponential distribution in Eq. (4.13) are $1/m+\theta$ and $1/m$, respectively. In the simulation of this random field, both m and θ vary along the x-axis to ensure a constant mean of 1, as explained in the next paragraph.

The shifted exponential random field simulated in this section is discretized into $N = 512$ points, leading to a complete RFS with a length of $N = 512$. In this example, a nonstationary covariance matrix $\mathbf{C}$ is adopted, which is formed from a stationary correlation coefficient matrix $\mathbf{C}^s$ and a linearly varying SD function K_x. $\mathbf{C}^s$ is defined as

$$C_{i,j}^s = \exp\left(-\frac{2\Delta_{i,j}}{\lambda_c}\right)\cdot\left(1+\frac{2\Delta_{i,j}}{\lambda_c}\right) \tag{4.14}$$

where $C_{i,j}^s$ represents the correlation coefficient between two f values, f_i and f_j, at coordinates x_i and x_j, respectively; $\Delta_{i,j}$ is the relative distance between x_i and x_j, and $\Delta_{i,j} =| x_i - x_j|$; and λ_c is the correlation length ($\lambda_c = 25$ in this example). Furthermore, the linearly varying SD function K_x is defined as

$$K_x = \left(-\frac{0.6}{511}(x-1)+1\right)\sigma \tag{4.15}$$

where σ is taken as 0.5 in this example. Based on the stationary correlation matrix $\mathbf{C}^s$ and the linearly varying K_x, the nonstationary covariance matrix $\mathbf{C}$ in this example is defined as

$$C_{i,j} = C_{i,j}^s \cdot K_{x_i} \cdot K_{x_j} \tag{4.16}$$

where $C_{i,j}$ is the element of $\mathbf{C}$ indexed by i and j, and K_{x_i} is the SD of f_i at x_i. Eq. (4.16) indicates that the $C_{i,j}$ among two points depends on the absolute locations x due to the incorporation of K_x, and therefore $\mathbf{C}$ is a nonstationary covariance matrix. In addition, as pointed out in the previous paragraph, the decay factor m and location parameter θ in Eq. (4.13) vary along the x-axis to ensure a constant mean of 1. In this example, m at each x location, i.e., x_i (i = 1, 2, …, 512), is taken as the square root of the ith element of the first row of matrix $\mathbf{C}$ and $\theta = 1\text{-}1/m$.

In this example, a truncated KL expansion in Eq. (4.2) with 200 KL terms is used to simulate this non-Gaussian and nonstationary random field. Note that 200 KL terms preserve 99.9% of the total variance of the random field using the sum of 200 eigenvalues and the sum of all eigenvalues, i.e., $\sum_{i=1}^{200} \lambda_i / \sum_{i=1}^{512} \lambda_i = 99.9\%$. Therefore, the 200 KL terms can accurately represent the prescribed random field. In this example, 1,000 sets of complete RFSs are generated using the KL expansion. Note that an exponential and stationary correlation function is used in Eq. (4.14) because it is widely used in the literature and previous studies (e.g., Fenton and Griffiths 2008). Although the correlation function is stationary, the covariance function (i.e., Eq. 4.16) is nonstationary due to the linearly varying K_x (see Eq. 4.15).

Figure 4.21 compares the characteristics of the theoretical random field and those estimated from the 1,000 RFSs simulated, including CDF at a specific location $x = 256$ (see Figure 4.21a), varying correlation coefficient (see Figure 4.21b), and linearly varying SD (see Figure 4.21c). Figure 4.21a shows a very good match between the theoretical and simulated CDFs at $x = 256$. The difference between theoretical and simulated CDF is below 0.5% for all x values. Additionally, the theoretical correlation coefficient is compared with the one simulated from KL expansion in Figure 4.21b. The theoretical correlation coefficient is the modified exponential term in Eq. (4.13); hence, it does not depend on K_x or the coordinate of x. A good match is achieved using the KL expansion. Figure 4.21c compares the theoretical K_x function with the one simulated from KL expansion. The latter is a square root of the diagonalization of the covariance function for all 1,000 RFSs, which is also the SD evaluated for all 1,000 values at each x coordinate. The theoretical function and the one obtained from KL expansion are almost identical. Figure 4.22 shows a direct comparison between the theoretical covariance matrix and the covariance matrix obtained from 1,000 RFSs generated by KL expansion. The two covariance matrices are in good agreement with each other. As shown by Phoon et al. (2005), KL expansion is capable of accurately

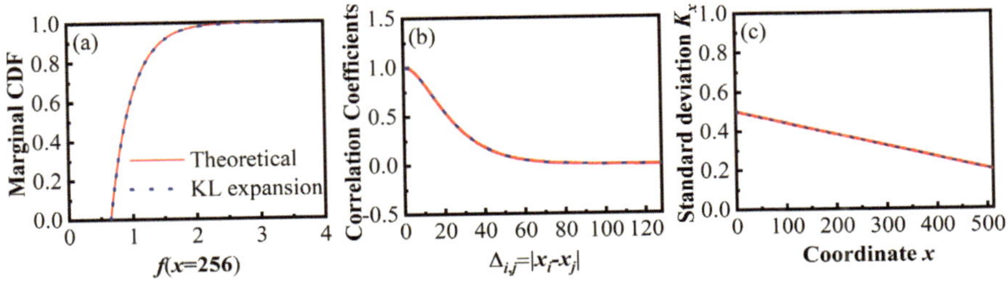

FIGURE 4.21 Comparison between the theoretical and KL expansion: (a) marginal CDF at $x = 256$; (b) correlation coefficient, i.e., the first row of matrix $\mathbf{C}^s$; and (c) standard deviation function K_x for the nonstationary shifted exponential random field (modified from Montoya-Noguera et al. 2019).

simulating strongly non-Gaussian marginal distributions and nonstationary covariance functions simultaneously.

4.4.1 Simulation of a Shifted Exponential Random Field from Sparse Measurements

The BCS-based generator is able to directly simulate a random field from a limited number (N_s) of sets of sparse measurement data. For demonstration, 100 out of the 1,000 RFSs above are used. In each RFS, only 50 out of the 512 data points are taken as measurements or input (about 1% of the original data) to the BCS-KL generator (i.e., $N_s = 100$ and M = 50). Using each set of sparse measurement data as input, the BCS-KL generator simulates $N_{\mathrm{RFS}} = 500$ complete RFSs. Hence, from the N_s sets of sparse measurement data, $N_s \times N_{\mathrm{RFS}}$ sets of complete RFSs in total are simulated by the BCS-KL generator. In this example, from a sparse data set consisting of a matrix of size 100 × 50, the BCS-KL generator results in a matrix with a size of 50,000 × 512.

Figure 4.23a shows an example of a complete RFS generated from the truncated KL expansion of the nonstationary shifted exponential, by a solid line, and the $M = 50$ data points on the RFS, by solid circles. These data points are measurements from the complete RFS and used as input for the BCS-KL generator. Examples of the $N_{\mathrm{RFS}} = 500$ complete RFSs simulated by the BCS-KL generator are shown in Figure 4.23b by gray lines. Figure 4.23b also shows the mean $\boldsymbol{f}$ vector obtained from the N_{RFS} RFSs by a red dashed line and the 95% CI by a pair of blue dotted lines.

The same procedure is performed for each of the $N_s = 100$ sets of sparse measurement data to generate 100 × 500 = 50,000 complete RFSs. The results of the generated RFSs are compared with the theoretical random field in Figure 4.24. The theoretical function of the marginal CDF of f at $x = 256$ is shown by a red solid line in Figure 4.24a, while the CDF obtained from the N_s sets of complete RFSs generated by KL expansion is shown by a blue dashed line. The CDF obtained from the RFSs

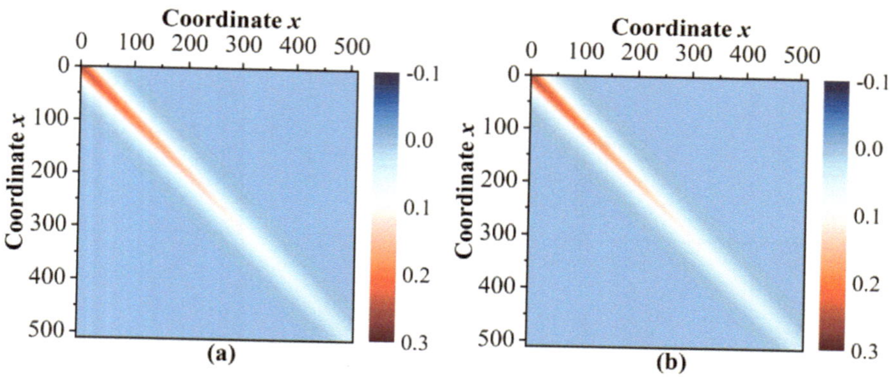

FIGURE 4.22 Comparison of covariance matrix for the nonstationary shifted exponential random field: (a) theoretical covariance matrix and (b) covariance matrix derived from 1,000 RFSs (modified from Montoya-Noguera et al. 2019).

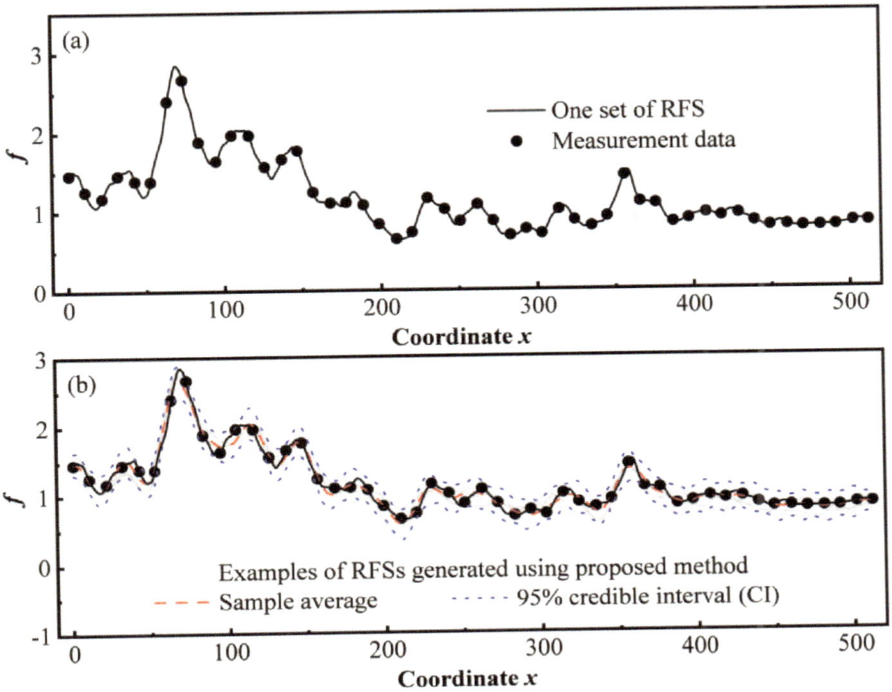

FIGURE 4.23 (a) One set of shifted exponential random field sample with M = 50 measurement data. (b) Examples of RFSs simulated by BCS-KL generator.

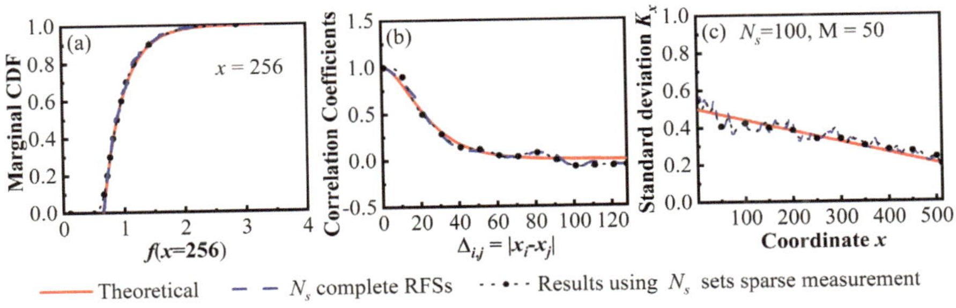

FIGURE 4.24 Simulation results from the BCS-KL generator for the nonstationary shifted exponential random field: (a) marginal CDF at x = 256, (b) correlation coefficient, and (c) standard deviation K_x.

generated by the BCS-KL generator using N_s sets of sparse measurement data as input is shown by a black dotted line with solid circles. The marginal CDF from the BCS-KL generator is highly consistent with the theoretical CDF and the one computed for the original N_s sets of complete RFSs.

Figure 4.24b shows the theoretical correlation coefficient function and those evaluated using the N_s sets of complete RFSs or from BCS-KL generator with the N_s sets of sparse measurement data. The correlation coefficient function from the BCS-KL results is in reasonable agreement with the theoretical function. Figure 4.24b also shows that the differences between the theoretical correlation coefficient function and that from the sparse measurement data are slightly larger than the difference between the correlation coefficient function for the N_s sets of complete RFSs and that from the BCS-KL results with the N_s sets of sparse measurement data.

Figure 4.24c shows the theoretical K_x function and the K_x functions obtained for the N_s sets of complete RFSs or from the BCS-KL generator with the N_s sets of sparse measurement data. The K_x function shows the nonstationarity of the random field generated in this example. As the number, N_s, of samples is relatively small, the K_x function obtained from the N_s sets of complete RFSs (i.e., the blue dashed line) does not match perfectly the theoretical function (i.e., the red solid line). Therefore, although the K_x function obtained from the BCS-KL generator matches reasonably well the one obtained for the N_s sets of complete RFSs (e.g., see a comparison between the black dotted line with solid circles and the blue dashed line in Figure 4.24c), both K_x functions obtained from simulations deviate slightly from the theoretical function (e.g., see a comparison between the black dotted line with solid circles and the red solid line in Figure 4.24c). In other words, the accuracy of the K_x function obtained from the BCS-KL generator depends on how well the N_s sets of complete RFSs represent the targeted nonstationary process.

Figure 4.25 compares the theoretical covariance matrix (see Figure 4.25a) with those obtained from the $N_s = 100$ sets of complete RFSs generated by KL expansion (see Figure 4.25b) and the BCS-KL generator using $N_s = 100$ sets of sparse measurement data (see Figure 4.25c). It is evident that the three covariance matrices contain similar autocorrelation patterns (i.e., see the similar distribution of color in Figure 4.25) and have similar magnitude for the autocorrelation (i.e., see the similar color in Figure 4.25), although some local variations are observed for off-diagonal terms in Figures 4.25b and 4.25c. In this case, only 10% of the complete RFSs (i.e., 100 out of 1,000 complete RFSs generated by KL expansion) from which less than

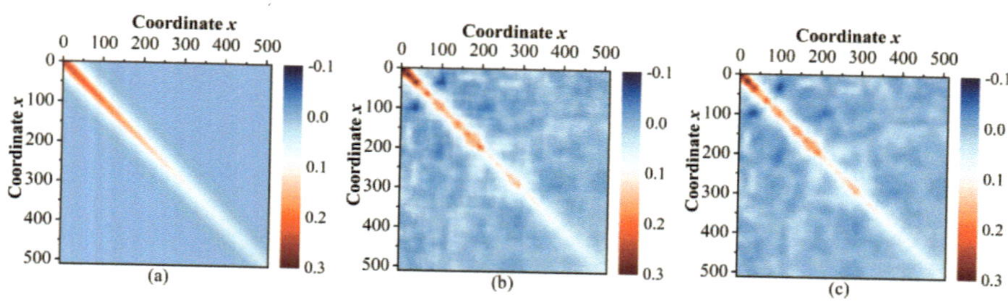

FIGURE 4.25 Comparison of covariance matrix for the nonstationary shifted exponential random fields: (a) theoretical covariance matrix, (b) covariance matrix derived from 100 complete RFSs, and (c) covariance matrix derived from $N_s \times 500$ samples by the BCS-KL generator (M = 50).

10% of data points (i.e., 50 data points out of 512 data points in each RFS, in total less than 10%×10% = 1% of the data) were taken as input to the BCS-KL generator. The results shown in Figures 4.24 and 4.25 suggest that the BCS-KL generator presented in this chapter is able to simulate reasonably the theoretical marginal CDF, the correlation coefficient function, the K_x function, and the covariance matrix simultaneously.

4.4.2 Effect of Data Quantity on the RFSs Simulated from the BCS-KL Generator

The accuracy of the marginal CDF, the correlation coefficient function, and the K_x function obtained from the BCS-KL generator are affected by the number, N_s, of sets of sparse measurement data and the number, M, of measurement data points within each data set. In this section, the effects of both N_s and M on the accuracy of the results are investigated below.

4.4.2.1 Effect of Number, N_s, of Sets of Measurements

To investigate the effect of N_s, three N_s scenarios were tested: i.e., N_s = 10, 50, and 200, in addition to the N_s = 100 in Section 4.4.1. From each of the N_s datasets, M = 50 measurement data points are taken from the same coordinates as those in Figure 4.23a. Then, for each set of sparse measurement data, N_{RFS} = 500 RFSs with a length N = 512 are simulated by the BCS-KL generator. Then, the RFSs obtained for each N_s case are represented by a matrix with 512 columns and 10 N_{RFS}, 50 N_{RFS}, 100 N_{RFS}, and 200 N_{RFS} rows for N_s = 10, 50, 100, and 200, respectively. For each N_s value, the marginal CDF at x = 256, correlation coefficient function, and the K_x function are estimated accordingly, as shown in Figures 4.26–4.28.

Figures 4.26a–d plot the estimated marginal CDFs at x = 256 for N_s = 10, 50, 100, and 200, respectively. Figures 4.26a and b (i.e., N_s = 10 and 50, respectively) clearly show that there are relatively large differences between the marginal CDF obtained from the BCS-KL generator and the theoretical marginal CDF at x = 256, as shown by black dotted lines with solid circles and red solid lines, respectively. However, the differences are smaller in Figure 4.26b than those shown in Figure 4.26a, when the number N_s increases from 10 to 50. This is reasonable as more information on the targeted random field is used in the simulation when the number of sets of sparse measurement data used increases from 10 to 50.

A comparison between Figures 4.26b and c (i.e., N_s = 50 and 100, respectively) shows that the differences mentioned above further reduce as more sets of sparse measurement data are used. When N_s increases from 100 to 200, the differences between the theoretical and the marginal CDF from the BCS-KL generator are similar, although the difference in Figure 4.26d is slightly smaller than those in Figure 4.26c, no significant improvement is gained when increasing the N_s values from 100 to 200 in this example. In addition, the marginal CDFs are evaluated from the original N_s sets of complete RFSs for x = 256, as shown in Figure 4.26 by blue dashed lines. As the N_s value increases, these blue dashed lines plot closer to the red solid lines, i.e., the theoretical marginal CDF. Since the results of the BCS-KL generator are dependent on the original N_s sets of complete RFSs used, the results of the BCS-KL generator

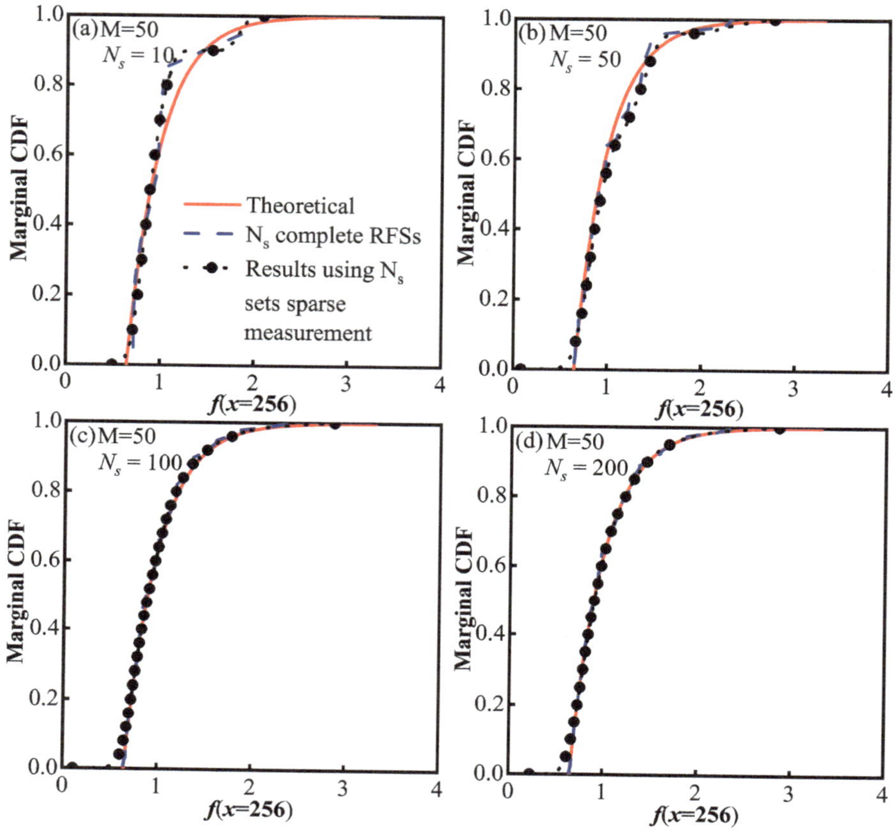

FIGURE 4.26 Effect of the number N_s set of sparse measurements used for BCS-KL generator on the marginal CDF at $x = 256$ with M = 50: (a) $N_s = 10$, (b) $N_s = 50$, (c) $N_s = 100$, and (d) $N_s = 200$.

match the marginal CDF prescribed by the N_s sets of complete RFSs, but they might deviate from the theoretical marginal CDF, when the N_s is small.

Figure 4.27 shows the effect of the number of sets of sparse measurement data on the correlation coefficient function obtained from the simulation results. As shown in Figures 4.27a and b, when N_s is small, relatively large differences are observed between the correlation coefficient function obtained from the BCS-KL generator and the theoretical one. A comparison between Figure 4.27c and d (i.e., $N_s = 100$ and 200, respectively) shows that the accuracy obtained in the correlation coefficient function only improves slightly as N_s increase. This suggests that, similar to the results in Figure 4.26 for the marginal CDF, $N_s = 100$ is able to achieve a reasonable accuracy in the correlation coefficient function.

Figure 4.28 shows the theoretical K_x function and those obtained from the BCS-KL generator for different N_s values and from the N_s sets of complete RFSs. As shown in Figure 4.28a (i.e., $N_s = 10$), the function obtained from the BCS-KL method oscillates

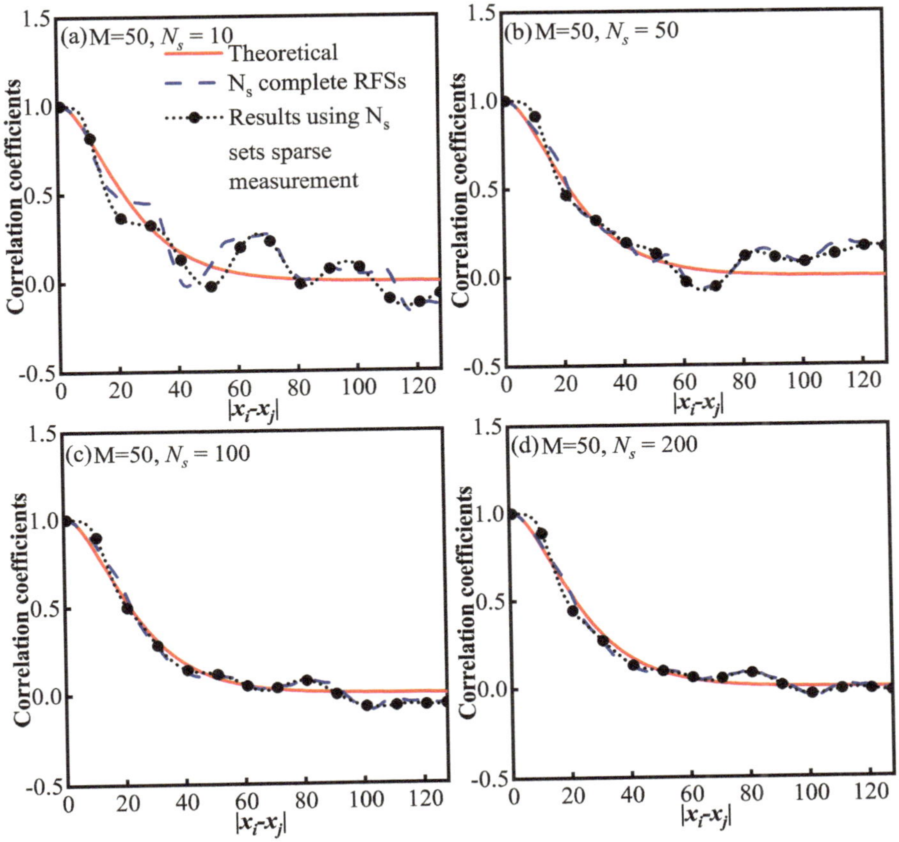

FIGURE 4.27 Effect of the number N_s of set of sparse measurements used for BCS-KL generator on the correlation coefficient with M = 50: (a) N_s = 10; (b) N_s = 50; (c) N_s = 100; (d) N_s = 200.

around the theoretical function, presenting big positive and negative differences with respect to the theoretical value. As the N_s value increases, the variations around the theoretical function reduce (see Figures 4.28b–d, for N_s = 50, 100, and 200, respectively). The differences between the theoretical K_x function and the one from the BCS-KL generator may be evaluated using the mean absolute error (MAE), which is defined as

$$\text{MAE} = \frac{1}{N} \mid K_{x,i} - \hat{K}_{x,i} \mid \tag{4.17}$$

where $K_{x,i}$ is the ith theoretical SD value computed by Eq. (4.15); $\hat{K}_{x,i}$ is the corresponding value re-estimated from the RFSs simulated by the BCS-KL generator.

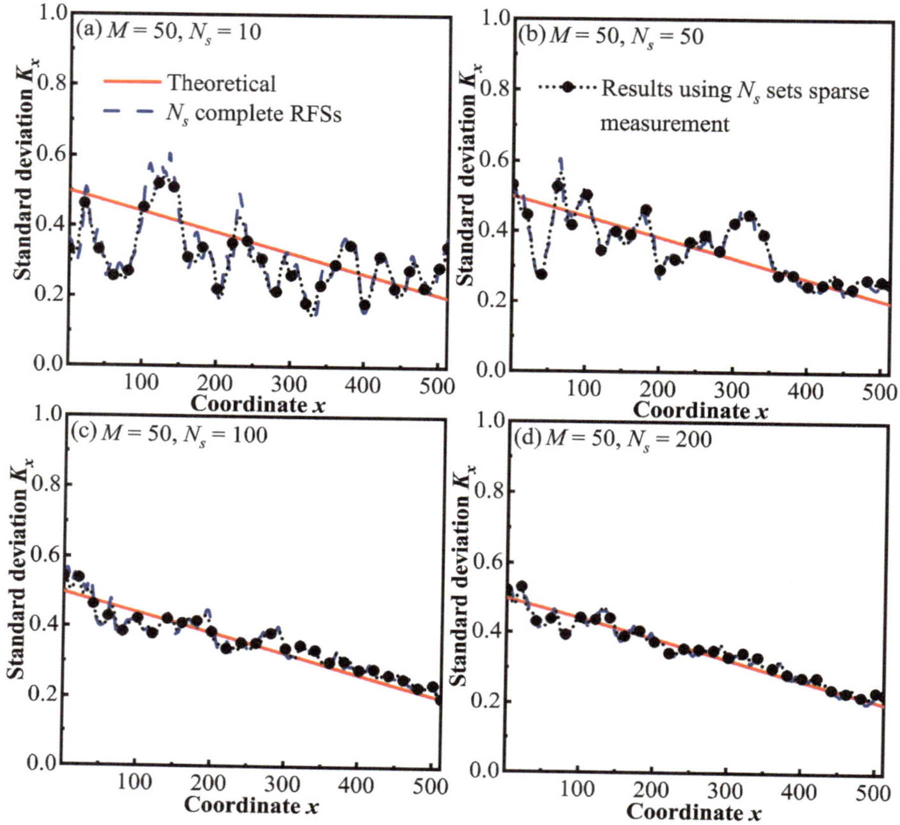

FIGURE 4.28 Effect of the number N_s of set of sparse measurements used for the BCS-KL generator on the correlation coefficient with M = 50: (a) N_s = 10, (b) N_s = 50, (c) N_s = 100, and (d) N_s = 200.

Based on Eq. (4.17), MAE are computed as 0.079, 0.048, 0.026, and 0.018, respectively, for N_s = 10, 50, 100, and 200. The MAE values are relatively small, especially when N_s is reasonably large, e.g., N_s = 100. Additionally, the K_x function obtained from the BCS-KL generator matches well the one obtained from the N_s sets of complete RFSs (i.e., the black dotted line with solid circles and the blue dashed line). All these observations suggest the BCS-KL generator is able to simulate the nonstationarity in the targeted random field using N_s sets of sparse measurement data.

4.4.2.2 Effect of Number, *M*, of Measurements within One Dataset

This section investigates the effect of the number, M, of measurement data points used in each of the N_s sets of sparse data on the marginal CDF, the correlation coefficient function, and the K_x function obtained from the BCS-KL generator. In addition to the M = 50 scenario used previously, three additional M scenarios are also tested, i.e., M = 10, 100, and 200. These values correspond to approximately 2%, 20%, and 40%

of all data points (i.e., 512 data points) in each set of complete RFS. $N_s = 100$ sets of incomplete RFSs are used as sparse measurement data for the BCS-KL generator for all M scenarios. Then, for each set of sparse measurement data, $N_{RFS} = 500$ RFSs with a length $N = 512$ are generated by the BCS-KL generator. Thus, the resulting RFSs for each M case are represented by a matrix with 512 columns and $100 \times N_{RFS}$ rows. For each case, the marginal CDF at a given x coordinate (e.g., $x = 256$), correlation coefficient function, and the K_x function can be estimated from the resulting RFSs.

Figure 4.29 shows the results of marginal CDF at $x = 256$ when M increases from 10 to 200. As M increases, the marginal CDF obtained from the BCS-KL generator gradually converges to the one obtained from $N_s = 100$ sets of complete RFSs (i.e., the black dotted lines with solid circles converge to the blue dashed lines). Note that there are slight differences between the theoretical marginal CDF (i.e., the red solid line) and the marginal CDF obtained from $N_s = 100$ sets of complete RFSs (i.e., the blue dashed lines), particularly at the tails of the CDF. This difference diminishes as N_s value increases, as discussed in the previous section. When

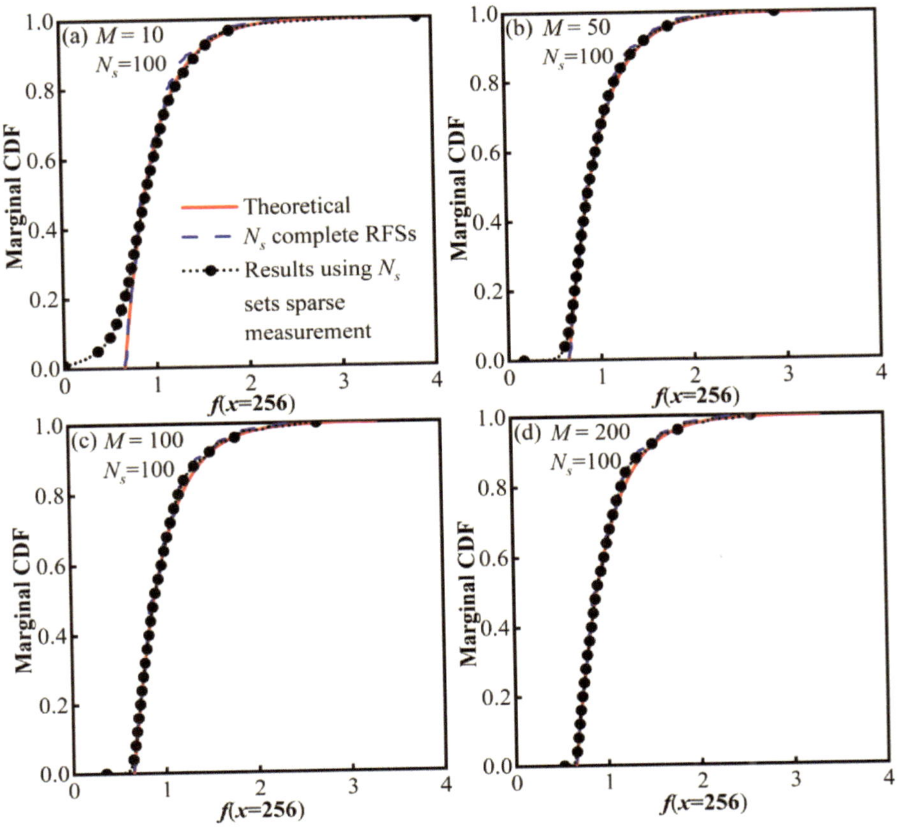

FIGURE 4.29 Effect of the number of data points M in each set of sparse measurement on the correlation coefficient: (a) $M = 10$, (b) $M = 50$, (c) $M = 100$, and (d) $M = 200$.

comparing Figures 4.29b–d, no significant difference can be observed, and no significant improvement is gained when increasing the M values from 50 to 200 in this example. This observation suggests that using $M = 50$ measurement data points in each of the $N_s = 100$ RFSs in this example is able to provide reasonable marginal CDF when using the BCS-KL generator.

Figure 4.30 shows the effect of M on the correlation coefficient function. As M increases, the differences decrease between the theoretical function and that obtained from the BCS-KL generator. A comparison between Figures 4.30a and b (i.e., $M = 10$ and 50, respectively) shows that the accuracy obtained in the correlation coefficient function significantly improves as the number of data points in each sparse measurement increases. However, the improvement gradually diminishes when M further increases to 50, 100, and 200 (see a comparison among Figures 4.30b–d). This suggests that $M = 50$ data points in each set of sparse measurements in this example

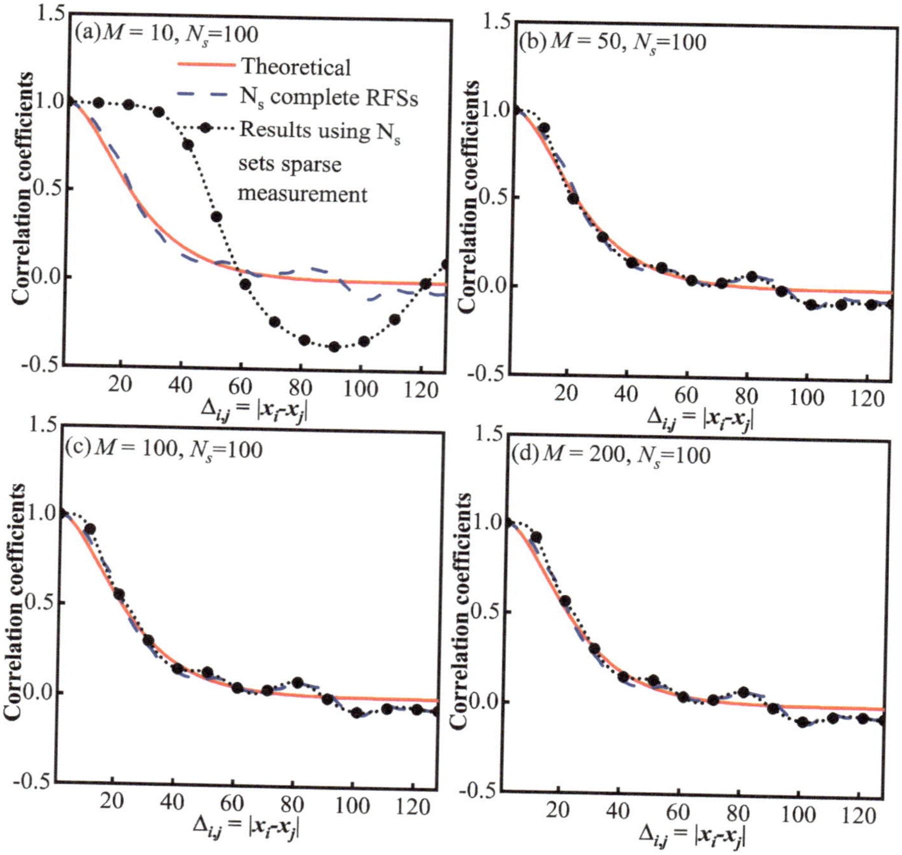

FIGURE 4.30 Effect of the number of data points M in each set of sparse measurement on the correlation coefficient (in terms of the first row of matrix $\mathbf{C}^s$): (a) $M = 10$, (b) $M = 50$, (c) $M = 100$, and (d) $M = 200$.

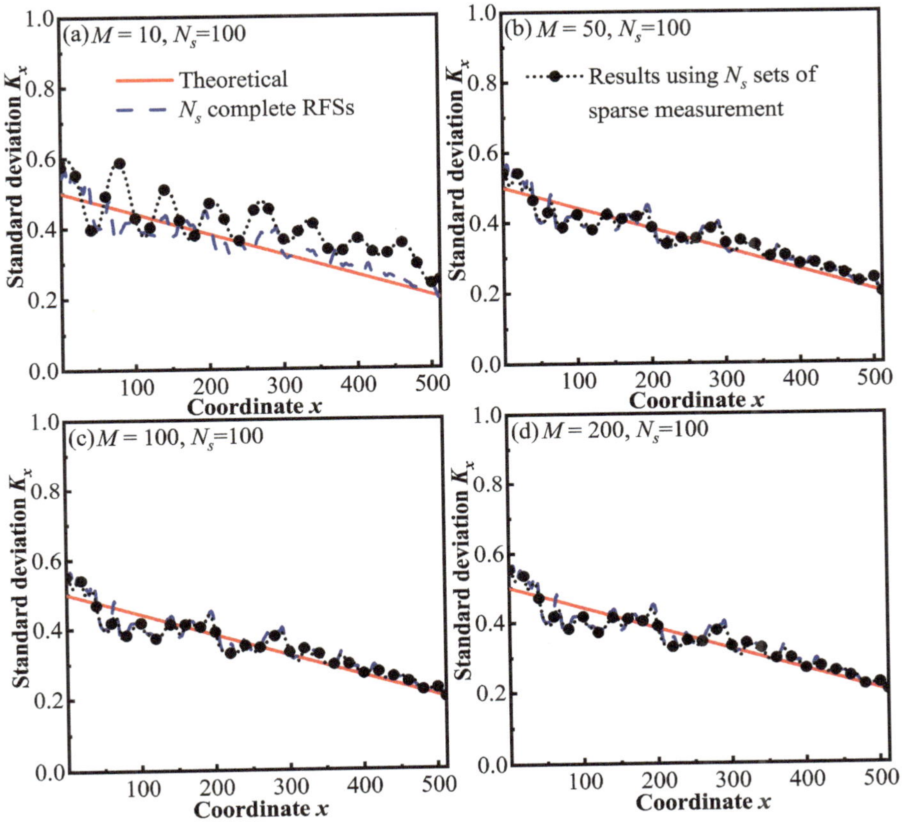

FIGURE 4.31 Effect of the number of data points M in each set of sparse measurement on the standard deviation K_x: (a) $M = 10$, (b) $M = 50$, (c) $M = 100$, and (d) $M = 200$.

are sufficient for BCS-KL to accurately simulate the correlation coefficient function. Note that the correlation coefficient function computed from the original N_s sets of complete RFSs is shown by blue dashed lines in Figure 4.30. It is evident that, as M increases, the function obtained from the BCS-KL generator converges to that obtained from the original N_s sets of complete RFSs.

The effect of the M value on the K_x function obtained from the RFSs generated by the BCS-KL generator is shown in Figure 4.31 for the four M values tested. As the M value increases from 10 to 50 (i.e., Figures 4.31a and b, respectively), the accuracy in the K_x function obtained from the BCS-KL generator improves significantly. However, as the M value further increases from 50 to 200, the accuracy is only improved slightly (see a comparison among Figures 4.31b–d). This suggests that $M = 50$ and $N_s = 100$ are sufficient for accurately simulating the marginal CDF, the correlation coefficient function, and the K_x function in this example. In other words,

50 data points per set of measurements and 100 sets of measurement data are sufficient to simulate properly the cumulative distribution function and the nonstationarity of the random field.

4.5 SIMULATION OF 2D OR 3D RANDOM FIELDS USING BCS-MCMC GENERATOR

The BCS-MCMC generator, which utilizes Gibbs sampling as detailed in Section 4.2.2, can also be used to generate 1D to 3D RFSs directly from sparse measurements. In this section, the BCS-MCMC generator is used to simulate 2D RFSs for geotechnical site characterization using the real-life example discussed in Section 3.6. In this example, $M = 30$ measurement points of electrical conductivity of soil are taken as input. Figure 3.15a in Chapter 3 shows the locations of these measurements by open circles. With these sparse measurements, 2D RFSs of electrical conductivity are simulated by two steps: (1) identification of nontrivial coefficients by OMP described in Section 3.2 and (2) estimation of these nontrivial coefficients with quantified uncertainty using MCMC described in Section 4.2.2.

As discussed in Chapter 3, the measurements $\mathbf{Y}$ in Figure 3.15a are first vectorized as a column vector $\boldsymbol{y}^{2D}$ with corresponding $\mathbf{A}_s^{2D}$. Then, OMP is adopted to identify the nontrivial coefficients and corresponding 2D basis functions. After that, MCMC is used to probabilistically estimate these nontrivial coefficients with quantified uncertainty. Some initial values are provided for $\boldsymbol{\alpha}$, τ, and γ, e.g., $\alpha_t = 10^{-4}$ ($t = 1, 2, \ldots, s$), $\tau = 100/\text{var}(\boldsymbol{y}^{2D})$, and $\gamma = 10^{-4}$, followed by updates of $\boldsymbol{\mu}_{\boldsymbol{\omega}_s^{2D}}$ and $\mathbf{COV}_{\boldsymbol{\omega}_s^{2D}}$ using Eq. (4.6) (**Step 1**). Note that var($\boldsymbol{y}^{2D}$) represents the variance of sparse measurements. Then, a set of $\boldsymbol{\omega}_s^{2D}$ samples can be simulated with the updated $\boldsymbol{\mu}_{\boldsymbol{\omega}_s^{2D}}$ and $\mathbf{COV}_{\boldsymbol{\omega}_s^{2D}}$. Only the nontrivial elements in $\boldsymbol{\omega}_s^{2D}$ are simulated, while the other elements are taken as zero (**Step 2**). With the newly generated $\boldsymbol{\omega}_s^{2D}$ sample, parameters of PDF $p(\boldsymbol{\alpha}|\boldsymbol{\omega}_s^{2D}, \gamma, \boldsymbol{y}^{2D})$ are updated with $a_t = \acute{E}_{s,t}^{2D}$ and $b_t = \gamma$ in Eq. (4.7), based on which a set of $\boldsymbol{\alpha}$ samples is simulated (**Step 3**). Update the parameter d_n in Eq. (4.8) using the updated $\boldsymbol{\mu}_{\boldsymbol{\omega}_s^{2D}}$, new $\boldsymbol{\omega}_s^{2D}$ sample, and $\boldsymbol{y}^{2D}$, and simulate a τ sample (**Step 4**). Update the parameter γ_b in Eq. (4.9) using the newly simulated $\boldsymbol{\alpha}$ sample, from which a new γ sample is simulated (**Step 5**). Collect newly simulated $\boldsymbol{\alpha}$, τ, and γ samples, using which to update $\boldsymbol{\mu}_{\boldsymbol{\omega}_s^{2D}}$ and $\mathbf{COV}_{\boldsymbol{\omega}_s^{2D}}$, and repeat Steps 2–5 until a prescribed number (e.g., 10,500) of MCMC samples is achieved.

In this example, the first 500 iterations of the Markov chains are considered as burn-in period and are discarded before further analysis. Then, a $\boldsymbol{\omega}_s^{2D}/\tau$ sample is picked up and recorded from every ten iterations to avoid correlations among the $\boldsymbol{\omega}_s^{2D}/\tau$ Markov chains. This leads to 1,000 sets of independent $\boldsymbol{\omega}_s^{2D}/\tau$ samples in total. Figures 4.32a–d, respectively, show 1,000 MCMC samples for th first elements of $\boldsymbol{\omega}_s^{2D}$, i.e., $\omega_{s,1}^{2D}$, the first element of $\boldsymbol{\alpha}$, i.e., α_1, γ, and τ as examples.

One set of $\boldsymbol{\omega}_s^{2D}$ and τ samples is then used to calculate one 2D RFS of electrical conductivity by Eq. (3.43). Residuals $\boldsymbol{\varepsilon}^{2D}$ in Eq. (3.43) are obtained by a simulation of $N_1 \times N_2$ random numbers from a Gaussian distribution with the zero mean and variance

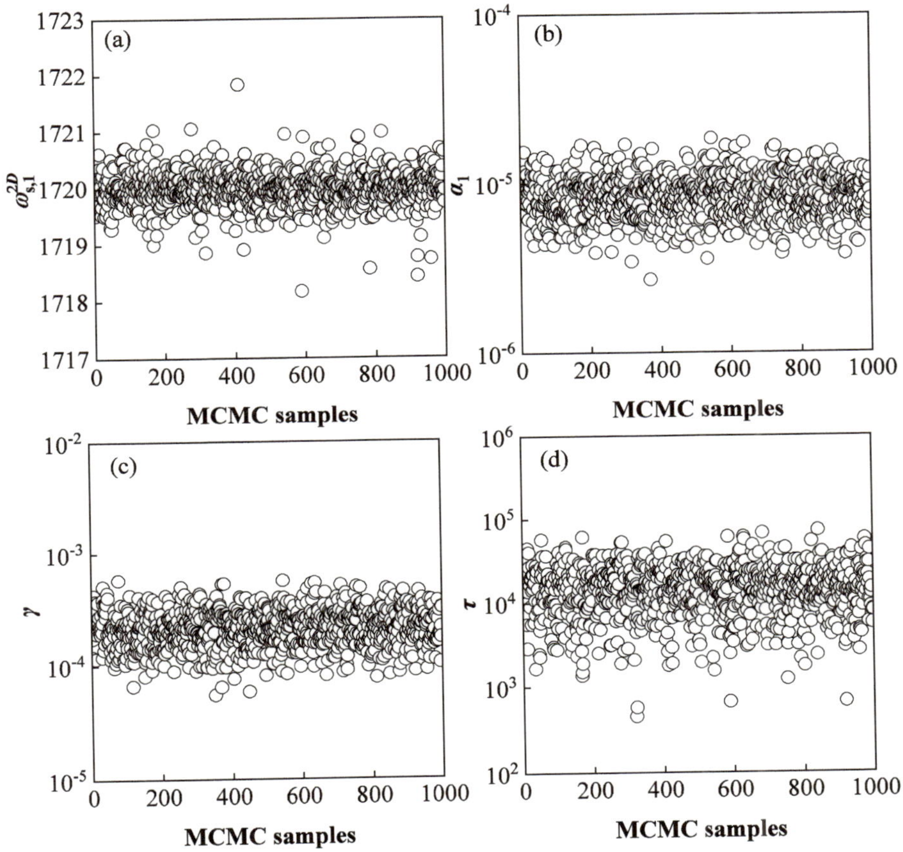

FIGURE 4.32 Generated MCMC samples of $\omega_{s,1}^{2D}$, α_1, γ, and τ.

of (τ^{-1}). Therefore, 1,000 sets of MCMC samples lead to 1,000 *2D* RFSs of interest. Figures 4.33a–d, respectively, show four examples of 2D RFS obtained. All four *2D* RFS exhibit a spatially varying trend that is similar to the original *2D* profile shown in Figure 3.15, indicating that the *2D* RFSs of electrical conductivity simulated by the BCS-MCMC generator are realistic and consistent with the true site conditions. The spatially varying mean and SD of electrical conductivity are further computed at each location based on 1,000 2D RFSs, and they are shown in Figures 4.34b and c, respectively. For comparison, Figure 4.34a plots the underlying 2D profile of electrical conductivity. It is seen from Figure 4.34 that distribution of electrical conductivity in Figure 4.34b is consistent with those in Figure 4.34a. This observation is further justified by the nearly zero residuals shown in Figure 4.34d and relatively small SD in Figure 4.34c. Although the BCS-MCMC generator is illustrated using 2D RFSs in this example, it is equally applicable to simulate 3D RFSs from sparse measurements for geotechnical site characterization.

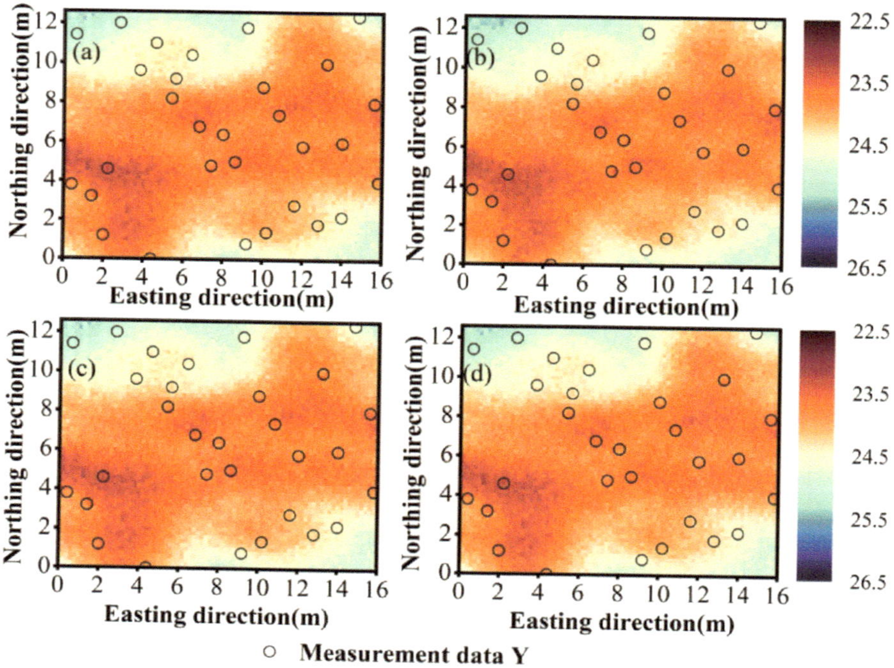

FIGURE 4.33 Four $2D$ RFSs simulated from sparse measurements **Y** by BCS-MCMC generator.

4.6 APPLICATION EXAMPLES IN GEOTECHNICAL SITE CHARACTERIZATION

In previous sections, both BCS-KL and BCS-MCMC generators were developed and illustrated for simulating RFSs directly from sparse measurements. In this section, these generators are applied to geotechnical site characterization using sparse measurements, which are often encountered in geotechnical practice. Two application examples are presented, including 2D soil stratification and characterization of 3D spatial variability from sparse CPT dataset, which are discussed in detail below.

4.6.1 2D Probabilistic Soil Stratification Using Sparse CPT Data

In this section, the BCS-KL generator is used for 2D soil stratification from a limited number of CPTs performed at an experimental site in the South Parklands of the city of Adelaide (e.g., Jaksa 1995). The CPT data, including the normalized resistance Q_t, and normalized friction ratio F_R, is firstly used to calculate soil behavior type (SBT) index I_c at each data points using the following equation (Robertson and Wride 1998):

$$I_c = \sqrt{(3.47 - \log Q_t)^2 + (\log F_R + 1.22)^2} \tag{4.18}$$

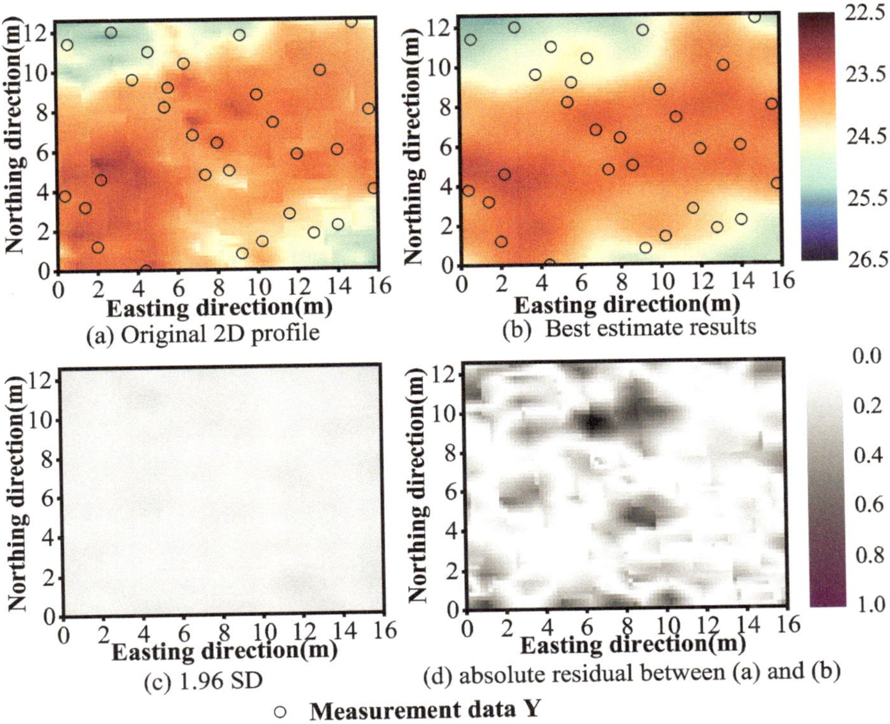

FIGURE 4.34 Comparison between the original data of soil electrical conductivity and statistical results of 2*D* RFSs simulated from 30 measurements using BCS-MCMC generator: (a) original 2*D* profile, (b) best estimate results, (c) 1.96 SD, and (d) absolute residual between (a) and (b).

TABLE 4.4
Soil Classification Based on SBT Index I_c (after Robertson and Wride 1998)

Range of I_c	SBT ID	SBT Description
$I_c < 1.31$	7	Gravelly sand to dense sand
$1.31 < I_c < 2.05$	6	Sands: clean sand to silty sand
$2.05 < I_c < 2.60$	5	Sand mixtures: silty sand to sandy silt
$2.60 < I_c < 2.95$	4	Silt mixtures: clayey silt to silty clay
$2.95 < I_c < 3.60$	3	Clays: silty clay to clay
$I_c > 3.60$	2	Organic soil: peats

Based on the I_c values, soils are classified into various soil behavior types, as summarized in Table 4.4.

In this site, ten sets of CPT-based I_c data within the 2D vertical cross-section SS-SS', as shown in Figure 4.35, are taken as input to the BCS-KL generator, while the

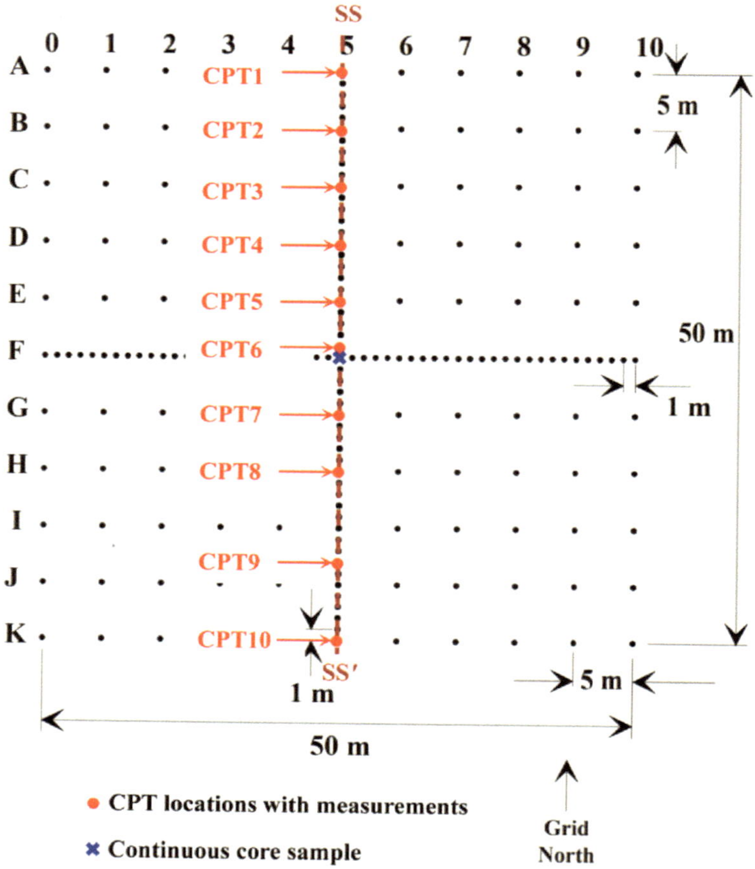

FIGURE 4.35 Layout of ten CPT soundings used as input to the BCS-KL generator and those used for comparison at the South Parklands site in Australia (Data from Jaksa 1995).

other sets of I_c data within this 2D vertical section are used for comparison and validation. Figure 4.35 shows a plan view of the study site (e.g., Jaksa 1995). The 2D vertical cross-section SS-SS' starts at the location of CPT sounding A5 and ends at the location of CPT sounding K5. Ten sets of I_c data used as input to the BCS-based generator are denoted as CPT1 to CPT10, respectively, as shown in Figure 4.35. Figure 4.36 shows the *1D* I_c profiles corresponding to these ten CPT soundings by solid lines. In this example, each CPT sounding covers a depth of 2.55 m with a resolution of 0.01 m, leading to 256 data points for each I_c dataset. In addition, horizontal spacing of the ten CPT soundings is around 5 m. Suppose that the target CPT sounding resolution along the horizontal direction is 0.1 m, the target *2D* RFS therefore has a dimension of $N_1 = 256$ by $N_2 = (50/0.1+1) = 501$.

In this example, measurements **Y** is a matrix with a dimension of 256×10. Following the method discussed in Section 3.6.1, s_1 and s_2 are computed as $s_1 = 256$

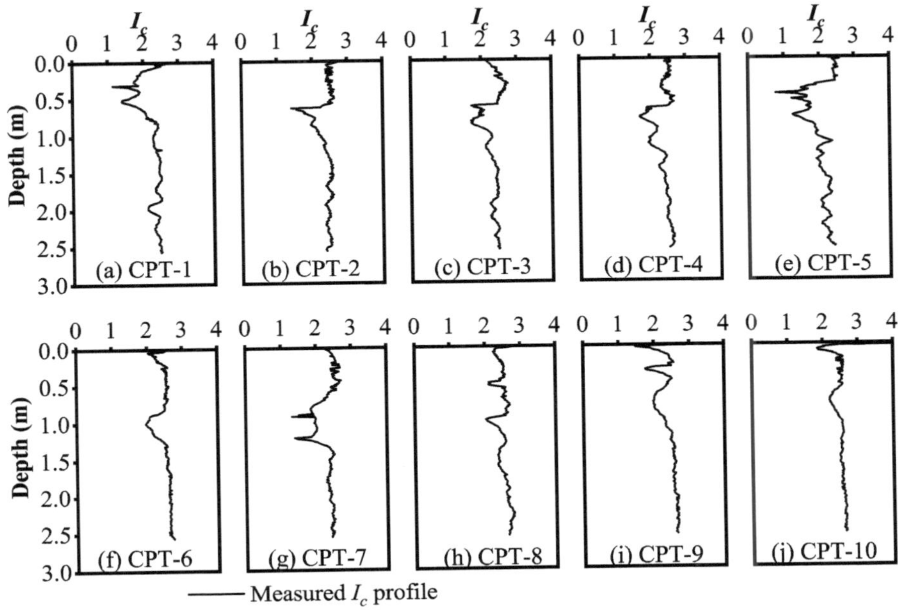

FIGURE 4.36 I_c profiles of ten CPT soundings at the South Parklands site.

and $s_2 = 10$. As a result, $\mathbf{B}_{1,s} = \mathbf{B}_1$; while only the first $s_2 = 10$ columns of matrix $\mathbf{B}_2$ are taken as $\mathbf{B}_{2,s}$. Then, $\mathbf{Y}$ is vectorized as $\boldsymbol{y}^{2D}$, and matrix $\mathbf{A}_s$ corresponding to $\boldsymbol{y}^{2D}$ is obtained from $\mathbf{B}_{1,s}$ and $\mathbf{B}_{2,s}$. After that, OMP discussed in Section 3.2 is used to identify nontrivial coefficients in $\boldsymbol{\omega}_s^{2D}$, followed by the BCS-KL generator to simulate 2D cross-section of I_c with a dimension of 256×501. Figures 4.37a–4.37d, respectively, show four 2*D* cross-sections of I_c generated at this site, each covers a width of 50 m and a depth of around 2.5 m.

Each 2D RFS of I_c may be used together with the SBT classifications summarized in Table 4.4 to produce a 2D cross-section of soil stratification. Therefore, 1,000 2D RFSs of I_c leads to 1,000 sets of 2D stratification cross-sections, from which the most probable SBT cross-section may be obtained. Note that the most probable SBT cross-section is obtained by setting the soil type as the one with the highest occurrence frequency at each point. Let $MP_{SBT}(x_1^i, x_2^j)$ denote occurrence frequency of the most probable soil type at a specific location (x_1^i, x_2^j) ($i = 1, 2, \ldots, 256$; $j = 1, 2, \ldots, 501$), and it is calculated as

$$MP_{SBT}(x_1^i, x_2^j) = \begin{cases} \max & \dfrac{\sum_{k=1}^{N_r} I\left(SBT_k(x_1^i, x_2^j) = ID\right)}{N_r} \\ subject\ \ to\ ID = \{2,3,4,5,6,7\} \end{cases} \tag{4.19}$$

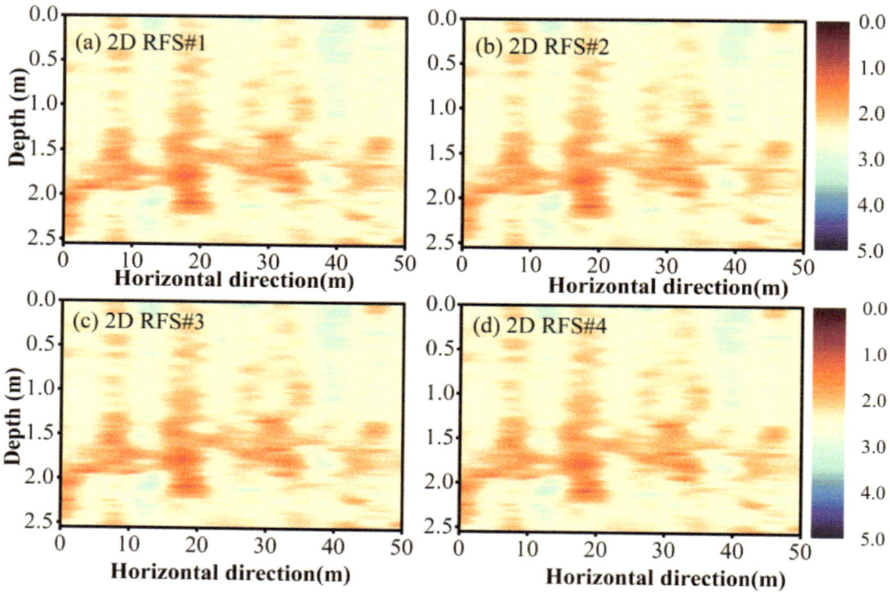

FIGURE 4.37 Four examples of I_c cross-sections simulated from the BCS-KL generator.

where x_1 and x_2 represent the depth and horizontal directions, respectively; $I(\cdot)$ is an indicator function. For a given soil type ID, $I(SBT_k(x_1^i, x_2^j) = ID) = 1$ when and only when $SBT(x_1^i, x_2^j) = ID$ is true, otherwise $I(SBT_k(x_1^i, x_2^j) = ID) = 0$. N_r represents the total number of 2D RFS of I_c, and it is taken as $N_r = 1{,}000$ in the example. It is worthwhile to note that the uncertainty associated with each location can also be quantified by a dispersion, DP_{SBT}, which is defined as (Shi and Wang 2021a and b)

$$DP_{SBT}(x_1^i, x_2^j) = \frac{\sum_{k=1}^{N_r} I\left(SBT_k(x_1^i, x_2^j) \neq MP_{SBT}(x_1^i, x_2^j)\right)}{N_r} \tag{4.20}$$

where SBT_{MP} denotes the most probable SBT. It is seen from Eq. (4.20) that DP_{SBT} generally represents the proportion of mismatching soil types at each location (x_1^i, x_2^j) ($i = 1, 2, \ldots, 255$; $j = 1, 2, \ldots, 501$) among 1,000 realizations, as compared with that of SBT_{MP}. Therefore, DP_{SBT} varies from 0 to 1 theoretically.

Figure 4.38a shows the most probable SBT obtained. The majority of the most probable SBT in the 2D vertical cross-section is sand mixtures (i.e., SBT5). Several zones of silt mixtures (SBT4) are identified at around horizontal coordinate of 10 m, 25 m, and 40 m. In addition, Figure 4.38b shows the dispersion map computed from Eq. (4.20), quantifying the uncertainty associated with the soil stratification. Yellow or orange zones indicate relatively large uncertainty. In contrast, blue or light

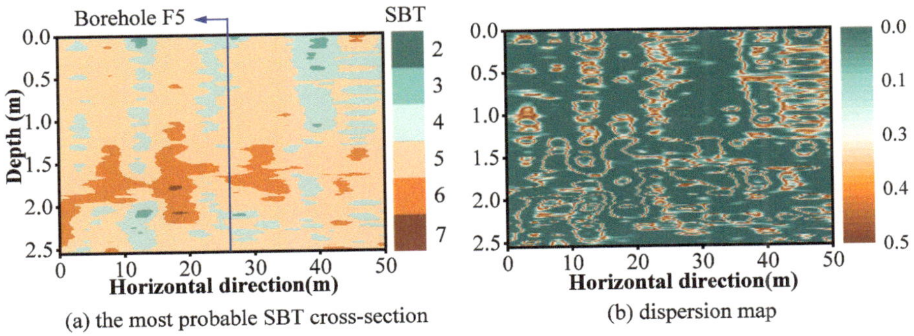

FIGURE 4.38 Statistics of 2D SBT cross-section at the South Parklands site: (a) the most probable SBT cross-section and (b) dispersion plot of 1,000 SBT cross-sections.

blue zones indicate relatively small uncertainty. Figure 4.38b shows that most yellow zones correspond to the soil layer boundaries in the most probable SBT cross-section.

To further evaluate the results obtained, soil logging from borehole F5 (see Figure 4.35 and Figure 4.38a) within the 2D vertical cross-section is compared with the results of the BCS-KL generator at the corresponding location. Figure 4.39a shows the borehole logging of F5, whereas Figure 4.39b shows the most probable SBT extracted from the corresponding location. It is seen from Figure 4.39a that the borehole logs at F5 indicate sandy clay/clayey sand, silty sandy clay with medium plasticity, silty sandy clay with high plasticity, and silty clay with high plasticity. The depths of boundaries for these four strata are 0.3 m, 0.9 m, and 1.6 m, respectively. In general, the borehole logs can be simplified as two strata, i.e., sandy soils from 0 to 1.6 m and silty soils from 1.6 m to 2.56 m. Figure 4.39b shows that soils above a depth of around 1.8 m are largely sand mixtures, and soils below around 1.8 m depth are mainly silt mixtures. Such observations are generally consistent with the logging results shown in Figure 4.39a, despite the inherent differences between the CPT-based SBT classification and composition-based soil classification for borehole logging. In addition, SBT results in Figure 4.39b may be interpreted probabilistically in terms of occurrence probability. Note that the occurrence probabilities of silt mixtures (SBT4) and sand mixtures (SBT5) versus depth at the location of borehole F5 can be readily computed as the corresponding identification frequency within the 1,000 simulations. The occurrence probability is calculated as

$$P_{r,SBT=ID}(x_1^i) = \frac{\sum_{k=1}^{N_r} I\left(SBT_k(x_1^i) = ID\right)}{N_r} \tag{4.21}$$

where $P_{r,SBT=ID}(x_1^i)$ represents the occurrence probability associated with silt mixtures (with ID = 4) or sand mixtures (with ID = 5) at the location of borehole F5 with varying depth x_1^i (i = 1, 2, …, 256). $SBT_k(x_1^i)$ represents the kth SBT profile at the

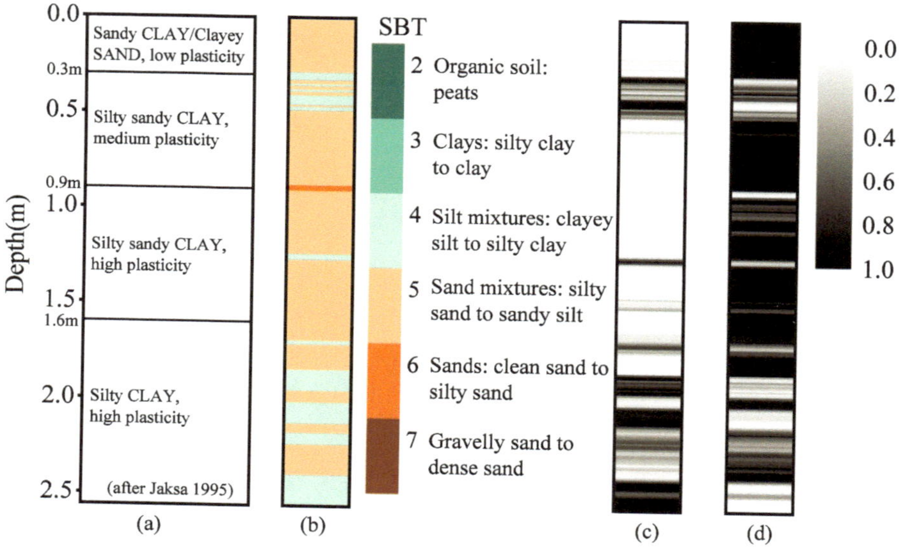

FIGURE 4.39 Comparisons between soil logging from borehole F5, most probable SBT results and probability distributions of SBTs 4&5 along borehole F5: (a) logging of borehole F5, (b) the most probable SBT results, (c) probability of SBT4 (silt mixtures), and (d) probability of SBT5 (sand mixtures).

location of borehole F5. Based on Eq. (4.21), the occurrence probabilities of silt mixtures (SBT4) and sand mixtures (SBT5) versus depth at the location of borehole F5 are obtained and, respectively, shown by gray scale columns in Figures 4.39c and 4.39d. The darker the grey scale bar, the higher the corresponding probability. It is observed in Figure 4.39c that areas below a depth of 1.8 m are apparently darker than areas above this level. In contrast, in Figure 4.39d, areas below a depth of 1.8 m are apparently lighter than areas above it, which implies that the soils above a depth of 1.8 m are likely to be sand mixtures, and soils below a depth of 1.8 m are likely to be silt mixtures. The results from the BCS-KL generator using a limited number of CPT soundings are generally consistent with the borehole logging data in the 2D vertical cross-section.

4.6.2 3D Probabilistic Characterization of Soil Spatial Variability Using Limited CPT Soundings

In this section, the BCS-MCMC generator is used to characterize 3D spatial variability of soil properties at the South Parklands site in Australia (after Jaksa 1995). As shown in Figure 4.40a, the rectangular shaded area of 20 m × 30 m is taken as the study area. Within the target area, 54 CPTs in total were carried out, as shown in Figure 4.40b. Most CPT soundings within the target area in Figure 4.40b cover a

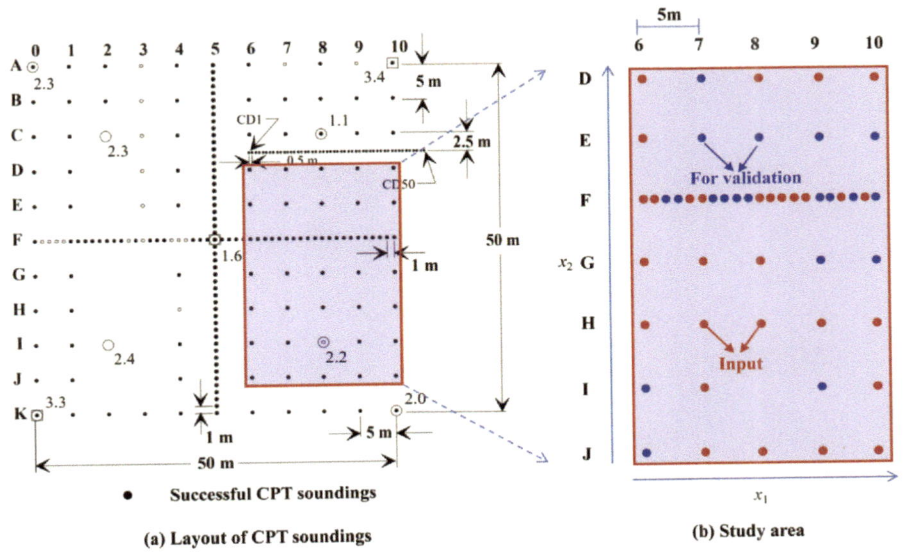

FIGURE 4.40 Plan view of the study area within Parklands site in Australia (after Jaksa 1995 and Lyu et al. 2023).

depth of around 5 m with a resolution of 0.01 m, leading to 501 data points for each CPT dataset. Some CPT soundings, however, have a depth of less than 5 m, and the number of data points along the depth is less than 500 in these cases.

In this example, a resolution of 0.2 m is used for two horizontal directions, and the number of cells along x_1 and x_2 directions is 100 and 150, respectively. Besides, a resolution of 0.01 m is used for the depth direction. Therefore, the study area to be characterized has a dimension of 101 × 151 × 501. n_b = 30 CPT datasets are used as input for the BCS-MCMC generator, as labeled by solid circles in Figure 4.40b. Based on the descriptions in Section 4.2.2, N_1 = 101, N_2 = 151, and N_3 = 501 are used to construct 1D basis function $\mathbf{B}_1$, $\mathbf{B}_2$, and $\mathbf{B}_3$, respectively. Then, following the method discussed in Section 3.7.1, sampling interval Δ_h and Δ_v are computed as Δ_h = 4.2 m, and Δ_v = 0.01 m. After that s_1, s_2 are taken as $s_1 \approx 20/\Delta_h+1 = 6$, $s_2 \approx 30/\Delta_h+1 = 8$, and s_3 = 501, leading to $\mathbf{B}_{1,s}$, $\mathbf{B}_{2,s}$, and $\mathbf{B}_{3,s}$. Subsequently, n_b = 30 sets of CPT q_c data are re-shaped as a column vector $\boldsymbol{y}^{3D}$. Matrix $\mathbf{A}_s$ corresponding to $\boldsymbol{y}^{3D}$ is also derived using their respective locations, together with $\mathbf{B}_{1,s}$, $\mathbf{B}_{2,s}$, and $\mathbf{B}_{3,s}$. Modified OMP is then used to identify nontrivial coefficients in the vector $\boldsymbol{\omega}_s^{3D}$, followed by Bayesian estimation with MCMC simulation. A total of 1,000 sets of representative $\boldsymbol{\omega}_s^{3D}$ samples are simulated in this example, based on which 1,000 3D RFSs of q_c are obtained. Figures 4.41a–d show four examples of 3D RFSs of q_c for the study area. Based on 1,000 3D profiles of q_c for the study area, the mean and SD of q_c at each location of the study area are obtained, as shown, respectively, in Figures 4.42a and c. To improve visualization of the results inside the 3D space, Figures 4.42b and d,

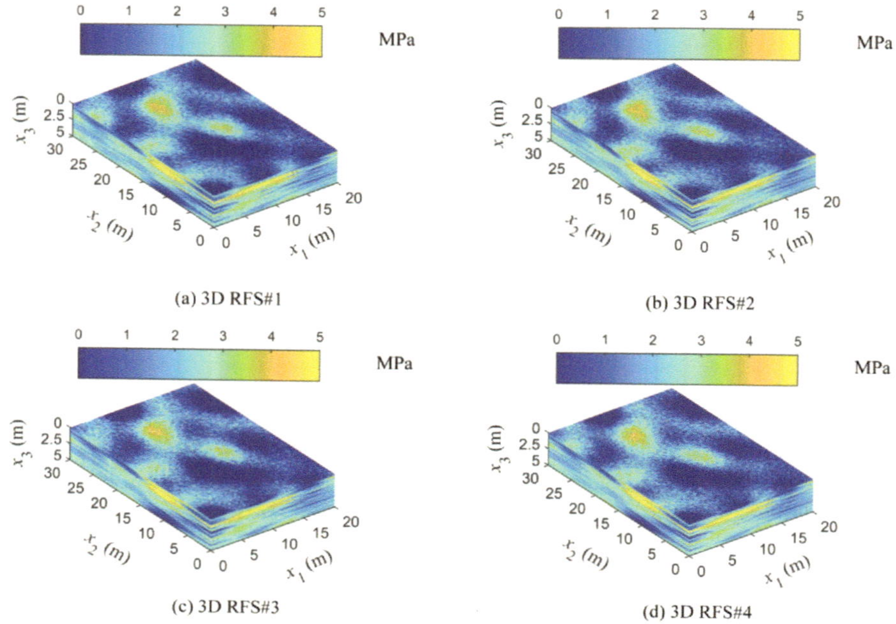

FIGURE 4.41 Examples of 3D profiles of q_c of the study area simulated from the BCS-MCMC generator.

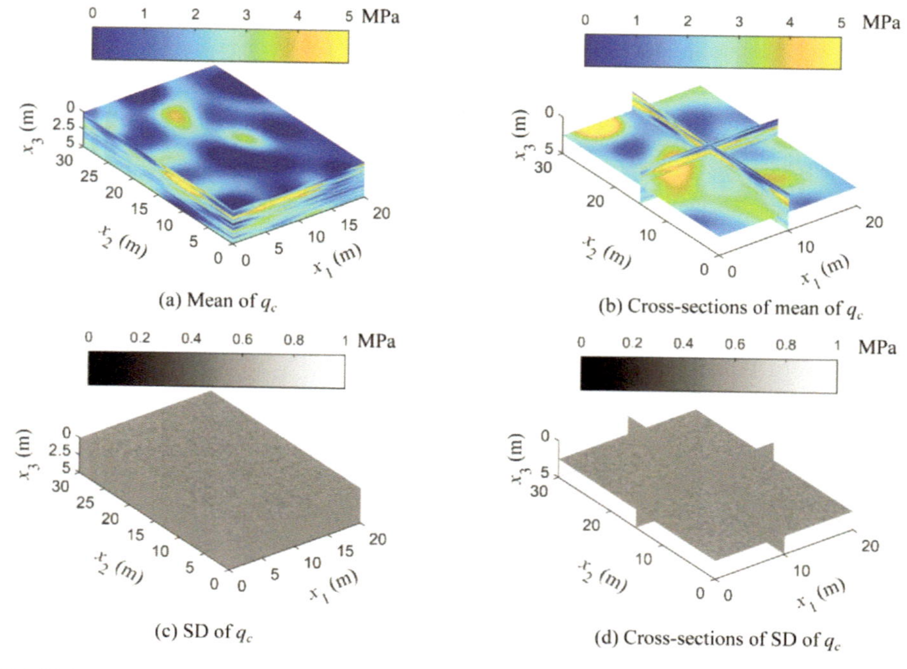

FIGURE 4.42 Statistics of 3D profiles of q_c at the study area from 1,000 3D RFS simulated by the BCS-MCMC generator.

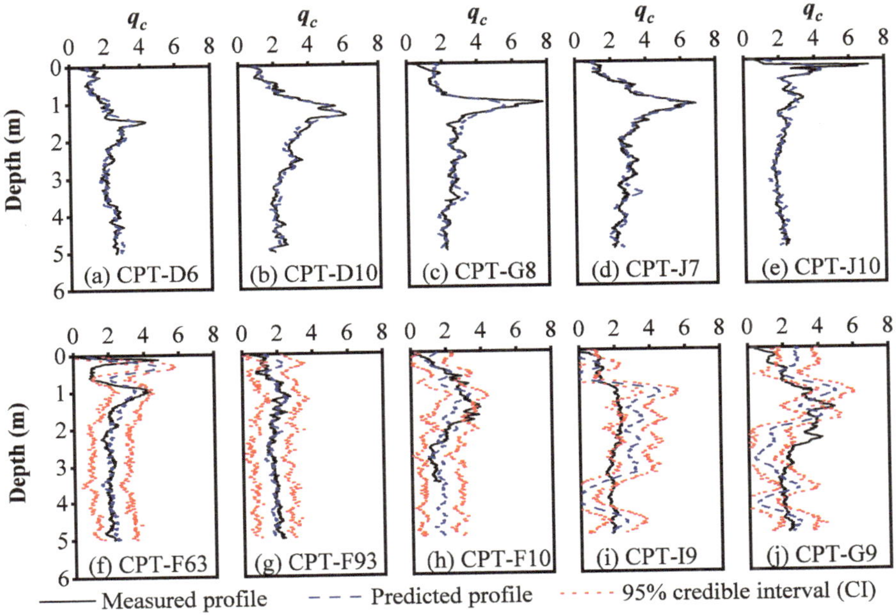

FIGURE 4.43 Comparison between original and predicted q_c profiles: (a)–(e) five CPT soundings (i.e., CPT-D6, CPT-D10, CPT-G8, CPT-J7, and CPT-J10) that are used as input; (f)–(j) five CPT soundings (i.e., CPT-F63, CPT-F93, CPT-F10, CPT-I9, and CPT-G9) that are not used as input.

respectively, show three cross-sections extracted from the mean and SD of the study area, each of which is located at the center of each coordinate.

To examine the q_c predicted from 1,000 3D RFSs for the study area, five sets of CPT q_c data that have been used as input for the BCS-MCMC generator for simulation of 3*D* RFS of q_c are, respectively, shown in Figures 4.43a–e, while five sets of CPT q_c data that are not used as input are, respectively, shown in Figures 4.43f–j. Figures 4.43a–e, respectively, show the original and predicted q_c at the location of CPT-D6, CPT-D10, CPT-G8, CPT-J7, and CPT-J10 (see Figure 4.40b for their locations) by solid and dashed lines, respectively. It is evident that the dashed lines almost overlap with the solid lines, indicating that q_c data are reasonably predicted at the locations where CPT data are used as input for the BCS-MCMC generator. Figures 4.43f–j, respectively, show the original and predicted q_c at the locations of CPT-F63, CPT-F93, CPT-F10, CPT-I9, and CPT-G9 (see Figure 4.40b for their locations). It is seen that the dashed lines are generally consistent with the respective trends of the solid lines, although there are some local differences among them, particularly for CPT-F63, CPT-F93, and CPT-F10. Besides, note that the measured data of CPT-F10 end at a depth of about 3.5 m (see Figure 4.43h) and contains less than 501 data points. However, BCS-MCMC generator is able to provide predictions up to 5 m depth.

REFERENCES

Andrieu, C., N. De Freitas, A. Doucet, and M. Jordan. 2003. An introduction to MCMC for machine learning. *Machine learning* 50 (1–2): 5–43.

Ang, A., and Wilson H. Tang. 2007. *Probability concepts in engineering: Emphasis on applications to civil & environmental engineering*. 2nd ed. New York: Wiley.

Au, S. K., and Y. Wang. 2014. *Engineering risk assessment with subset simulation*. Singapore: John Wiley & Sons.

Beer, M., Y. Zhang, S. T. Quek, and K. K. Phoon. 2013. Reliability analysis with scarce information: Comparing alternative approaches in a geotechnical engineering context. *Structural Safety* 41: 1–10. doi: org/10.1016/j.strusafe.2012.10.003

Cami, B., Sina Javankhoshdel, K.-K. Phoon, and J. Ching. 2020. Scale of fluctuation for spatially varying soils: Estimation methods and values. *ASCE-ASME Journal of Risk and Uncertainty in Engineering Systems, Part A: Civil Engineering* 6 (4): 03120002. doi: 10.1061/AJRUA6.0001083.

Cao, Z., and Y. Wang. 2014. Bayesian model comparison and selection of spatial correlation functions for soil parameters. *Structural Safety* 49: 10–17. doi: org/10.1016/j.strusafe.2013.06.003

Ching, J., and K. K. Phoon. 2017. Characterizing uncertain site-specific trend function by sparse Bayesian learning. *Journal of Engineering Mechanics* 143 (7): 04017028. doi: doi:10.1061/(ASCE)EM.1943-7889.0001240

Ching, J., S. Wu, and K. K. Phoon. 2016. Statistical characterization of random field parameters using frequentist and Bayesian approaches. *Canadian Geotechnical Journal* 53 (2): 285–298. doi: 10.1139/cgj-2015-0094

Christian, J. T., C. C. Ladd, and G. B. Baecher. 1994. Reliability applied to slope stability analysis. *Journal of Geotechnical Engineering* 120 (12):2180–2207.

Fenton, G. A. 1999a. Estimation for stochastic soil models. *Journal of Geotechnical and Geoenvironmental Engineering* 125 (6): 470–485.

Fenton, G. A. 1999b. Random field modeling of CPT data. *Journal of Geotechnical and Geoenvironmental Engineering* 125 (6): 486–498.

Fenton, G. A., and D. V. Griffiths. 2002. Probabilistic foundation settlement on spatially random soil. *Journal of Geotechnical and Geoenvironmental Engineering* 128 (5): 381–390. doi: doi:10.1061/(ASCE)1090-0241(2002)128:5(381)

Fenton, G. A., and D. V. Griffiths. 2005. Three-dimensional probabilistic foundation settlement. *Journal of Geotechnical and Geoenvironmental Engineering* 131 (2): 232–239.

Fenton, G. A., and D. V. Griffiths. 2008. *Risk assessment in geotechnical engineering*. New York: Wiley.

Gelman, A., J. B. Carlin, H. S. Stern, and D. B. Rubin. 2004. *Bayesian data analysis*. Vol. 2. Boca Raton, FL: Chapman & Hall/CRC.

Geweke, J 1992. Evaluating the accuracy of sampling-based approaches to the calculation of posterior moments. *Bayesian Statistics* 4: 169–193.

Griffiths, D. V., and G. A. Fenton. 2009. Probabilistic settlement analysis by stochastic and random finite-element methods. *Journal of Geotechnical and Geoenvironmental Engineering* 135 (11): 1629–1637. doi: doi:10.1061/(ASCE)GT.1943-5606.0000126

Griffiths, D. V., J. Huang, and G. A. Fenton. 2009. Influence of spatial variability on slope reliability using 2-D random fields. *Journal of Geotechnical and Geoenvironmental Engineering* 135 (10): 1367–1378.

Hicks, M. A., and W. A. Spencer. 2010. Influence of heterogeneity on the reliability and failure of a long 3D slope. *Computers and Geotechnics* 37 (7–8): 948–955. doi: org/10.1016/j.compgeo.2010.08.001

Hu, K., P. Ch Ivanov, Z. Chen, P. Carpena, and H. E. Stanley. 2001. Effect of trends on detrended fluctuation analysis. *Physical Review E* 64 (1): 011114.

Jaksa, M. B. 1995. The influence of spatial variability on the geotechnical design properties of a stiff, overconsolidated clay. PhD, Faculty of Engineering, University of Adelaide.

Jaksa, M. B., P. I. Brooker, and W. S. Kaggwa. 1997. Inaccuracies associated with estimating random measurement errors. *Journal of Geotechnical and Geoenvironmental Engineering* 123 (5): 393–401.

Johnson, R. A., and D. W. Wichern. 2002. *Applied multivariate statistical analysis*. 5th ed., Vol. 4. New Jersey, NJ: Prentice Hall.

Kulatilake, P. H. S. 1991. Discussion of "Probabilistic Potentiometric Surface Mapping" by Pinnaduwa HSW Kulatilake (November, 1989, Vol. 115, No. 11). *Journal of Geotechnical Engineering* 117 (9): 1457–1458.

Lloret-Cabot, M., G. A. Fenton, and M. A. Hicks. 2014. On the estimation of scale of fluctuation in geostatistics. *Georisk: Assessment and Management of Risk for Engineered Systems and Geohazards* 8 (2): 129–140. doi: 10.1080/17499518.2013.871189

Lyu, B., Y. Hu, and Y. Wang. 2023. Data-driven development of three-dimensional subsurface models from sparse measurements using Bayesian compressive sampling: A benchmarking study. *ASCE-ASME Journal of Risk and Uncertainty in Engineering Systems, Part A: Civil Engineering* 9 (2): 04023010.

Mara, T. A., N. Fajraoui, A. Guadagnini, and A. Younes. 2016. Dimensionality reduction for efficient Bayesian estimation of groundwater flow in strongly heterogeneous aquifers. *Stochastic Environmental Research and Risk Assessment* 31(9): 2313–72316 doi: 10.1007/s00477-016-1344-1

Montoya-Noguera, S., T. Zhao, Y. Hu, Y. Wang, and K. K. Phoon. 2019. Simulation of non-stationary non-Gaussian random fields from sparse measurements using Bayesian compressive sampling and Karhunen-Loève expansion. *Structural Safety* 79: 66–79.

Pérez, C. J., Jacinto Martín, María Jesús Rufo. 2006. Sensitivity estimations for Bayesian inference models solved by MCMC methods. *Reliability Engineering System Safety* 91 (10–11): 1310–1314.

Phoon, K. K., H. W. Huang, and S. T. Quek. 2005. Simulation of strongly non-Gaussian processes using Karhunen–Loeve expansion. *Probabilistic Engineering Mechanics* 20 (2): 188–198. doi: org/10.1016/j.probengmech.2005.05.007

Phoon, K. K., S. P. Huang, and S. T. Quek. 2002a. Implementation of Karhunen–Loeve expansion for simulation using a wavelet-Galerkin scheme. *Probabilistic Engineering Mechanics* 17 (3): 293–303. doi: org/10.1016/S0266-8920(02)00013-9.

Phoon, K. K., S. P. Huang, and S. T. Quek. 2002b. Simulation of second-order processes using Karhunen–Loeve expansion. *Computers & Structures* 80 (12): 1049–1060. doi: org/10.1016/S0045-7949(02)00064-0

Phoon, K. K., S. T. Quek, and H. W. Huang. 2004. Simulation of non-Gaussian processes using fractile correlation. *Probabilistic Engineering Mechanics* 19 (4): 287–292. doi: 10.1016/j.probengmech.2003.09.001

Qi, X. H., and H. X. Liu. 2019. Estimation of autocorrelation distances for in-situ geotechnical properties using limited data. *Structural Safety* 79: 26–38.

Robertson, Peter K., and C. E. Wride. 1998. Evaluating cyclic liquefaction potential using the cone penetration test. *Canadian Geotechnical Journal* 35 (3): 442–459.

Shi, C., and Wang, Y. 2021a. Development of subsurface geological cross-section from limited site-specific boreholes and prior geological knowledge using iterative convolution XGBoost. *Journal of Geotechnical and Geoenvironmental Engineering*, 147(9): 04021082.

Shi, C., & Wang, Y. 2021b. Nonparametric and data-driven interpolation of subsurface soil stratigraphy from limited data using multiple point statistics. *Canadian Geotechnical Journal*, 58(2): 261–280.

Spanos, P. D., and Roger Ghanem. 1989. Stochastic finite element expansion for random media. *Journal of Engineering Mechanics* 115 (5): 1035–1053. doi: doi:10.1061/(ASCE)0733-9399(1989)115:5(1035)

Strang, Gilbert. 2006. *Linear algebra and its applications*. New York: Academic.

Stuedlein, Armin W, Steven L. Kramer, Pedro Arduino, and Robert D. Holtz. 2012. Geotechnical characterization and random field modeling of desiccated clay. *Journal of Geotechnical and Geoenvironmental Engineering* 138 (11): 1301–1313.

Van Trees, Harry L., K. L. Bell, and Z. Tian. 2013. *Detection, estimation, and modulation theory, Part I: Detection, estimation, and filtering theory*. 2nd ed. New York: John Wiley & Sons.

Vanmarcke, E., M. Shinozuka, S. Nakagiri, G. I. Schuëller, and M. Grigoriu. 1986. Random fields and stochastic finite elements. *Structural Safety* 3 (3): 143–166. doi: org/10.1016/0167-4730(86)90002-0

Vanmarcke, Erik H. 1977. Probabilistic modeling of soil profiles. *Journal of the geotechnical engineering division* 103 (11): 1227–1246.

Wang, Y., and Z. Cao. 2013. Probabilistic characterization of Young's modulus of soil using equivalent samples. *Engineering Geology* 159: 106–118. doi: org/10.1016/j.enggeo.2013.03.017

Wang, Y., and T. Zhao. 2017. Statistical interpretation of soil property profiles from sparse data using Bayesian compressive sampling. *Géotechnique* 67 (6): 523–536. doi: 10.1680/jgeot.16.P.143.

Wang, Y., T. Zhao, and K. K. Phoon. 2018. Direct simulation of random field samples from sparsely measured geotechnical data with consideration of uncertainty in interpretation. *Canadian Geotechnical Journal* 55 (6): 862–880. doi: org/10.1139/cgj-2017-0254

Wang, Y., T. Zhao, Y. Hu, and K-K Phoon. 2019. Simulation of random fields with trend from sparse measurements without detrending. *Journal of Engineering Mechanics* 145 (2):04018130. doi: doi:10.1061/(ASCE)EM.1943-7889.0001560

Yuen, K. V., and G. A. Ortiz. 2018. Multiresolution Bayesian nonparametric general regression for structural model updating. *Structural Control and Health Monitoring* 25 (2): e2077. doi: 10.1002/stc.2077

Yuen, K. V., G. A. Ortiz, and K. Huang. 2016. Novel nonparametric modeling of seismic attenuation and directivity relationship. *Computer Methods in Applied Mechanics and Engineering* 311: 537–555.

Zevgolis, Ioannis E., and Philippe L. Bourdeau. 2010. Probabilistic analysis of retaining walls. *Computers and Geotechnics* 37 (3): 359–373. doi: org/10.1016/j.compgeo.2009.12.003

Zhang, J., and B. Ellingwood. 1994. Orthogonal series expansions of random fields in reliability analysis. *Journal of Engineering Mechanics* 120 (12): 2660–2677. doi: doi:10.1061/(ASCE)0733-9399(1994)120:12(2660)

Zhao, T., and Y. Wang. 2020. Non-parametric simulation of non-stationary non-gaussian 3D random field samples directly from sparse measurements using signal decomposition and Markov Chain Monte Carlo (MCMC) simulation. *Reliability Engineering & System Safety* 203: 107087. doi: org/10.1016/j.ress.2020.107087

5 Sample Size Determination for Site Characterization

Zheng Guan and Yu Wang

ABSTRACT

Site investigation is a fundamental aspect of geotechnical engineering practice. However, only a small fraction of geo-materials is typically sampled and tested, leading to considerable uncertainty and risk in the interpretation of ground conditions. This raises a critical question: How many samples (such as test specimens, borings, and in-situ test soundings) are required for a given site? Existing geotechnical guidelines and design codes provide conceptual principles for site investigation, offering general rules of thumb for the minimum number of exploration points. However, no scientific or quantitative method explicitly considers the spatial variation and correlation of geotechnical properties in sample size determination. To address this challenge, this chapter presents an extensive parametric study to develop quantitative methods and statistical charts for sample size determination using Bayesian compressive sampling.

5.1 INTRODUCTION

Geotechnical engineers use measurement data from in-situ or laboratory tests to infer ground conditions. However, only a small portion (e.g., 1/100,000 or less) of the total volume of soil is examined in engineering practice, resulting in significant uncertainty and risk in the interpreted ground conditions (e.g., Baecher and Christian 2003; Fenton and Griffiths 2008; Zhao et al. 2018). The interpretation uncertainty is significantly affected by the amount of measurement data (i.e., sample size). Generally speaking, a large sample size provides detailed information about the ground conditions, leading to improved accuracy and confidence in the interpretation of ground conditions and reduced risk for engineering projects.

The importance of an effective site investigation has long been recognized by geotechnical engineers, but the number of exploration points is selected mainly based on engineering judgment and local experience (e.g., Christian 2004; Wang et al. 2015). Overreliance on engineering judgment might lead to inconsistent or even improperly interpreted geotechnical models, inconsistent scopes in site investigations, and undesirable reliability in designed geotechnical structures. Currently, there is no scientific or quantitative method for planning geotechnical site investigations in engineering practice. Existing geotechnical design manuals and codes only provide conceptual

DOI: 10.1201/9781003381235-5

principles for the planning of site investigations. For example, Clause 2.4.2.4 in Eurocode 7 (CEN 2007) states that "the necessary number of specimens to be tested shall be established depending on the homogeneity of the ground, the quality and amount of comparable experience with the ground, and the geotechnical category of the problem." The National Cooperative Highway Research Program provides guidelines for planning subsurface investigations (Rix et al. 2018) and states that "the number of investigation locations may be increased or decreased depending on the expected variability of subsurface conditions and other project-specific considerations."

This chapter aims to develop a rational method for determining the sample size required to achieve a target level of accuracy or confidence in the interpretation of a geospatial model of subsurface soil. The proposed method is developed based on extensive parametric studies for investigating a relationship between sample size and the target level of accuracy or confidence. In the parametric studies, the Bayesian compressive sampling (BCS) method is used for the interpretation of a geospatial model of subsurface soil, because it not only provides the mean estimate of a geotechnical model from sparse measurements but also quantifies the corresponding uncertainties.

The remainder of this chapter is organized as follows. Section 5.2 provides an overview of the parametric study. Section 5.3 describes the development of a relationship between sample size and accuracy of the one-dimensional (1D) spatial model, followed by the relationship between sample size and confidence of the 1D spatial model in Section 5.4. Sections 5.5 and 5.6 further extend the proposed method to a two-dimensional (2D) geotechnical cross-section.

5.2 PARAMETRIC STUDIES FOR SAMPLE SIZE DETERMINATION

To determine the sample size required to achieve a desired level of accuracy or confidence in the interpreted geotechnical models, a relationship between sample size and the target level of accuracy or confidence should be established first. An extensive parametric study is performed to develop such a relationship. Figure 5.1 shows a conceptual framework for developing such a relationship in geotechnical site investigation. The framework mainly involves three stages: (1) the relationship between the desired level of accuracy or confidence and sample size for a given geotechnical property model (e.g., spatial variation of soil properties along depth), (2) the relationship for a specific site condition, and finally (3) the relationship for a wide variety of typical sites commonly encountered in geotechnical site investigation.

In Stage 1, it is instructive to first consider the required sample size for one soil property profile (i.e., Steps 1–6 in Figure 5.1). Suppose that there is a soil property profile that is completely known (i.e., all data points in the profile are known). To mimic the data measurement and interpretation in geotechnical site investigation, a number, M (e.g., $M = 5$), of data points is obtained from the soil property profile and used as measurement data points for interpreting a complete soil property profile using BCS. Because the original soil property profile is completely known, the soil property profile interpreted from the M measurement data points may be compared with the original soil property profile for evaluating the accuracy of the interpreted profile (i.e., Steps 2–4 in Figure 5.1). Let $\boldsymbol{f} = \left[f_1, f_2, \ldots, f_N\right]^{\mathrm{T}}$ be a vector representing the original soil property profile along depths with a total number of data points, N. In

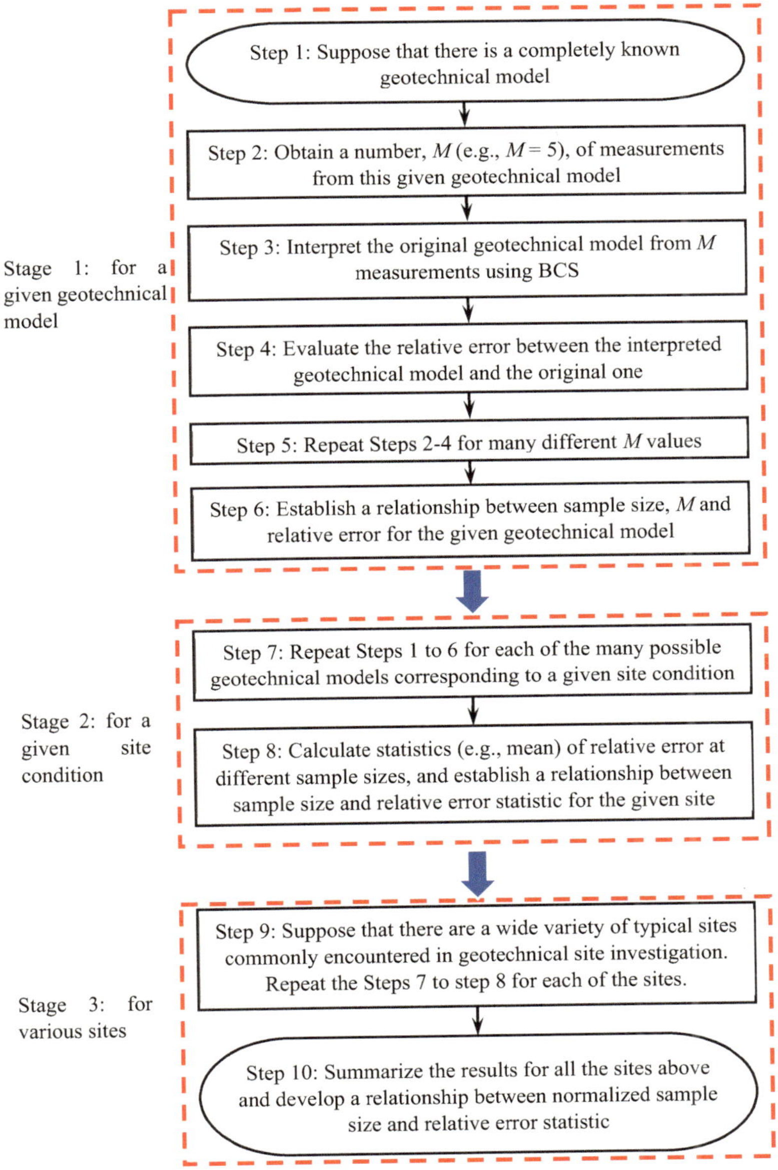

FIGURE 5.1 Framework for establishing a relationship between sample size and accuracy in geotechnical site investigation (after Wang et al. 2019).

the context of compressive sensing (CS), the soil property profile might be expressed as $\boldsymbol{f} = \mathbf{B}^1\boldsymbol{\omega}$, where $\mathbf{B}^1$ is an orthonormal matrix with a dimension of $N \times N$, each column of which represents a basis function, and $\boldsymbol{\omega}$ is a weight column vector with a length of N, corresponding to N basis functions in $\mathbf{B}^1$. Note that the matrix $\mathbf{B}^1$ is independent of the soil property profile $\boldsymbol{f}$, and that $\mathbf{B}^1$ can be constructed using discrete cosine transform. Using the Bayesian framework, the soil property profile interpreted from sparse measurement data $\boldsymbol{y}$, denoted as $\hat{\boldsymbol{f}} = \left[\hat{f}_1, \hat{f}_2, ..., \hat{f}_N\right]^{\mathrm{T}}$, can be obtained (e.g., Ji et al. 2008; Wang and Zhao 2017). The accuracy of the interpreted soil property profile, $\hat{\boldsymbol{f}}$, can be quantified by comparing a relative error, RE, between $\hat{\boldsymbol{f}}$ and $\boldsymbol{f}$, which is expressed as:

$$RE = \frac{\left\|f-\hat{f}\right\|}{\left\|f\right\|} \times 100\% = \sqrt{\frac{\sum_{i=1}^{N}\left(f_i-\hat{f}_i\right)^2}{\sum_{i=1}^{N} f_i^2}} \times 100\% \tag{5.1}$$

where the symbol "$\|\cdot\|$" denotes the Euclidian norm of a vector, which is defined as the square root of the sum of the squared value of all elements, as shown in the right-hand side of Eq. (5.1). Note that the definition of RE in Eq. (5.1) is similar to the normalized root mean square error in statistics if the original soil property profile is treated as the mean. To investigate the influence of difference sample sizes (i.e., different M values) on the accuracy of the interpreted soil property profile, Steps 2–4 described above are repeated for many different M values (e.g., $M = 10$), and the corresponding REs are calculated (i.e., Step 5 in Figure 5.1).

When the sample size M is small, the soil property profile interpreted from such sparse measurement data might not be accurate, and the RE is relatively large. As M increases, the accuracy of the interpreted soil property profile improves, and the RE gradually decreases. When all data points in the original soil property profile are treated as measurement data points (i.e., $M = N$), the interpreted soil property profile converges to the original soil property profile, and the RE approaches to zero. The relationship between RE and M for a given soil property profile can be established after Steps 2–4 are repeated for many different M values (i.e., Step 6). This relationship quantifies the variation of the estimation uncertainty as the sample size changes for a given soil property profile.

In Stage 2 (i.e., Steps 7–8 in Figure 5.1), the study is extended from a given soil property profile (i.e., Stage 1) to a given site. Suppose that there is a given site where many (e.g., 100) sets of complete soil property profiles are available. The procedure developed in Stage 1 (i.e., Steps 1–6) may be used repeatedly for each set of the complete soil property profiles in this site, resulting in many sets of relationships between RE and M. Statistical analysis can be performed to obtain the mean of RE, μ_{RE}, at different M values for all the relationships obtained. This establishes a relationship between the interpretation accuracy (e.g., using μ_{RE}) and sample sizes for a given site.

In Stage 3 (i.e., Steps 9–10 in Figure 5.1), the study is further extended from a given site (i.e., Stage 2) to a wide variety of typical sites commonly encountered in geotechnical site investigation. The procedure developed in Stage 2 (i.e., Steps 7–8) may be used repeatedly for each typical site, resulting in many sets of relationships between μ_{RE} and M. Because well-investigated and well-documented sites are rare in reality, random field simulation (e.g., Vanmarcke 1977) is adopted in this study to generate a large number of typical sites as a parametric study in Stage 3. Three statistical parameters are commonly needed in a random field simulation of spatial variation of a soil property: (1) mean μ_{RE} of the soil property, (2) variance or coefficient of variation, COV, and (3) correlation length, which reflects quantitatively spatial correlation of the soil property at different depths. Typical ranges of μ_{RE}, COV, and correlation length for a variety of soil properties have been reported in the literature (e.g., Phoon and Kulhawy 1999; Cao et al. 2016), as shown in Tables 5.1 and 5.2, and these typical ranges are used in the random field simulation in the parametric study to generate a large number of typical sites that mimic spatially varying and correlated soil properties in real site conditions.

The procedure developed in Stage 2 (i.e., Steps 7–8) is used repeatedly for each typical site in the parametric study, resulting in many sets of relationship between μ_{RE} and M for different site conditions. The large number of μ_{RE} and M relationships obtained in the parametric study will be summarized in a dimensionless plot, and a relationship between the normalized μ_{RE} and normalized M will be developed from the dimensionless plot. The normalized μ_{RE} and normalized M relationship provides a link between the accuracy of the interpretation results and sample size, and this relationship will be further used to develop statistical chart for sample size determination with consideration of spatial variation and correlation of soil properties (Tables 5.1 and 5.2).

TABLE 5.1
Typical Ranges of the Mean, Coefficient of Variation of Soil Properties Reported in the Literature (after Phoon and Kulhawy 1999; Cao et al. 2016)

Soil Property	Soil Type	Range of Mean	Range of COV (%)
Total unit weight, γ (kN/m^3)	Fine grained	13~24	3~20
	Coarse grained	17~26	4~6
Dry unit weight, γ_d (kN/m^3)	Fine grained	9~18	2~21
Young's modulus, E (MPa)	Coarse grained	5.2~15.6	26~68
Liquid limit, w_L (%)	Fine grained	27~89	3~39
Undrained shear strength, S_u (kPa)	Fine grained	6~713	4~84
Natural water content, w_n (%)	Fine grained	13~105	7~46
Relative density, D_r (%)	Coarse grained	11~36	30~70
Plastic limit, w_P (%)	Fine grained	13~35	3~34
Plastic index, PI (%)	Fine grained	11~52	9~57
Liquidity index, LI	Fine grained	0.5~2.5	5~88
Effective stress friction angle, ϕ'(°)	Fine grained	9~41	4~50
	Coarse grained	30~42	2~17
q_c (MN/m^2) (CPT test)	Sand, silty clay	0.5~29.2	5~81
N (SPT test)	Sand, clay	7~74	19~62

TABLE 5.2
Summary of Correlation Length for Different Soils (after Cami et al. 2020)

Soil Type	Horizontal Correlation Length: λ_h (m)		Vertical Correlation Length: λ_v (m)	
	Range	Mean	Range	Mean
Ankara clay	—	—	1–6.2	3.63
Chicago clay	—	—	0.79–1.25	0.91
Hangzhou clay	40.4–45.4	42.9	0.49–0.77	0.63
Alluvial	1.07–49	14.2	0.07–1.1	0.36
Marine clay	8.37–66	30.9	0.11–6.1	1.55
Clay	0.14–163.8	31.9	0.05–3.62	1.29
Marine sand	15	15	0.07–7.2	1.43
Silty clay	9.65–45.4	29.8	0.095–6.47	1.40
Offshore soil	24.6–66.5	45.6	0.48–1.62	1.04
Overconsolidated clay	0.14	0.14	0.063–0.255	0.15
Sand	1.69–80	24.5	0.1–4	1.17
Sensitive clay	—	—	1.1–2	1.55
Silt	12.7–45.5	33.2	0.14–7.19	2.08
Soft clay	22.2–80	47.6	0.14–6.2	1.70
Undrained engineered soil	—	—	0.3–2.7	1.42

5.3 RELATIONSHIP BETWEEN SAMPLE SIZE AND ACCURACY OF SPATIAL MODEL

5.3.1 Random Field Simulation of Soil Property Profiles

Random fields have been successfully used to model spatially correlated geotechnical properties during the last several decades (e.g., Vanmarcke 1977; Vanmarcke et al. 1986; Phoon and Kulhawy 1999; Fenton and Griffiths 2008; Bong and Stuedlein 2018; Wang et al. 2018; Zhao and Wang 2018). In this study, soil property profiles from a wide variety of statistically homogeneous and stationary soil layers are simulated using random fields. Consider, for example, a 1D stationary Gaussian random field X. X is a random variable representing a soil property of interest with a mean μ and a standard deviation σ. The spatial correlation between soil properties at different depths is modeled by a correlation function $\rho(h)$. For a single exponential correlation function (SECF), the correlation function may be expressed as

$$\rho(h) = \exp(-2|h| / \lambda_v) \tag{5.2}$$

where λ_v is the vertical correlation length and h is the separation distance between any two points in a random field. Using a random field generator, such as Karhunen–Loève expansion, random field samples of X may be generated (e.g., Zhang and Ellingwood 1994; Huang et al. 2001; Au and Wang 2014; Wang et al. 2018).

TABLE 5.3
Summary of Parameters Used in the Parametric Study

Parameter	Variation
Thickness (m)	2.0, 5.0, 10.2, 20.4, 30.7, 40.9, 51.1
Mean, μ	0.5, 20, 40, 50, 60, 80, 100, 1000
Correlation length, λ_v (m)	0.5, 1, 2, 3, 5, 7, 9, 11, 13, 15, 18, 20, 50, 80
COV (%)	1, 5, 10, 20, 30, 40, 50, 100
Number of measurement data, M	5, 10, 15, 20, 25, 30, 35, 40, 45, 50, 60, 80

5.3.2 Parametric Study Using Random Field Simulation Data

To simulate a wide variety of typical sites, typical ranges of the three statistical parameters (i.e., μ, COV, and λ_v) for random field modeling of soil properties are considered and used in the parametric study. For example, Phoon and Kulhawy (1999) and Cao et al. (2016) summarized the typical ranges of μ and COV for a wide variety of soil properties, including undrained shear strength, friction angle, natural water content, plastic and liquid limits, unit weight, relative density, and field measurements (e.g., cone penetration test [CPT], vane shear test, standard penetration test [SPT]). Generally speaking, the mean values range from 0.5 to 700, while the COV varies from 2% to 90%. The typical values of vertical correlation length λ_v for different soil properties range from 0.8 m to 12.7 m. Hence, these typical ranges of random field parameters are used in the parametric study. Table 5.3 summarizes the parameters used in the parametric study. The μ values range from 0.5 to 1,000, while the COV values range from 1% to 100%. The values of vertical correlation length λ_v vary from 0.5 m to 80 m. The thicknesses of the soil layers vary from 2.0 m to 51.1 m, which covers the typical ranges of thickness for a homogeneous soil layer in geotechnical practice. In total, 504 random fields are simulated. For each random field, 100 random field samples are generated, and the total number of data points for each random field sample is $N = 512$. Then, 12 different sample sizes (i.e., $M = 5$, 10, 15, 20, 25, 30, 35, 40, 45, 50, 60, 80) are adopted for each random field sample. A number of measurement data points ranging from 5 to 80 are used as input to BCS for interpreting the complete soil property profiles.

Steps 1–6 are repeatedly performed for each random field sample at 12 different sample sizes (i.e., Step 7), resulting in $504 \times 100 \times 12 = 604{,}800$ relative errors. Then, Step 8 is repeatedly performed to estimate the mean relative error for each M value and random field, leading to $504 \times 12 = 6{,}048$ μ_{RE}. In the next section, the parametric study results (i.e., 6,048 data pairs of μ_{RE} and M) are summarized for developing a dimensionless relationship between the normalized sample size and μ_{RE}.

5.3.3 Dimensionless Relationship between Sample Size and Relative Error

The parametric study results are summarized in Figure 5.2 with two dimensionless parameters, normalized mean relative error μ_{RE} / COV, and normalized sampling

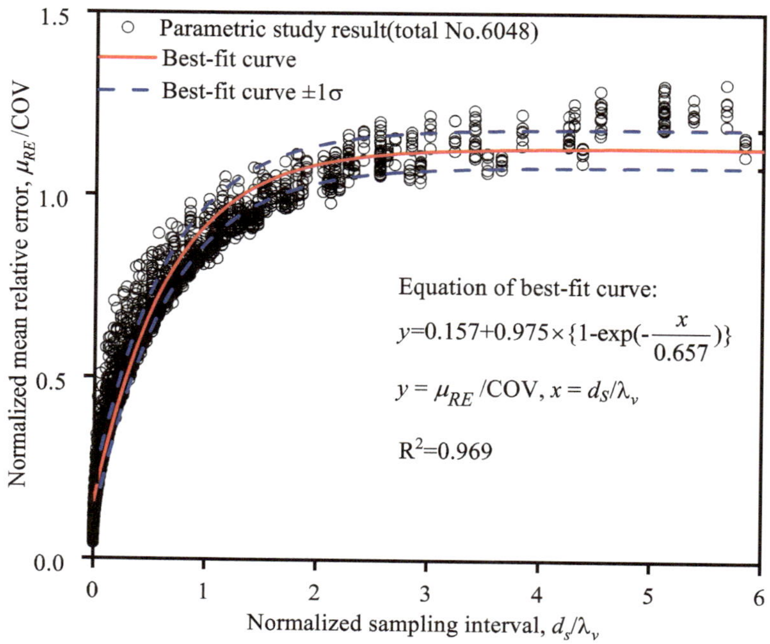

FIGURE 5.2 Dimensionless plot of normalized mean relative error versus normalized sampling interval (after Wang et al. 2019).

interval d_s / λ_v. The vertical axis in Figure 5.2 (i.e., μ_{RE} / COV) represents the interpretation accuracy (i.e., mean *RE*) normalized with respect to the variability (i.e., COV) of the soil property concerned. The horizontal axis in Figure 5.2 (i.e., d_s / λ_v) represents the sample size (i.e., sampling interval) normalized with respect to the spatial correlation (i.e., correlation length). Figure 5.2 shows the 6,048 data pairs of μ_{RE} / COV and d_s / λ_v by open circles. The data pairs in Figure 5.2 reveal a relationship that the value of μ_{RE} / COV increases as the value of d_s / λ_v increases from 0 to about 3, and that the μ_{RE} / COV value approaches to an asymptotic value when d_s / λ_v is larger than 3.

When the normalized sampling interval d_s / λ_v is close to 0, a small positive intercept is observed along the vertical axis of μ_{RE} / COV. There are two possible scenarios leading to this small intercept. In the first scenario, the sampling interval d_s approaches to 0, which means that the soil properties are sampled and measured at every location. The small intercept in the vertical axis of μ_{RE} / COV reflects the nugget effect or the difference in soil properties at neighboring locations (e.g., Matheron 1963). In the second scenario, the correlation length, λ_v is very large, which indicates that the soils are uniform within the layer and that the soil properties at different locations within the soil layer are perfectly correlated with each other. Therefore, the entire soil property profile can be represented by a single sample and measurement. The small intercept in the vertical axis of μ_{RE} / COV in this scenario represents the measurement error, which is also often referred to as the nugget effect in geostatistics. Note that the measurement error has been probabilistically modeled in BCS as a Gaussian random variable with a zero mean and unknown variance, which is subsequently integrated through a marginalization in the Bayesian framework. More discussion on the nugget effect is provided later in this section.

On the contrary, when the normalized sampling interval d_s / λ_v is large (e.g., larger than 3), the sample size is quite small and the sampling interval is much larger than the distance within which the measurement data are significantly correlated. In this case, the d_s / λ_v has negligible effect on the μ_{RE} / COV, which plots more or less horizontally over different d_s / λ_v values in Figure 5.2. When d_s / λ_v is larger than 3, reducing the sampling interval (i.e., d_s) or increasing the sample size (i.e., *M*) does not improve the level of accuracy for the interpretation results. This suggests that there is a minimum sample size threshold or maximum sampling interval threshold for geotechnical site investigation. When the sample size is smaller than the minimum sample size threshold or the sampling interval is larger than the maximum sampling interval threshold (e.g., $3\lambda_v$), changing the sample size or sampling interval has negligible influence on the level of accuracy for the interpretation results of spatial variation.

Note that the data pattern in Figure 5.2 is very similar to a semi-variogram in geostatistics, which shows variation of a half of auto-covariance with lag distance. Figure 5.3 shows a schematic of a typical semi-variogram where the experimental semi-variogram data (i.e., open circles) are fitted to an exponential function with a nugget effect component (i.e., solid line). As shown in Figure 5.3, the semi-variogram monotonically increases as lag distance increases, and it approaches to an asymptotic

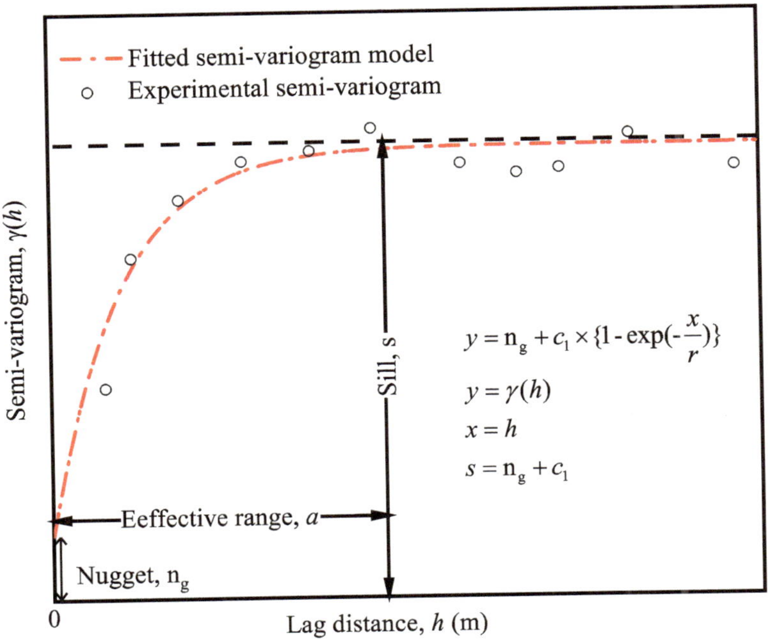

FIGURE 5.3 Illustration of an exponential semi-variogram model with the fitting parameters (after Wang et al. 2019).

value, defined as sill, s, in geostatistics, when the lag distance is larger than an effective range, a (e.g., Webster and Oliver 2007). The effective range is usually taken as the lag distance where the value of the semi-variogram reaches 95% of its sill value. The semi-variogram contains a small positive intercept, n_g, at the vertical axis, which is referred to as the nugget effect in geostatistics for modeling measurement errors and spatial variation within the shortest sampling interval (e.g., Matheron 1963).

Since Figure 5.2 exhibits a pattern that is similar to an exponential semi-variogram, an exponential function is used in a regression analysis to best fit the 6,048 data points. The best-fit line is shown by a solid line in Figure 5.2, and the corresponding best-fit equation, with a coefficient of determination $R^2 = 0.969$, is expressed as:

$$y=0.157+0.975\times\{1-\exp(-\frac{x}{0.657})\} \tag{5.3}$$

where $y = \mu_{RE} / \text{COV}$ and $x = d_s / \lambda_v$. The best-fit n_g value in Eq. (5.3) is 0.157, and the sill $s = 0.157 + 0.975 = 1.132$. The effective range $a = 1.8$ can be estimated from the best-fit line in Figure 5.2. Two dashed lines are also included in Figure 5.2 for representing the best-fit curve plus or minus one standard deviation. The equations for the two dashed lines are expressed, respectively, as:

$$y=0.209+0.975\times\{1-\exp(-\frac{x}{0.657})\} \tag{5.4}$$

$$y=0.105+0.975\times\{1-\exp(-\frac{x}{0.657})\} \tag{5.5}$$

5.3.4 Effect of Spatial Correlation Function

The spatial correlation between the soil property values at different locations in random fields is represented by the correlation function. Several theoretical correlation functions have been successfully used in the literature to model the variation of geotechnical properties in statistical analysis, and the most suitable correlation function for a certain site should be based on site conditions (e.g., Cao and Wang 2014). Four correlation functions are commonly used in geotechnical engineering, including SECF, binary noise correlation function (BNCF), second-order Markov correlation function (SMCF), and squared exponential correlation function (SQECF). For SECF, BNCF, SMCF, and SQECF, the correlation function $\rho(h)$ can be expressed as

$$\text{SECF: } \rho(h) = \exp(-2|h|/\lambda) \tag{5.6}$$

$$\text{BNCF}: \rho(h) = \begin{cases} 1-|h|/\lambda, \; for \; |h|<\lambda \\ 0, \; otherwise \end{cases} \tag{5.7}$$

$$\text{SMCF: } \rho(h) = (1+4|h|/\lambda)\exp(-4|h|/\lambda) \tag{5.8}$$

$$\text{SQECF: } \rho(h) = \exp\left(-\pi\,(h/\lambda)^2\right) \tag{5.9}$$

where λ is the correlation length and h is the separation distance between any two points in a random field.

In order to investigate the influence of different correlation functions on the parametric study results for degree of accuracy and confidence level, the procedures illustrated in the last two sections are repeated for normal distribution random fields with four different correlation functions shown in Eqs (5.6)–(5.9), together with the parameters in Table 5.3. The parametric study results for different correlation functions are summarized in Figure 5.4.

Figure 5.4 shows the parametric study results for the level of accuracy with different correlation function forms, and they are in a format similar to Figure 5.2. Their vertical axis is normalized mean relative error (i.e., μ_{RE} / COV), and their horizontal axis is normalized sampling interval (i.e., d_s / λ_v). Open circles in each figure indicate 6,048 data pairs of μ_{RE} / COV and d_s / λ_v. The data pairs in these figures reveal a similar relationship that the value of μ_{RE} / COV increases as the value of

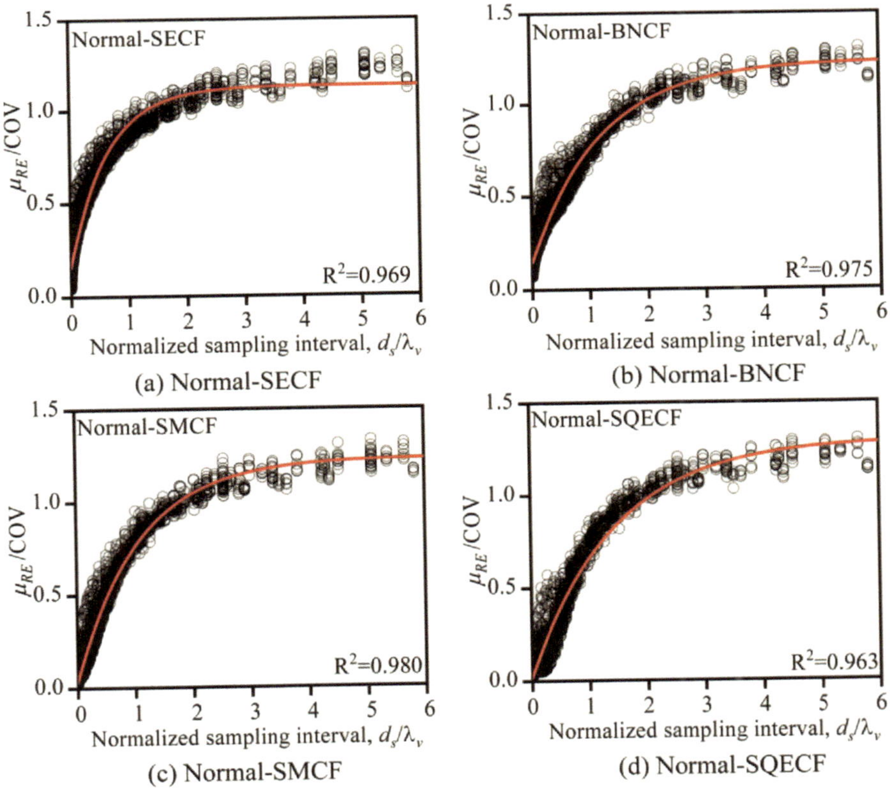

FIGURE 5.4 Parametric study results for normally distributed random fields with four correlation functions (after Guan and Wang 2020).

TABLE 5.4
Parametric Study Results for Normally Distributed Random Fields with Four Common Correlation Functions

Correlation Function	Best-Fit Equation
SECF	$y=0.157+0.975\times\{1-\exp(-\frac{x}{0.657})\}$
SMCF	$y=0.037+1.197\times\{1-\exp(-\frac{x}{1.085})\}$
BNCF	$y=0.143+1.090\times\{1-\exp(-\frac{x}{1.223})\}$
SQECF	$y=0.020+1.283\times\{1-\exp(-\frac{x}{1.454})\}$

Note: $y=\mu_{RE}/\text{COV}$, $x=d_s/\lambda_v$.

d_s/λ_v increases, and that the μ_{RE}/COV value approaches to an asymptotic value when d_s/λ_v is larger than a threshold value. The asymptotic values for SECF, BNCF, SMCF, and SQECF are 1.13, 1.23, 1.23, and 1.3, respectively. The solid lines in Figure 5.4 represent the best-fit curves obtained from regression analysis with a coefficient of determination $R^2 = 0.963 \sim 0.980$, respectively. The corresponding best-fit equations are summarized in the second column of Table 5.4. Although the regression equations share the same equation form, the values of the regression constants are different for the four different correlation function types. It is evident that the correlation function type has a significant effect on the relationship between sample size and the level of accuracy.

5.3.5 Effect of Random Field Distribution Types

Although normal distribution has been widely used to model geotechnical properties in the literature (e.g., Wang et al. 2010; Guan and Wang 2021c) and previous sections, it is well recognized that soil properties in a specific site might follow different probability distributions (e.g., Wang et al. 2015). For example, some soil properties (e.g., friction angle) are strictly nonnegative, and it is probably more suitable to model them by a lognormal distribution than a normal distribution. To explore the effect of random field distribution types on the relationships between sample size and levels of accuracy, the parametric study described in the previous sections is repeated for lognormally distributed random fields with four correlation function types using the parameters summarized in Table 5.3. The parametric study results for lognormally distributed random fields are shown in Figure 5.5 using legends similar to those in Figure 5.4.

Figure 5.5 shows the parametric study results for the level of accuracy using lognormal random fields with four correlation function types. Similar to Figure 5.4, the value of μ_{RE}/COV increases as the value of d_s/λ_v increases, and it approaches to an asymptotic value when d_s/λ_v is larger than a threshold value. The best-fit curves are

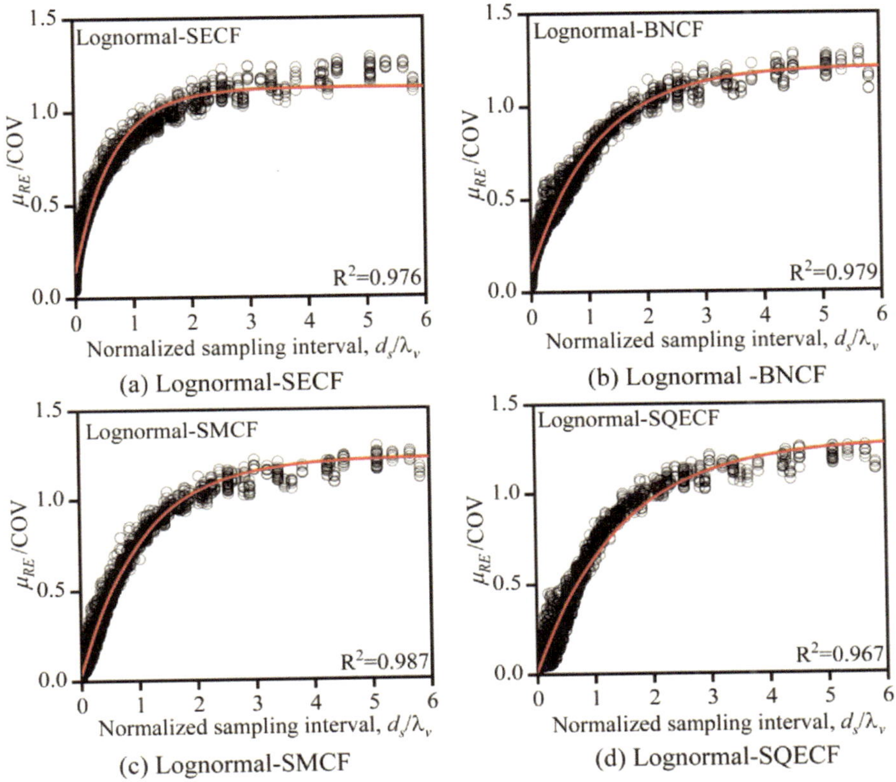

FIGURE 5.5 Parametric study results for lognormally distributed random fields with four correlation functions (after Guan and Wang 2020).

TABLE 5.5
Parametric Study Results for Lognormally Distributed Random Fields with Four Common Correlation Functions

Correlation Function	Best-Fit Equation
SECF	$y=0.143+0.983\times\{1-\exp(-\frac{x}{0.656})\}$
BNCF	$y=0.118+1.094\times\{1-\exp(-\frac{x}{1.148})\}$
SMCF	$y=0.028+1.198\times\{1-\exp(-\frac{x}{1.088})\}$
SQECF	$y=0.017+1.281\times\{1-\exp(-\frac{x}{1.476})\}$

Note: $y=\mu_{RE}/$ COV, $x=d_s/\lambda_v$.

obtained from regression analysis and shown by solid lines. The corresponding best-fit equations are summarized in Table 5.5, with $R^2 = 0.967{\sim}0.987$.

The relationship between the normalized mean relative error and the normalized sampling interval shown in Tables 5.4 and 5.5 provides a link between the sample size and the level of accuracy for the interpretation results. This relationship may be further used to develop a statistical chart for sample size determination with consideration of spatial variation and correlation of soil properties, as described in the next section.

5.3.6 Development of Statistical Chart for the Level of Accuracy

Figure 5.6 summarizes the best-fit curves for the level of accuracy obtained from the parametric studies described above. The value of μ_{RE} / COV increases as the value of d_s / λ_v increases, and it approaches to an asymptotic value when d_s / λ_v is larger than a threshold value. When the normalized sampling interval d_s / λ_v is close to 0, a small positive intercept is observed at the vertical axis of μ_{RE} / COV. The small intercept in the vertical axis of μ_{RE} / COV is similar to the nugget effect in geostatistics for modeling measurement errors and spatial variation within the shortest sampling interval (e.g., Matheron 1963; Wang et al. 2019). On the contrary, when the normalized sampling interval d_s / λ_v is large (i.e., the sampling interval is much larger than the distance within which the data are significantly correlated), the d_s / λ_v has a negligible effect on the μ_{RE} / COV, leading to the asymptotic behavior of best-fit curves shown in Figure 5.6. In other words, when d_s / λ_v is larger than a threshold value, reducing the sampling interval (i.e., d_s) or increasing the sample size (i.e., M) does not improve the

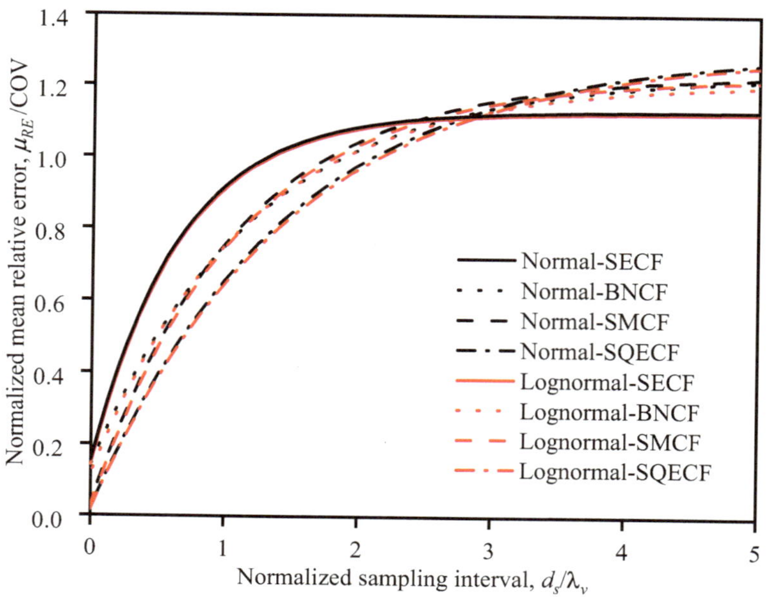

FIGURE 5.6 Summary of parametric study results for the level of accuracy (after Guan and Wang 2020).

level of accuracy for the interpretation results of spatial variation. This suggests that there is a minimum sample size threshold or maximum sampling interval threshold for geotechnical site investigation. When the sample size is smaller than the minimum sample size threshold or the sampling interval is larger than the maximum sampling interval threshold (e.g., $3\lambda_v$ for Normal-SECF and Lognormal-SECF), changing the sample size or sampling interval has negligible influence on the level of accuracy for the interpretation results of spatial variation.

In addition, the μ_{RE} / COV vs d_s / λ_v relationships for normally and lognormally distributed random fields are almost overlapped in Figure 5.6. The probability distribution types only have minimal effect on the relationship between sample size and accuracy level. On the other hand, the effect of spatial correlation function types is considerable, and different spatial correlation function types lead to different μ_{RE} / COV versus d_s / λ_v relationships. It is also interesting to note that the commonly used Gaussian random fields with single exponential correlation functions have the largest values of μ_{RE} / COV, when d_s / λ_v is smaller than about 3. In other words, modeling a specific site with unknown random field types or unknown spatial correlation function types as a Gaussian random field with a single exponential correlation function leads to overestimation of the normalized mean relative error, and hence, it is conservative.

Figure 5.7 shows statistical charts developed from Figure 5.6 for the determination of sample size along a vertical direction within a single statistically homogeneous soil layer at various levels of accuracy. The statistical chart for the level of accuracy in Figure 5.7 is developed from Figure 5.6 by switching the horizontal and vertical axes. Similar to Figure 5.6, Figure 5.7 is plotted in accordance with two normalized dimensionless parameters: a target normalized relative error (i.e., RE_T / COV) in the horizontal axis and a normalized sampling interval in the vertical axis (i.e., d_s / λ_v). The target relative error, RE_T, is the desired level of accuracy for the interpretation results that a user would like to have. When replacing μ_{RE} by RE_T and switching the vertical and horizontal axis of Figure 5.6, a statistical chart for the level of accuracy is generated, as shown in Figure 5.7. This figure shows that as the RE_T value decreases, or the desired level of accuracy for the interpretation results becomes stringent, the RE_T / COV value decreases, leading to a decreasing normalized sampling interval. As the sampling interval decreases, the required number of samples increases. To cater for the often-unknown spatial correlation function types or random field distribution types for a specific site in geotechnical practice, lower bound envelopes of all best-fit curves in Figure 5.7 are proposed and shown by the solid lines. The corresponding equation for the level of accuracy is expressed as:

$$y=0.00022e^{7.8x}+1.3e^{0.55x}-1.43 \tag{5.10}$$

where $y = d_s / \lambda_v$ and $x = RE_T$ / COV.

Figure 5.7 provides a quantitative tool for geotechnical practitioners to rationally determine sample size with consideration of spatial correlation and the desired levels of accuracy on the interpretation result. To use these statistical charts, a user shall first specify his/her target relative error RE_T. RE_T shall be determined depending on the importance and risk of the project (e.g., failure consequences), as well as the

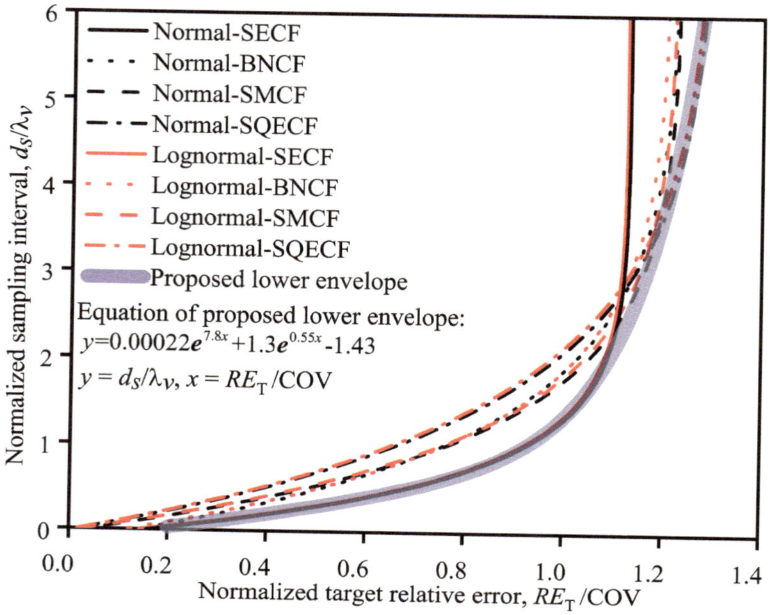

FIGURE 5.7 Statistical chart for sampling interval determination considering the degree of accuracy (after Guan and Wang 2020).

complexity of local geology (e.g., Baecher and Christian 2003; Wang et al. 2019). Then, the coefficient of variation, COV, and vertical correlation length, λ_v, of the soil properties concerned shall be estimated based on existing knowledge of the site and soil properties concerned, such as local experience on soil properties and sites under similar conditions, geologic and topographic maps, and the typical range of COV and λ_v for different soil properties reported in the literature (e.g., Phoon and Kulhawy 1999; Cao et al. 2016). After the COV value and RE_T value are determined, the RE_T / COV in the horizontal axis of Figure 5.7 is calculated accordingly. Then, the most suitable correlation function and soil property distribution type shall be determined and used, based on the site-specific information available. After that, the corresponding d_s / λ_v values in the vertical axis can be estimated from Figure 5.7. Finally, the required sampling interval d_s in geotechnical site investigation is the smaller value of the results obtained from Figure 5.7. The required sample size M is subsequently calculated from the d_s value and the thickness of the soil layer. In geotechnical engineering practice, the types of correlation function or probability distribution are often unknown during sample size determination. In these cases, the lower bound envelope in Figure 5.7 may be used as a conservative alternative to bypass the need for probability distribution types or spatial correlation function types. As the COV and λ_v of the soil properties concerned can be obtained from prior knowledge about the site, the proposed chart may be used even without any existing site-specific measurement data.

5.3.7 Application Example

To demonstrate the application of the proposed statistical chart, real CPT q_c data are used as illustrative examples in this section. Note that CPT data are used because CPT is performed in a nearly continuous manner. The CPT data can then be used for validating the interpretation results of measurement data, the sample size of which is determined from the statistical chart. The proposed quantitative method and statistical chart are equally applicable to measurements of other soil properties.

A CPT (i.e., test number CPT-1) was carried out in a cohesive soil layer in eastern Houston, Texas, USA, and its q_c data were downloaded online from ISSMGE TC304dB database website (http://140.112.12.21/issmge/Database_2011.htm) and used in this illustrative example, as shown in Figure 5.8. According to Stuedlein et al. (2012), the water table is about 1.8 m below the ground surface, and the site contains a thick layer of stiff, slightly silty clay. The CPT q_c data ranging from 5.10 m to 15.32 m are used in this example. The q_c data was recorded at 20 mm intervals, so the total number of data points for this CPT q_c data profile is 512.

Suppose that the target relative error, RE_T, in this application is 10%. The COV and λ_v of the q_c data in this site are estimated as 15% and 2.0 m, respectively, based

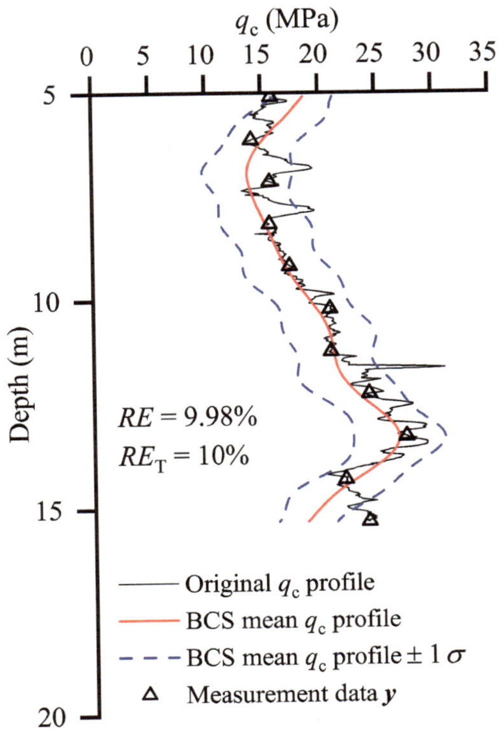

FIGURE 5.8 Comparison between the original q_c profile and the q_c profile interpreted from the BCS method for a real CPT data example (after Guan and Wang 2020).

on local experience (e.g., nearby CPT data). These ranges are also within the typical ranges of 5 ~ 40% for COV and 0.1 ~ 2.2 m for λ_v, respectively, for the clay CPT q_c data reported in the literature (e.g., Phoon and Kulhawy 1999). Then, the value of RE_T/COV is calculated as 0.67. Using the best-fit curve in the statistical chart, the normalized sampling interval $d_s / \lambda_v = 0.50$ is calculated. Because λ_v is estimated as 2.0 m, the sampling interval is then calculated as $d_s = 1.00$ m.

The thickness of the soil property profile is 10.22 m, and the number of measurement data points is estimated as $M = 10.22 / 1.00 + 1 \approx 11$. Subsequently, 11 data points with an equal interval are taken from the CPT data, and they are used as measurement data and input to the BCS method for interpreting the complete soil property profile. As shown in Figure 5.8, the measurement data are shown by open triangles. The solid line represents the original CPT q_c profile, and the dashed line indicates the mean q_c profile obtained from BCS. The statistical uncertainty of the interpreted q_c profiles is indicated by the interval within two dashed lines which are the mean q_c profile plus or minus one standard deviation. The *RE* between the original and mean q_c profiles is 9.98%, which is quite close to the target relative error of 10%. In other words, the statistical chart provides reasonable results and achieves the target level of accuracy for this site.

5.4 RELATIONSHIP BETWEEN SAMPLE SIZE AND CONFIDENCE OF THE SPATIAL MODEL

When a soil property profile is interpreted from sparse and limited measurements, the interpreted soil property profile unavoidably contains uncertainty. Confidence level is commonly used to quantitatively represent the uncertainty of an estimate in statistical analysis. In general, the confidence level improves (e.g., the confidence interval reduces) as the number of measurement data points (i.e., sample size) increases. To determine the sample size required to achieve a desired confidence level, a relationship between confidence level and sample size is needed. An extensive parametric study is performed in this section to establish a relationship between confidence level and sample size (e.g., Guan and Wang 2020).

5.4.1 Parametric Study for Confidence Level

Similar to the parametric study for accuracy level described in the previous section, the basic idea of the parametric study is to quantify the evolution of the confidence level of the interpreted soil property profile as the increase of sample size for a single homogenous soil layer in geotechnical site investigation. Based on the parametric study results, a relationship is established between the confidence level of interpretation results and the sample size for these typical sites.

Firstly, a number, M (e.g., $M = 5$), of data points are obtained from a synthetic soil property profile simulated from a random field X and used as measurement data points for interpreting a complete soil property profile by BCS to mimic the data measurement and interpretation procedures in geotechnical site investigation. Note that the BCS method provides not only the mean of the estimated soil property at each location but also the full marginal probability distribution function including standard deviation and quantiles of the estimated soil property. BCS method is used in this study because it can estimate standard deviation σ at each data point of an X profile, or each depth of an interpreted soil property profile in this study, from sparse measurements (e.g., Candès et al. 2006; Donoho 2006; Tropp and Gilbert 2007; Ji et al. 2008, 2009; Wang and Zhao 2016, 2017). The confidence level is measured by the relative standard deviation, *RSD*, which is defined as a ratio of standard deviation, σ to the mean, μ. It is the average of the coefficients of variation for interpreted soil properties along depth. Let a vector $\hat{\boldsymbol{f}} = \left[\hat{f}_1, \hat{f}_2, \ldots, \hat{f}_N\right]^{\mathrm{T}}$ with a total number, N, of data points denote the best estimate (i.e., mean) soil property profile along depths interpreted from the M measurement data points, and $\boldsymbol{\sigma} = \left[\sigma_{\hat{f}_1}, \sigma_{\hat{f}_2}, \ldots, \sigma_{\hat{f}_N}\right]^{\mathrm{T}}$ is a vector representing the standard deviation of interpreted soil property profile at the corresponding depths. The *RSD* for this interpreted soil property profile is defined as the arithmetic mean of all elements:

$$RSD = \frac{1}{N}\sum_{i=1}^{N}\frac{\sigma_{\hat{f}_i}}{\hat{f}_i} \tag{5.11}$$

Then, this procedure is repeated for many different M values (i.e., $M = 5$, 10, 15, 20, 25, 30, 35, 40, 45, 50, 60, 80), leading to many *RSD* values. In general, the confidence level improves as the sample size M increases. When all data points of a soil property profile are sampled, the confidence level converges to 100% (i.e., *RSD* approaches to zero). Therefore, the relationship between *RSD* and M for a given soil property profile can be established by plotting the variation of *RSD* as a function of M. In order to establish the relationship between *RSD* and M for a typical site, the above procedure is repeatedly performed for many soil property profiles (e.g., 100 synthetic soil property profiles) from a given typical site (e.g., a random field simulation), resulting in many (e.g., 100) sets of relationship between *RSD* and M. Median of these relative standard deviations (i.e., M_{RSD}) is calculated to quantify the confidence level of interpreted results at different M (e.g., $M = 5$, 10, 15, 20, 25, 30, 35, 40, 45, 50, 60, and 80) values. A relationship between M_{RSD} and M for this given typical site is established accordingly.

Then, the procedure is repeatedly performed for each of the 504 random fields (i.e., typical site conditions) at 12 different sample sizes (i.e., $M = 5$, 10, 15, 20, 25, 30, 35, 40, 45, 50, 60, 80), resulting in $504 \times 12 = 6{,}048$ M_{RSD}. Then, the parametric study results (i.e., 6,048 data pairs of M_{RSD} and M) are summarized for developing a dimensionless relationship between the normalized sample size (i.e., sampling interval) and M_{RSD}, as shown in Figure 5.9 using two dimensionless parameters, normalized median relative standard deviation M_{RSD} / COV and normalized sampling interval d_s / λ_v. The vertical axis in Figure 5.9 (i.e., M_{RSD} / COV) reflects the confidence level (i.e., median relative standard deviation) normalized with respect to the variability (i.e., coefficient of variation) of the soil property concerned. The horizontal axis in Figure 5.9 (i.e., d_s / λ_v) represents the sample size (i.e., sampling interval) normalized with respect to the spatial correlation (i.e., correlation length). Figure 5.9 shows the 6,048 data pairs

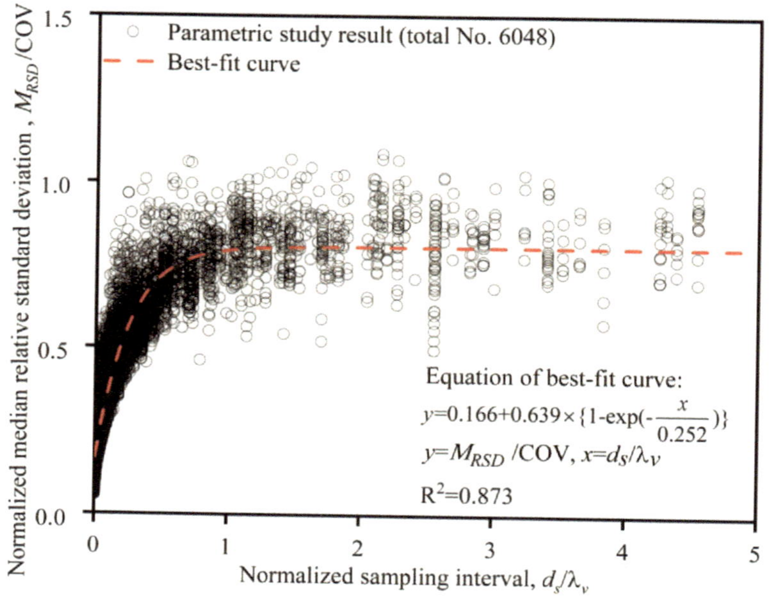

FIGURE 5.9 The dimensionless plot of normalized median relative standard deviation versus normalized sampling interval (after Guan and Wang 2020).

of M_{RSD} / COV and d_s / λ_v by open circles. The value of M_{RSD} / COV increases as the value of d_s / λ_v increases from 0 to about 1.3, and it approaches to an asymptotic value when d_s / λ_v is larger than 1.3. The best-fit curve to the 6,048 data pairs is shown by a solid line in Figure 5.9, and the corresponding best-fit equation, with a coefficient of determination $R^2 = 0.873$, is expressed as:

$$y=0.166+0.639\times\{1\text{-exp}(-\frac{x}{0.252})\} \tag{5.12}$$

where $y = M_{RSD}$ / COV and $x = d_s / \lambda_v$. Eq. (5.12) establishes a relationship between sample size M and confidence level (i.e., M_{RSD}).

A relationship between the sample size and confidence level has been established in Figure 5.9, based on the parametric study using normally distributed random fields with a single exponential correlation function. Figure 5.10 shows the parametric study results for the confidence level with different correlation function forms. The corresponding best-fit equations are summarized in the second column of Table 5.6. The parametric study results for lognormally distributed random fields are shown in Figure 5.11. The corresponding best-fit equations are summarized in the second column of Table 5.7.

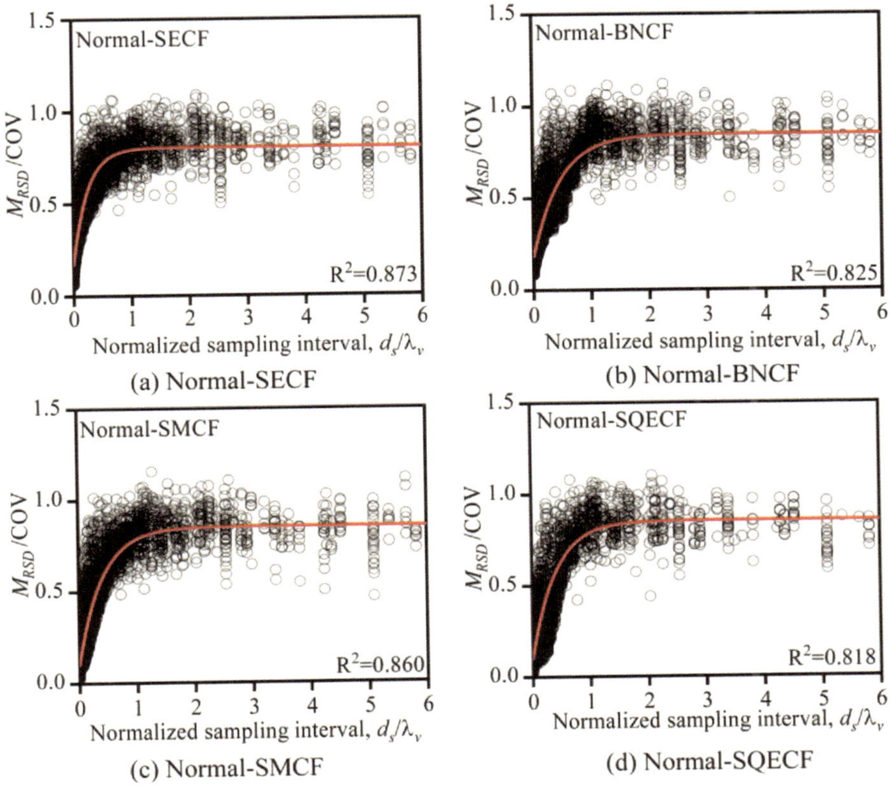

FIGURE 5.10 Parametric study results for normally distributed random fields with four correlation functions (after Guan and Wang 2020).

TABLE 5.6
Parametric Study Results for Normally Distributed Random Fields with Four Common Correlation Functions

Correlation Function	Best-Fit Equation
SECF	$y=0.166+0.639\times\{1-\exp(-\frac{x}{0.252})\}$
SMCF	$y=0.100+0.754\times\{1-\exp(-\frac{x}{0.407})\}$
BNCF	$y=0.177+0.671\times\{1-\exp(-\frac{x}{0.471})\}$
SQECF	$y=0.097+0.777\times\{1-\exp(-\frac{x}{0.519})\}$

Note: $y = M_{RSD}$ / COV, $x = d_s / \lambda_v$.

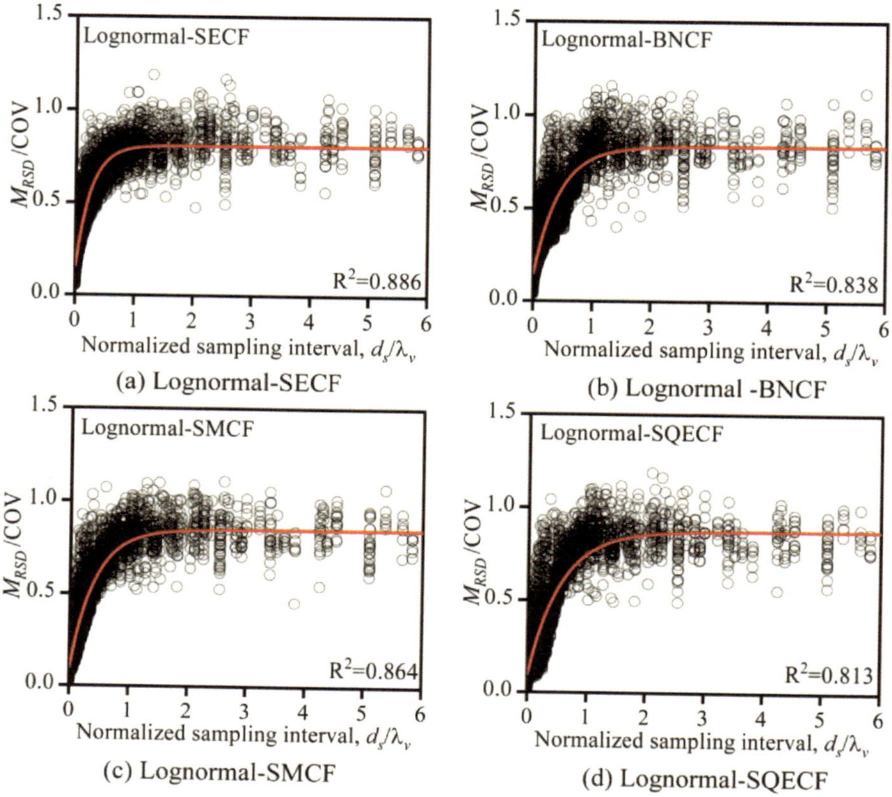

FIGURE 5.11 Parametric study results for lognormally distributed random fields with four correlation functions (after Guan and Wang 2020).

TABLE 5.7

Parametric Study Results for Lognormally Distributed Random Fields with Four Common Correlation Functions

Correlation Function	Best-Fit Equation
SECF	$y=0.155+0.654\times\{1-\exp(-\frac{x}{0.258})\}$
SMCF	$y=0.095+0.755\times\{1-\exp(-\frac{x}{0.415})\}$
BNCF	$y=0.148+0.688\times\{1-\exp(-\frac{x}{0.428})\}$
SQECF	$y=0.096+0.780\times\{1-\exp(-\frac{x}{0.545})\}$

Note: $y = M_{RSD} / \text{COV}$, $x = d_s / \lambda_v$.

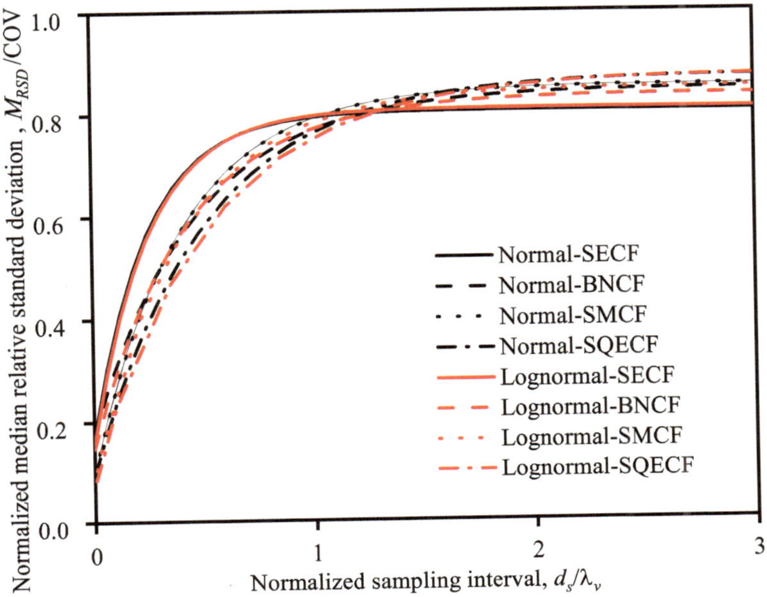

FIGURE 5.12 Summary of parametric study results for confidence level (after Guan and Wang 2020).

5.4.2 Development of Statistical Chart for the Confidence Level

Figure 5.12 summarizes the best-fit curves for the confidence level obtained from the parametric studies described above. The value of M_{RSD} / COV increases as the value of d_s / λ_v increases, and it approaches to an asymptotic value when d_s / λ_v is larger than a threshold value. When the normalized sampling interval d_s / λ_v is close to 0, a small positive intercept is also observed at the vertical axis of M_{RSD} / COV. On the contrary, when the sampling interval d_s is much larger than λ_v, the d_s / λ_v has a negligible effect on the M_{RSD} / COV. Similar to the result for the level of accuracy, a minimum sample size threshold or maximum sampling interval threshold is observed in Figure 5.12. When the sample size is smaller than the minimum sample size threshold or the sampling interval is larger than the maximum sampling interval threshold (e.g., 1.3λ_v for Normal-SECF and Lognormal-SECF), increasing the sample size does not improve the confidence level of the interpretation result. The results for normally and lognormally distributed random fields are plotted very closely in Figure 5.12. When d_s / λ_v < 1.3, the commonly used normally distributed random fields with a single exponential correlation function lead to the largest M_{RSD} / COV. However, note that the most suitable function types of the spatial correlation function and probability distribution might be different for different sites (e.g., Wang et al. 2015), and they should be determined based on the site-specific information available (e.g., DeGroot and Baecher 1993; Stuedlein et al. 2012; Cao and Wang 2014). The relationships between the normalized median relative standard deviation and normalized sampling interval

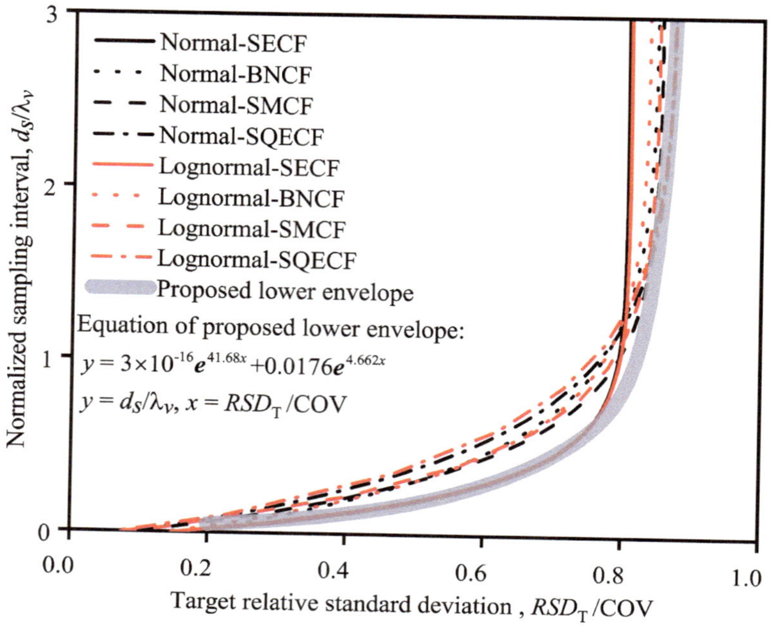

FIGURE 5.13 Statistical chart for sampling interval determination considering confidence level (after Guan and Wang 2020).

for different random field distribution types and correlation function types have been established in Figure 5.12. These relationships can be further used to develop statistical charts for sample size determination at various levels of confidence.

The statistical chart for the confidence level in Figure 5.13 is developed from Figure 5.12. Figure 5.13 is plotted with two normalized dimensionless parameters: a target normalized relative standard deviation (i.e., RSD_T / COV) in the horizontal axis and a normalized sampling interval in the vertical axis (i.e., d_s / λ_v). The target relative standard deviation, RSD_T, is the desired confidence level for the interpretation result that a user would like to have. By replacing M_{RSD} with RSD_T and switching the vertical and horizontal axis of Figure 5.12, a statistical chart for the confidence level is developed, as shown in Figure 5.13.

5.4.3 Application Example

In this subsection, natural water content, w_n, measured by laboratory tests is used to illustrate the application of the proposed statistical charts. Consider, for example, the determination of sample size for quantifying spatial variation of natural water content along depth within a silty clay layer in the construction site of Taipei City Hall Station, Taiwan (Chin et al. 1994). The site contains a thick layer of silty clay from a depth of about 2.9 m to 30.0 m, as shown in Figure 5.14.

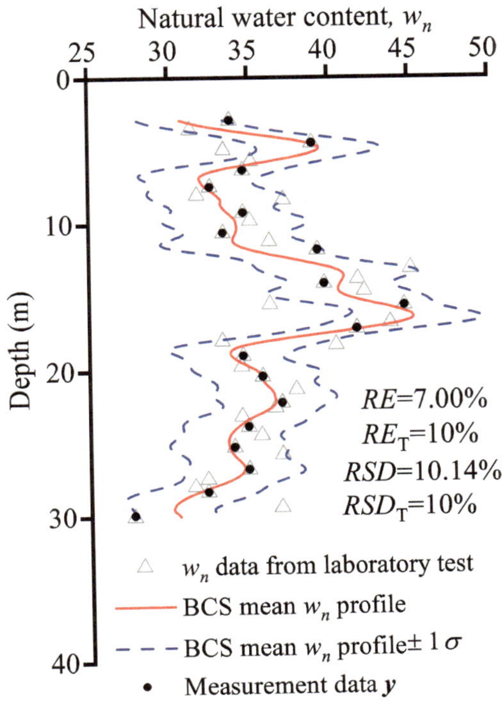

FIGURE 5.14 Comparison between the original profile and the profile interpreted from the BCS method in the natural water content data example (after Guan and Wang 2020).

Suppose that the target relative error and the target relative standard deviation are both 10%. The COV and λ_v of natural water content in this silty clay layer are estimated as 17% and 5.7 m, respectively. These two values are approximately the mean of the typical ranges of 8 ~ 30% for COV and 1.6 ~ 12.7 m for λ_v, respectively, of the clay's natural water content reported in the literature (e.g., Phoon and Kulhawy 1999). Then, the RE_T / COV = 0.59 and RSD_T / COV = 0.59 are calculated. Using the lower bound envelopes in Figures 5.7 and 5.13, $d_s / \lambda_v = 0.39$ and $d_s / \lambda_v = 0.28$ are obtained for the target relative error and the target relative standard deviation, respectively. Because λ_v is estimated as 5.7 m, sampling intervals required for RE_T and RSD_T are then calculated, respectively, as d_s = 2.22 m and d_s = 1.60 m. The thickness of the silty clay layer is 27.1m, and the numbers of measurement data points for RE_T and RSD_T are estimated as $M = 27.1/2.17+1 \approx 13$ and $M = 27.1/1.60+1 \approx 18$, respectively. In order to achieve both the target level of accuracy and confidence, a sample size M = 18 is adopted for this site. Subsequently, 18 specimens within the silty clay layer (i.e., from a depth of about 2.9 m to 30.0 m) shall be obtained from a borehole with an equal sampling interval for carrying out natural water content tests in the laboratory and quantifying spatial variation of the natural water content along the depth.

Chin et al. (1994) reported that a total number of 39 specimens (i.e., T1-T39) was obtained from a borehole within this silty clay layer for carrying out natural water content tests in the laboratory. Figure 5.14 shows the 39 w_n values measured along depth by open triangles. For validating the performance of the proposed statistical charts in this illustrative example, 18 w_n values are taken from the 39 w_n values with an approximately equal sampling interval, and they are shown in Figure 5.14 by solid points. The 18 w_n values obtained are used as the measurement data, i.e., input to BCS, for interpolating a high-resolution natural water content profile with 256 data points. Figure 5.14 also shows the mean natural water content profile interpreted from BCS by a solid line and the mean plus or minus one standard deviation profiles by two dashed lines, respectively. It is observed from Figure 5.14 that the dashed line follows a trend quite similar to that of the open triangles (i.e., all 39 w_n values), and that most of the 39 w_n data points (i.e., the open triangles) fall within the dashed lines. The relative standard deviation is 10.14%, which is very consistent with the target relative standard deviation of 10%. The relative error between the 39 original natural water content values and mean natural water content values at the corresponding depths obtained from BCS is calculated as 7%, which is smaller than the target relative error of 10%. It is reasonable because the sample size (i.e., 18) adopted in this site is larger than the sample size (i.e., 13) required to achieve the target level of accuracy. The consistent results suggest that the sample size determined by the statistical charts (i.e., Figures 5.7 and 5.13) achieves the target levels of accuracy and confidence in the results interpreted from the given number of measurements. In summary, the proposed statistical charts perform well for this illustrative example.

5.5 CPT SAMPLE SIZE IN A 2D VERTICAL CROSS-SECTION

In-situ tests, such as the CPT, are often performed during site investigation for characterization of subsurface soil properties and stratification, which are often simplified as a vertical cross-section for subsequent geotechnical design and analysis in practice (e.g., Lunne et al. 1997; Baecher and Christian 2003; Robertson 2009; Gharahbagh et al. 2013; Wang et al. 2019; Zhang et al. 2019; Shiau and Al-Asadi 2020; Guan and Wang 2021d; Guan et al. 2021). It is widely recognized that the number of CPTs performed has a significant effect on the accuracy of site investigation results (e.g., Rix et al. 2018; Wang et al. 2019). However, no rational method is available to determine the minimum CPT number for achieving a target level of accuracy in an objective and quantitative manner. Existing geotechnical design codes or guidelines only provide conceptual principles for the determination of CPT number (e.g., accuracy of site investigation improves as the CPT number increases) and recommend some rule of thumb minimal numbers (e.g., at least one CPT every 100 m).

This section aims to develop a rational method for addressing this issue. The method developed is based on an extensive parametric study that uses random field simulation to cover a wide variety of CPT data encountered in geotechnical practice and BCS to interpolate the spatially varying, but correlated, CPT data along both depth and horizontal directions simultaneously (e.g., Guan and Wang 2021a, 2021b).

Real CPT data are used to illustrate and validate the proposed method, which is shown to perform well.

5.5.1 Parametric Study for CPT Sample Size

In this study, 2D random fields are used in an extensive parametric study for generating many 2D high-resolution vertical cross-sections of CPT data to mimic various typical site conditions commonly encountered in engineering practice. Consider, for example, the normalized cone tip resistance, Q_t, obtained from CPT and expressed below (e.g., Robertson 2009):

$$Q_t = \frac{q_t - \sigma_{vo}}{\sigma_{vo}'} \tag{5.13}$$

where σ_{vo} and σ_{vo}'represent the total and effective overburden stresses, respectively; and q_t indicates the corrected cone tip resistance. Q_t has been widely used for the estimation of various soil properties (e.g., over-consolidation ratio, undrained shear strength ratio, and friction angle) and identification of soil behavior type in engineering practice (e.g., Robertson 2016).

Three parameters, (1) mean μ, (2) coefficient of variation, COV, and (3) correlation length (i.e., vertical correlation length, λ_v and horizontal correlation length, λ_h), of Q_t data together with its marginal distribution type (e.g., Gaussian distribution) and autocorrelation function form (e.g., an exponential correlation function) are required to specify a 2D Q_t random field. Typical ranges of μ, COV, and correlation length for CPT Q_t data have been reported for different site conditions in the literature (e.g., Robertson 1990; Firouzianbandpey et al. 2014), and these typical values are used in random field simulations for modeling various typical site conditions. Consider, for example, a 2D Gaussian random field with an exponential autocorrelation function that is widely used in geotechnical engineering (e.g., Vanmarcke 2010; Luo et al. 2012):

$$\rho(\Delta x_1, \Delta x_2) = \exp\left(-2\sqrt{\frac{(\Delta x_1)^2}{\lambda_v^2} + \frac{(\Delta x_2)^2}{\lambda_h^2}}\right) \tag{5.14}$$

where Δx_1 and Δx_2 represent the absolute vertical and horizontal distance between any two points in a random field, respectively; λ_v and λ_h represent the correlation length of Q_t data along depth and horizontal direction, respectively.

Table 5.8 summarizes typical ranges of μ, COV, λ_v, and λ_h for Q_t data covering a wide variety of site conditions (e.g., Robertson 1990; Phoon and Kulhawy 1999; Uzielli set al. 2005; Firouzianbandpey et al. 2014; Oguz et al. 2019). The μ and COV of Q_t data vary from 1 to 1,000 and 1% to 100%, respectively. The typical ranges of vertical correlation length λ_v and horizontal correlation length λ_h are from 0.1 m to 20 m and 0.5 m to 100 m, respectively. In the parametric study reported in

TABLE 5.8
Typical Ranges of the Mean, Coefficient of Variation, and Correlation Length of CPT Data for a Variety of Site Conditions Considered in the Parametric Study

CPT Data	Mean	Coefficient of Variation, COV (%)	Correlation Length (m)	
			Vertical Direction: λ_v	Horizontal Direction: λ_h
Normalized cone tip resistance, Q_t.	0.1–1000	1–100	0.1–20	0.5–100

Note: $Q_t = (q_t\text{-}\sigma_{vo})/\sigma'_{vo}$, where σ_{vo} and σ'_{vo} represent total and effective overburden stresses, respectively; q_t is corrected cone tip resistance.

TABLE 5.9
Summary of Parameters Used in the Parametric Study

Parameter	Variation
Mean, μ	0.1, 10, 100, 1000
Coefficient of variation, COV (%)	1, 5, 10, 20, 30, 40, 50, 100
Vertical correlation length, λ_v (m)	0.1, 0.5, 1, 2, 3, 5, 10, 20
Horizontal correlation length, λ_h (m)	0.5, 1, 5, 10, 15, 20, 25, 30, 35, 40, 50, 100
CPT number, M_c	5, 8, 10, 12, 15, 18, 20, 22, 25, 28, 30, 35, 40, 50
Depth of vertical cross-section (m)	1, 10, 20, 40, 100
Width of vertical cross-section (m)	10, 100, 200, 400, 1000

this section, the dimension of the 2D vertical cross-section varies from 1 m to 100 m along depth (i.e., x_1 direction) and from 10 m to 1,000 m for horizontal direction (i.e., x_2 direction). This range of dimension covers the typical range of vertical cross-sections commonly encountered in geotechnical practice. Table 5.9 summarizes various random field parameters used for the simulation of Q_t data. Different combinations of the parameters in Table 5.8 lead to 350 Q_t random fields for modeling various typical site conditions in engineering practice. Samples of each random field are generated and stored in a 2D matrix with up to 160,000 elements (e.g., 800 × 200).

Mathematically, let a matrix **F** with a dimension of $N_{x_1} \times N_{x_2}$ denote a true 2D high-resolution vertical cross-section of CPT data, while the corresponding high-resolution cross-section interpreted from M_c CPT soundings is denoted by a matrix $\hat{\mathbf{F}}$. Then, relative error, *RE*, between **F** and $\hat{\mathbf{F}}$ is used to quantify the accuracy of the interpretation results, which is expressed as:

$$\text{Relative error}: RE = \frac{\|\hat{\mathbf{F}} - \mathbf{F}\|}{\|\mathbf{F}\|} = \frac{\sqrt{\sum_{i=1}^{N_{x_2}} \sum_{j=1}^{N_{x_1}} \left(\hat{x}_{i,j} - x_{i,j}\right)^2}}{\sqrt{\sum_{i=1}^{N_{x_1}} \sum_{j=1}^{N_{x_2}} \left(x_{i,j}\right)^2}} \tag{5.15}$$

where $\hat{x}_{i,j}$ and $x_{i,j}$ indicate the entry in the ith row and jth column of the matrix $\hat{\mathbf{F}}$ and $\mathbf{F}$, respectively; $\|\bullet\|$ represents the Frobenius norm operation which is defined as a square root of the sum of all square of all elements in a matrix.

Using the similar procedure described in Section 5.2, a statistical analysis is performed to explore the effect of different site conditions (i.e., different values of random field parameters such as mean μ, COV, and correlation length) on the relationship between μ_{RE} and M_c. The statistical analysis results provide a quantitative link for the development of a relationship between the accuracy of the interpretation result (i.e., μ_{RE}) and the CPT number, M_c, for various typical site conditions. Based on the quantitative relationship developed, a statistical chart is further developed for determining the minimal number of CPTs required for achieving a target level of accuracy on site investigation results in geotechnical practice.

The parametric study results from the 350 random fields are summarized in a dimensionless plot, as shown in Figure 5.15. The vertical axis of Figure 5.15, i.e., the normalized mean relative error, μ_{RE} / COV, represents the interpretation accuracy μ_{RE} normalized by variability (i.e., COV) of Q_t data within a site. The horizontal axis of Figure 5.15, i.e., the normalized CPT spacing, d_c / λ_h, is the CPT horizontal spacing d_c

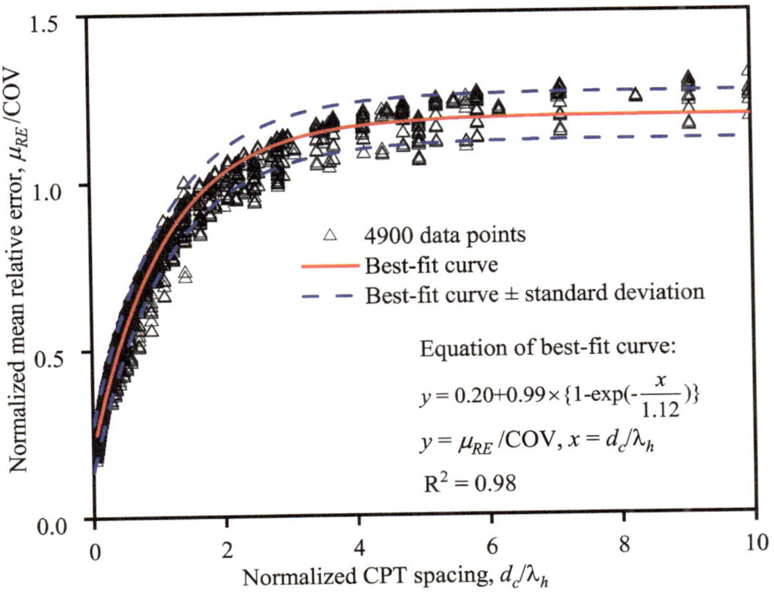

FIGURE 5.15 Dimensionless plot of normalized mean relative error versus normalized CPT spacing (after Guan and Wang 2021b).

normalized by its horizontal correlation length λ_h of Q_t data within a site. Figure 5.15 shows that the normalized mean relative error increases from about 0.2 to about 1.2 as the normalized CPT spacing increases from 0 to 5. When d_c / λ_h is larger than about 5, the μ_{RE} / COV reaches an asymptotic value. This suggests that when the CPT number is too small or the CPT spacing is significantly larger than the horizontal distance within which Q_t data are highly correlated (i.e., $d_c \geq 5\lambda_h$), increasing the CPT number or decreasing CPT spacing has a negligible effect on the improvement of the accuracy of interpretation result. In other words, there is a maximum CPT spacing threshold (i.e., $5\lambda_h$) in geotechnical site investigation. On the other hand, a small positive intercept is found on the vertical axis. Such a small positive intercept represents the measurement errors within the shortest CPT spacing (e.g., Webster and Oliver 2007; Wang et al. 2019).

The effect of COV and λ_h on the relationship between μ_{RE} and d_c (or CPT number, M_c) is also shown in Figure 5.15. For a given COV and μ_{RE} (or a target level of accuracy), d_c / λ_h is more or less a constant, and therefore, as the horizontal correlation length λ_h of Q_t data increases, the required CPT spacing increases, or the required CPT number decreases. Note that the horizontal correlation length represents the horizontal distance within which Q_t data are highly correlated. A small value of λ_h indicates that Q_t data highly fluctuate, and thus a large CPT number or small CPT spacing is required to capture the variation of Q_t data within a site. In contrast, a large value of λ_h suggests that Q_t data vary smoothly along the horizontal direction, and therefore, a relatively small CPT number or relatively large CPT spacing is sufficient to interpolate a Q_t vertical cross-section with a desired level of accuracy. Similarly, for a given λ_h and μ_{RE}, the required CPT spacing decreases as the COV of Q_t data increases. This is reasonable because COV represents the variability of Q_t data within a site, and a Q_t vertical cross-section with large COV indicates the high variability of Q_t data, leading to the difficulty in capturing the variation of Q_t data using sparse CPT measurements. Hence, relatively small CPT spacing (i.e., relatively large CPT number) is needed for interpreting Q_t vertical cross-section with large COV, and vice versa.

The pattern of data pairs in Figure 5.15 is similar to an exponential semi-variogram model in geostatistics (e.g., McBratney and Webster 1986; Webster and Oliver 2007), and thus, an exponential function is fitted to the data pairs in Figure 5.15. In Figure 5.15, a solid line represents the best-fit exponential function with a coefficient of determination $R^2 = 0.98$, which indicates that the solid line fits quite well the data pairs. The corresponding equation for the best-fit curve is expressed as:

$$y=0.20+0.99\times\{1\text{-}\exp(-\frac{x}{1.12})\} \tag{5.16}$$

where $y = \mu_{RE}$ / COV and $x = d_c / \lambda_h$. A confidence interval of the best-fit curve is shown in Figure 5.15 by two dashed lines which are expressed, respectively, as:

$$y=0.27+0.99\times\{1\text{-}\exp(-\frac{x}{1.12})\}$$

$$y=0.13+0.99\times\{1\text{-exp}(-\frac{x}{1.12})\} \quad (5.17)$$

Eqs. (5.16) and (5.17) provide a quantitative link for developing a relationship between CPT number, M_c, and the level of accuracy, μ_{RE}. Based on the quantitative relationship developed, a statistical chart is further developed for determining the minimal number of CPTs required for achieving a target level of accuracy in geotechnical site investigation, as described in the following section.

5.5.2 Development of Statistical Chart for CPT Sample Size

The statistical chart developed from the quantitative relationship between M_c and μ_{RE} is shown in Figure 5.15. By swapping the vertical and horizontal axis of Figure 5.15 and replacing the mean relative error, μ_{RE}, by a target level of relative error, RE_T, which represents the target level of accuracy on the interpreted Q_t vertical cross-section, a statistical chart is obtained for determining the preliminary number of required CPTs. The vertical axis and horizontal axis of the statistical chart represent the target relative error RE_T normalized by the coefficient of variation, COV (i.e., RE_T / COV), and the CPT spacing, d_c, normalized by horizontal correlation length, λ_h (i.e., d_c / λ_h), respectively.

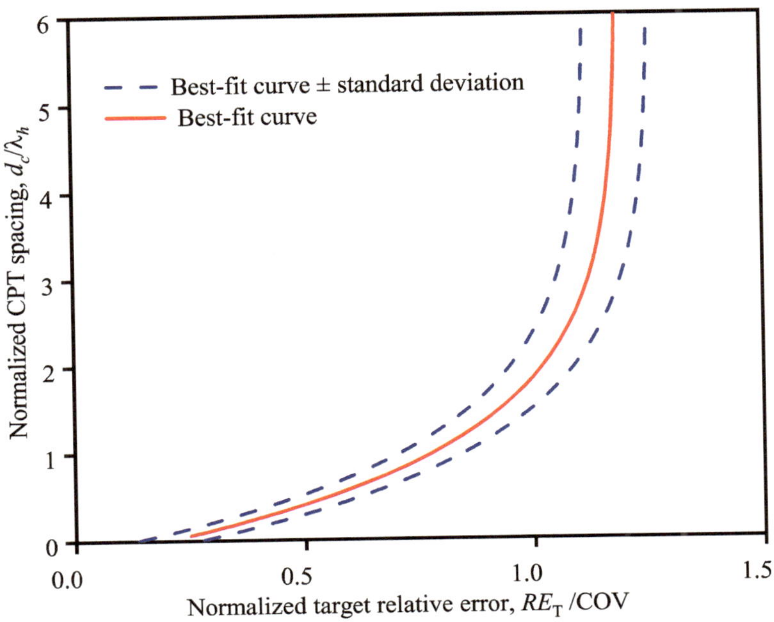

FIGURE 5.16 Statistical chart for determining the preliminary number of CPTs required for achieving a target level of accuracy in geotechnical site investigation (after Guan and Wang 2021b).

To use this statistical chart and determine the required number of CPTs for achieving a target accuracy level, RE_T, in engineering practice, COV and λ_h of CPT data for the site of interest are needed, and they may be estimated from existing knowledge about the site concerned. Such knowledge might be obtained from relevant literature or databases, geological maps, previous investigations of nearby sites, etc. Note that RE_T denotes the desired level of accuracy that is determined by engineers and owners of the projects and based on the risk level they can tolerate, and geological complexity of site conditions, etc. After the RE_T / COV and λ_h are specified, the required number of CPTs can be obtained directly from the statistical chart or Eqs. (5.16) and (5.17). In the following section, a set of real CPT data from Australia is used to illustrate and validate the proposed method and statistical chart.

5.5.3 Real Data Application Example

In this section, the proposed method is applied to determine the minimal number of CPTs required for the interpretation of a vertical cross-section of Q_t data within Keswick clay at a site in Adelaide, Australia. Jaksa (1995) reported a large number of CPT soundings carried out in the Keswick clay layer at this site, as shown in Figure 5.17. To illustrate and verify the proposed method, 51 sets of normalized cone resistance Q_t data (i.e., CPT-01 to CPT-51 shown in Figure 5.17) with a horizontal spacing of 1.0 m are used in this illustration example. The CPT data were obtained from the TC304 database (http://140.112.12.21/issmge/tc304.htm?=6). 51 sets of Q_t data are considered as the true high-resolution vertical cross-section of Q_t data which has a dimension of 2 m × 50 m (depth × width) with vertical and horizontal resolution of 0.005 m and 1 m, respectively. This leads to a true Q_t vertical cross-section with 400 × 51 data points in total, as shown in Figure 5.18a.

Suppose that 24% of relative errors on site investigation results is considered allowable for a project based on the site condition and importance of the project. In other words, the target relative error RE_T is determined as 24%. Based on existing knowledge such as nearby CPT data (Jaksa 1995), COV, and λ_h of Q_t data on Keswick clay are estimated as 21% and 2.0 m, respectively. The RE_T / COV = 1.14 is calculated. Then, the required minimal number of CPTs can be determined using Figure 5.16. Using the best-fit curve in the statistical chart, d_c / λ_h = 3.4 is obtained for RE_T / COV = 1.14. Because λ_h = 2.0 m, the CPT spacing, d_c, is calculated as 6.8 m. The width of the site is 50 m, and then the required minimal number of CPTs is estimated as $M_c = 50 / 6.8 + 1 \approx 8$. Therefore, 8 CPT soundings shall be carried out with an equal horizontal spacing for obtaining Q_t vertical cross-section at this site.

To mimic the measurements and interpretation of 8 CPT soundings in engineering practice, 8 CPT Q_t data sets (400 × 8 data points) are taken from the original 51 Q_t data sets with approximately equal spacing and used as input to interpolate vertical cross-section of Q_t data with a dimension of 400 × 51 using 2D BCS. Figure 5.18(b) shows the Q_t cross-section interpreted from M_c = 8 CPTs by a color

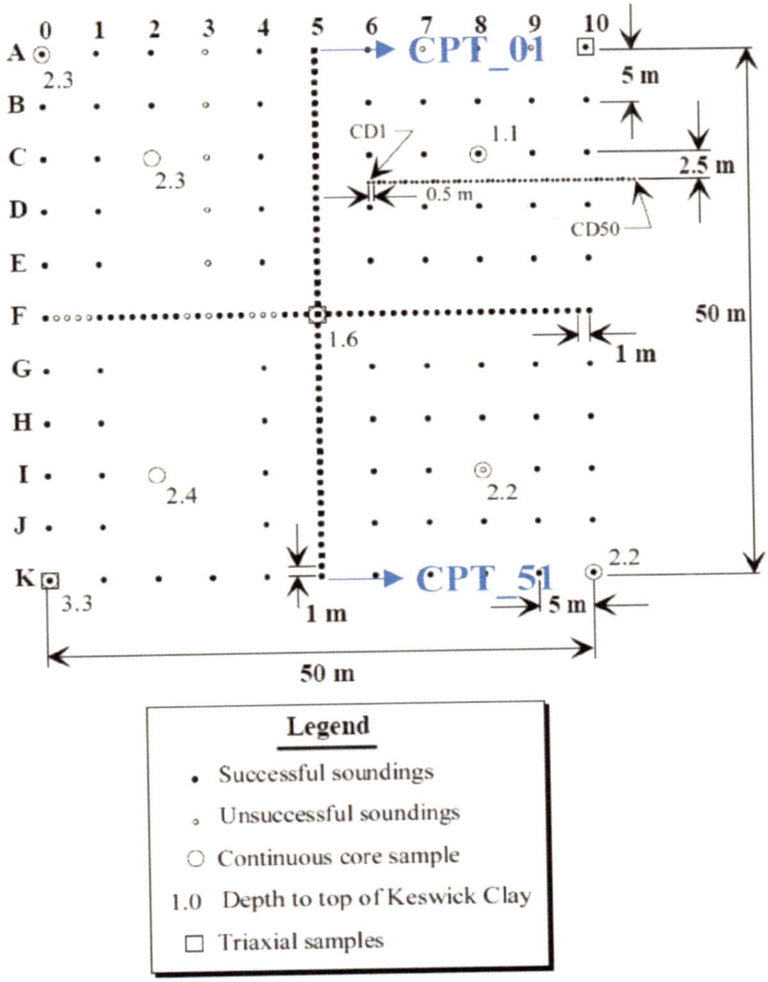

FIGURE 5.17 Layout of 51 CPT soundings at the South Parklands Site used in illustration example (after Jaksa 1995).

map. In Figure 5.18(b), dashed lines represent 8 CPT soundings. The patterns of the color map in Figure 5.18(b) are quite similar to those in Figure 5.18a, indicating that the interpreted Q_t cross-section is generally consistent with the true Q_t cross-section. The relative error, *RE*, between the true Q_t cross-section and the one interpreted from $M_c = 8$ CPTs is calculated as 24.5%, which is close to $RE_T = 24.0\%$. Such agreement indicates that the Q_t vertical cross-section interpreted from 8 CPTs, as determined by the developed statistical chart, achieves the target accuracy level. The result shows that the proposed method performs well for this real CPT Q_t data example.

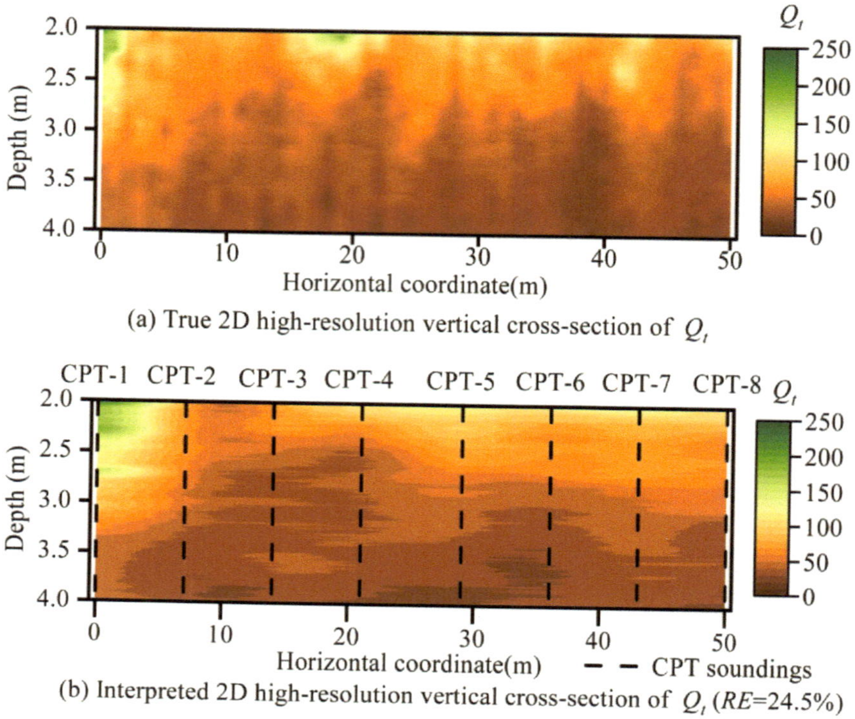

FIGURE 5.18 Comparison between the true 2D high-resolution vertical cross-section of Q_t and the one interpreted from eight CPTs using Bayesian compressive sensing (after Guan and Wang 2021b).

5.6 SAMPLE SIZE OF STANDARD PENETRATION TESTS (SPT) IN A 2D VERTICAL CROSS-SECTION

In geotechnical site investigation, standard penetration test (SPT) is probably the most widely used in-situ test for characterizing horizontal and vertical variation of soil properties, and the results of site investigation are usually simplified as a vertical cross-section for subsequent planning, design, and construction of engineering projects (e.g., Skempton 1986; Clayton 1995; Wang et al. 2019; Guan and Wang 2021a). It is widely recognized that the accuracy of site investigation results is greatly affected by the number of investigation points (e.g., SPT) (e.g., AASHTO 2018). However, no quantitative guideline is available for determining the minimum number of SPTs needed for satisfying an accuracy requirement of site investigation. Current geotechnical design codes only provide general recommendations for selecting the appropriate number of in-situ tests (e.g., SPT) (e.g., the greater variability of subsurface conditions, the larger number of SPTs required to obtain sufficient underground information) (e.g., CEN 2007; New York State DOT 2013; FHWA 2017; Michigan DOT 2019). No quantitative or rational method is available for selecting the appropriate number of SPTs considering spatial variability and auto-correlation in subsurface conditions. This section aims to address this issue and proposes a rational method for quantitatively determining the minimum number of SPTs required for satisfying an accuracy requirement of site investigation in a vertical cross-section.

5.6.1 Parametric Study for SPT Sample Size

It is well-recognized that the soil properties are spatially correlated and varying along both horizontal and vertical directions owing to the complex geological processes that soils have gone through during their formation (e.g., Fenton and Griffiths 2008; Wang et al. 2016). In geotechnical engineering, spatial variability of soil properties may be modeled explicitly using random fields (e.g., Vanmarcke 1977, 2010; Stuedlein et al. 2012; Guan and Wang 2023). Therefore, in this study, a wide variety of typical site conditions commonly encountered in engineering practice are modeled using 2D SPT data random fields. Consider, for example, the normalized SPT value $(N_1)_{60}$, which is usually expressed as (e.g., Boulanger and Idriss 2014):

$$(N_1)_{60} = C_N N_{60} \tag{5.18}$$

where N_{60} is the corrected SPT N value; $C_N = (\sigma_{atm} / \sigma_{vo}')^{0.5} \le 2$, $\sigma_{atm} = 100$ kPa, σ_{vo}' = effective vertical overburden stress. $(N_1)_{60}$ has been widely used to estimate various soil properties such as friction angle, pre-consolidation stress, and cyclic resistance ratio (e.g., Mayne 2006).

The $(N_1)_{60}$ data is often taken to be normally distributed since the normal distribution is commonly used for the modeling probability distribution of soil properties (e.g., Fenton and Griffiths 2008). To model a 2D Gaussian random field for $(N_1)_{60}$ data, three parameters are needed, including mean $\mu_{(N_1)_{60}}$, standard deviation $\sigma_{(N_1)_{60}}$, and spatial correlation length (i.e., vertical correlation length, λ_v and horizontal correlation length, λ_h). Numerous studies have been carried out to evaluate the spatial variability of SPT data, and the typical ranges of $\mu_{(N_1)_{60}}$, $COV_{(N_1)_{60}}$, λ_v, and λ_h have been reported in the literature, as summarized in Table 5.10 (e.g., Phoon and Kulhawy 1999; Zhang and Chen 2012; Oguz et al. 2019). In general, $\mu_{(N_1)_{60}}$ ranges from 0.1 to 100, while $COV_{(N_1)_{60}}$ ranges from 1% to 100%; λ_v and λ_h range from 0.1m to 10m and 1m to 100m, respectively. Using these typical values, a wide variety of site conditions can be simulated using 2D random fields. Table 5.11 summarizes the random field parameters used in the parametric study, and it covers typical ranges reported in the literature. In addition, different sizes of cross-sections (e.g., Depth × Width, $D \times W$ = 9 m × 50 m, 9 m × 100 m, 15 m × 250 m, ..., 90 m × 1000 m) are also used in random field simulation. To consider many possible 2D vertical cross-sections of $(N_1)_{60}$ data within a given site condition, 100 random field samples (i.e., 2D high-resolution vertical cross-sections of $(N_1)_{60}$ data) are generated for each random field using the circulant embedding algorithm (e.g., Dietrich and Newsam 1993).

After the cross-section of $(N_1)_{60}$ data is interpreted from sparse SPT measurements, the accuracy of the interpretation results may be quantified by calculating the relative error, *RE*, between the interpreted cross-section of $(N_1)_{60}$, $\hat{\mathbf{F}}$ and the true one, $\mathbf{F}$ using

TABLE 5.10
Typical Ranges of the Mean, Coefficient of Variation and Correlation Length of SPT Data for a Variety of Site Conditions

SPT Data	Mean	Coefficient of Variation, COV (%)	Correlation Length (m)	
			Vertical Direction: λ_v	Horizontal Direction: λ_h
Normalized SPT value $(N_1)_{60}$	0.1–100	1–100	0.1–10	1–100

Note: $(N_1)_{60} = C_N N_{60}$, where N_{60} is the corrected SPT value; $C_N = (\sigma_{atm} / \sigma_{vo}')^{0.5} \leq 2$, $\sigma_{atm} = 100$ kPa, σ_{vo}'= effective vertical overburden stress.

TABLE 5.11
Summary of Parameters Used in Parametric Study

Parameter	Variation
Mean, $\mu_{(N_1)_{60}}$	0.1, 10, 100
Coefficient of variation, $COV_{(N_1)_{60}}$ (%)	5, 10, 20, 30, 40, 50, 80, 100
Vertical correlation length, λ_v (m)	0.5, 1, 2, 4, 5, 6, 8, 10
Horizontal correlation length, λ_h (m)	1, 5, 10, 15, 20, 30, 40, 50, 75, 100
The number of boreholes with SPTs	5, 8, 10, 12, 15, 18, 20, 22, 25, 28, 30
Depth interval of SPT (m)	1.0, 1.5, 2.5, 3.0, 5.0
Depth of vertical cross-section (m)	9, 15, 30, 45, 90
Width of vertical cross-section (m)	50, 100, 200, 1000

Eq. (5.15). Using simulated typical site conditions and 2D BCS, a comprehensive parametric study is carried out to quantify the influence of spatial variability on the relationship among the number, M_S, of boreholes with SPTs, SPT depth interval, d_v, and μ_{RE}.

Results of the parametric study (i.e., 4,088 μ_{RE} at different combinations of d_v and M_S) are plotted in a dimensionless plot in Figure 5.19. Note that a given M_S value represents different borehole spacings along a horizontal direction for a given width of a cross-section. For consistency between horizontal and vertical directions, borehole spacing, d_h, rather than M_S is used in Figure 5.19, which can be readily calculated by the width of the site divided by ($M_S - 1$). The horizontal axis of Figure 5.19 is the normalized borehole spacing, $\tilde{d}$, which is defined as:

$$\tilde{d} = \frac{d_h}{\lambda_h} \times \frac{d_v}{\lambda_v} \tag{5.19}$$

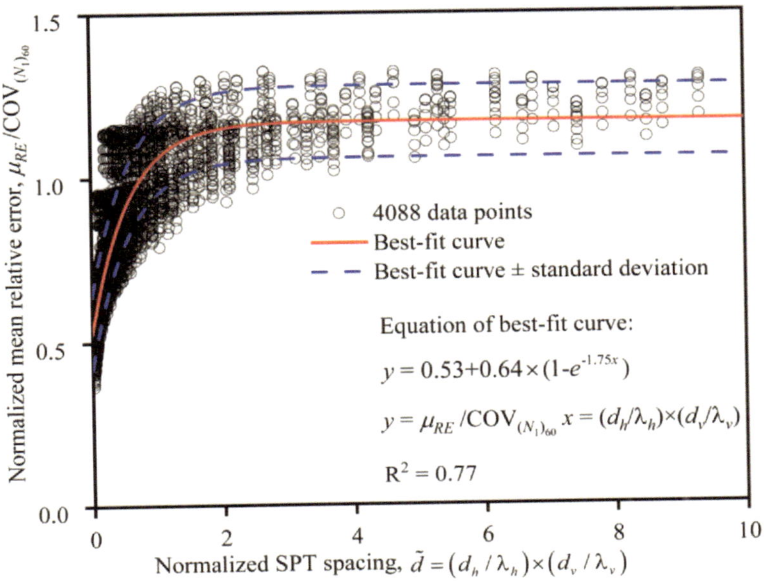

FIGURE 5.19 Dimensionless plot of normalized mean relative error versus normalized SPT spacing (after Guan and Wang 2021a).

where d_h is the borehole spacing along horizontal direction; d_v represents the vertical interval of SPT data along depth; and λ_v and λ_h are vertical and horizontal correlation length, respectively. The vertical axis of Figure 5.19 is the mean relative error μ_{RE} normalized by the coefficient of variation, $\mathrm{COV}_{(N_1)_{60}}$. It can be seen from Figure 5.19 that 4,088 data pairs of $\tilde{d}$ and $\mu_{RE}/\mathrm{COV}_{(N_1)_{60}}$ exhibit an obvious trend between $\tilde{d}$ and $\mu_{RE}/\mathrm{COV}_{(N_1)_{60}}$. The value of $\mu_{RE}/\mathrm{COV}_{(N_1)_{60}}$ increases rapidly from about 0.4 to about 1.3 as $\tilde{d}$ increases from about 0 to about 2.5, and it converges to an asymptotic value of about 1.3 when $\tilde{d}$ is larger than 2.5.

The effect of spatial variability and correlation in soil properties (i.e.,$\mathrm{COV}_{(N_1)_{60}}$, λ_v, and λ_h) on the relationship among μ_{RE}, d_v, and d_h is also shown in Figure 5.19. The relationship between normalized relative error, $\mu_{RE}/\mathrm{COV}_{(N_1)_{60}}$, and normalized SPT spacing is nonlinear. $\mu_{RE}/\mathrm{COV}_{(N_1)_{60}}$ increases monotonically as $\tilde{d}$ increases, and it approaches to an asymptotic value of about 1.3 when $\tilde{d}$ is larger than 2.5. One of the simple models for such an asymptotic growth is an exponential model (e.g., Ott 1994) which is adopted herein to best fit the data. In this case, the asymptotic exponential form might be written as:

$$y=a+b\times(1-e^{-cx}) \tag{5.20}$$

where $x = \tilde{d}$, $y = \mu_{RE}/\mathrm{COV}_{(N_1)_{60}}$, a is the initial value at $x = 0$, and the asymptote is $y = a + b$ when x is larger than 2.5. A regression analysis is then performed for all data pairs shown in Figure 5.19. The best-fit curve for the 4,088 data pairs is shown in Figure 5.19 by the solid line with the equation given below with $R^2 = 0.77$:

$$y=0.53+0.64\times(1-e^{-1.75x}) \tag{5.21}$$

The corresponding confidence interval (e.g., best-fit curve ± standard deviation) is also shown in Figure 5.19 by two dashed lines which are expressed as:

$$y=0.64+0.64\times(1-e^{-1.75x})$$

$$y=0.42+0.64\times(1-e^{-1.75x}) \tag{5.22}$$

The quantitative relationship between the SPT spacing and level of accuracy, μ_{RE}, is obtained from Eqs. (5.21) and (5.22). Such a relationship provides a rational and quantitative way for geotechnical engineers to select the minimum number of SPTs needed to meet the accuracy requirement of site investigation, considering spatial variability and correlation within a specific site. To facilitate the application of such a relationship in engineering practice, a statistical chart is further developed, as described in the next section.

5.6.2 Statistical Chart for Determination of the Minimum SPT Number

In this section, a statistical chart for selecting the minimum number of SPTs is developed based on the quantitative relationship between M_S and μ_{RE}. Note that μ_{RE} may be replaced by a target relative error, RE_{T}, which represents an accuracy requirement of site investigation. By switching the vertical and horizontal axis of Figure 5.19 and replacing μ_{RE} with RE_{T}, a statistical chart for selecting the minimum number of SPTs is generated, as shown in Figure 5.20.

To use the statistical chart in Figure 5.20 for selecting the minimum number of SPTs in engineering practice, accuracy requirement (i.e., RE_{T}) and anticipated site conditions (i.e., $\mathrm{COV}_{(N_1)_{60}}$, λ_v, and λ_h) should be determined first. Accuracy requirement (i.e., RE_{T}) represents the desired accuracy level of the interpreted cross-section of SPT data, which shall be determined based on the acceptable and tolerable risk thresholds of a project, anticipated site conditions, and the importance of geotechnical structures. For example, when the type of structure is sensitive to differential settlement or the tolerable risk is set at a stringent level, a stringent accuracy requirement (i.e., a relatively small RE_{T}) should be adopted in engineering practice. $\mathrm{COV}_{(N_1)_{60}}$, λ_v, and λ_h reflect the spatial variability of site conditions, which may be obtained from different sources, such as aerial photographs, topographic maps, soil survey maps and data, historical data from previous investigations at the site or in the surroundings, previous experiences from the area, and published reports and literature about similar soil properties and site conditions (e.g., Phoon and Kulhawy 1999; FHWA 2017).

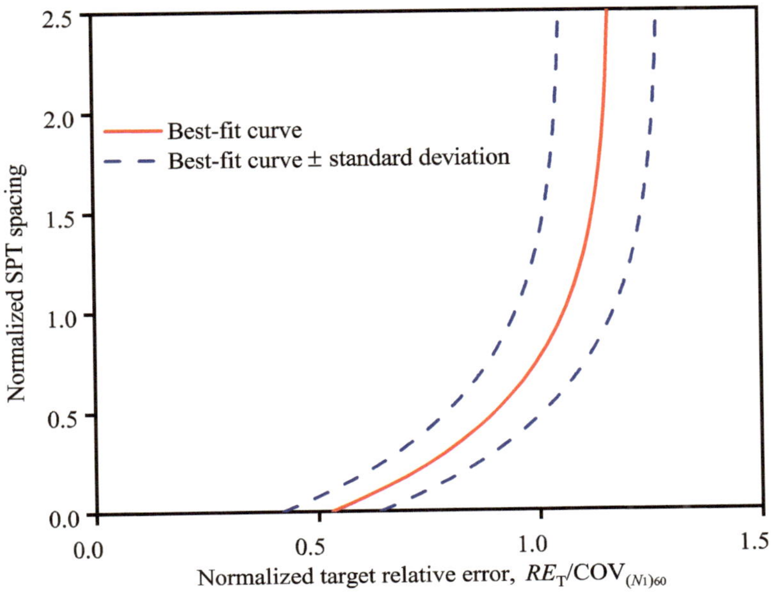

FIGURE 5.20 Statistical chart for selecting the minimum number of SPTs needed to meet an accuracy requirement of site investigation (after Guan and Wang 2021a).

After RE_T, $COV_{(N_1)_{60}}$, λ_v, and λ_h are determined, normalized target relative error, $RE_T / COV_{(N_1)_{60}}$, can be calculated. Then, the corresponding normalized SPT spacing, $\tilde{d}$, can be obtained directly using the proposed statistical chart (e.g., the best-fit curve in Figure 5.19). When the depth interval, d_v, of SPT is determined, the required SPT spacing along a horizontal direction can be obtained as $d_h = (\tilde{d} \times \lambda_h \times \lambda_v) / d_v$ for a specific site. In the next section, a set of real SPT data from New Zealand is used to illustrate and validate the proposed method and statistical chart.

5.6.3 Real SPT Data Example

In this section, the proposed method and statistical chart are used to determine the minimum number of SPTs for characterizing spatial variation of SPT data (e.g., $(N_1)_{60}$) within a cross-section A-A' at a site in Wellington, New Zealand, as shown by adashed line in Figure 5.21. Based on the records from previous boreholes conducted at this site, the upper layer of the site ranging from a depth of 0 m to about 10 m mainly consists of silty sand. The saturated unit weight of silty sand γ_{sat} is about 20 kN/m^3, and the water table at the site ranges from 0.1 m to 5.2 m below the ground surface.

The width of cross-section A-A' is about 354 m, while the depth of cross-section A-A' is 7.5 m, i.e., from a depth of about 1.5 m to about 9.0 m. Suppose that the accuracy requirement for the interpreted cross-section of SPT data, RE_T, is set as 45%.

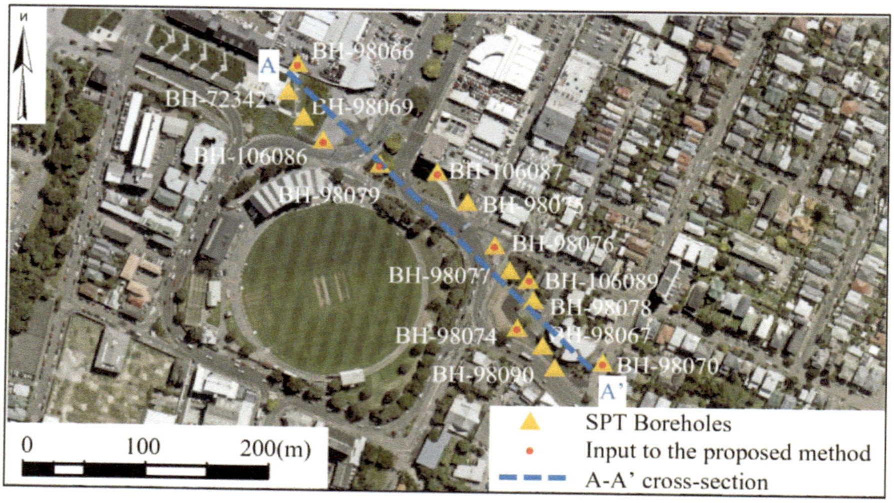

FIGURE 5.21 Locations of 15 boreholes with SPTs and a cross-section at Wellington, New Zealand (data from NZGD 2012).

$COV_{(N_1)_{60}}$, λ_v, and λ_h of this site are estimated as 50%, 2.5 m, and 60 m, respectively, based on SPT data performed nearby previously. They are also within the typical ranges of these parameters reported in the literature (e.g., Vanmarcke 1977; Phoon and Kulhawy 1999). Then, the normalized target relative error, $RE_T \,/\, COV_{(N_1)_{60}}$, can be calculated as 0.9. Using the best-fit curve in Figure 5.20, the corresponding normalized SPT spacing, $\tilde{d}$, can be obtained as 0.49. The depth interval d_v of SPT is set as 1.5 m, which is commonly used in geotechnical practice. Then, the number of SPT from a single borehole, denoted as M_{x_2}, can be calculated as $D/d_v + 1 =$ 7.5 m / 1.5 m + 1 = 6. As $d_h = (\tilde{d} \times \lambda_h \times \lambda_v)/d_v$, the required borehole spacing, d_h, is estimated as 49 m. Therefore, the number of boreholes, $M_S = 354/49 + 1 \approx 8$, with an equal horizontal spacing should be conducted for characterizing variation of SPT data within the cross-section A-A'. The total number of SPT measurement data needed is $M = M_{x_2} \times M_S = 6 \times 8 = 48$.

To examine the performance of the proposed method for this site, a set of real SPT data along cross-section A-A' is obtained from the New Zealand Geotechnical Database (NZGD) (NZGD 2012). In total, 94 $(N_1)_{60}$ data points are obtained from 15 boreholes with SPTs performed at this site, i.e., BH-98066, BH-72342, ..., BH-98070, as shown in Figure 5.21 by solid triangles. The measurement data of these SPTs are summarized in Figure 5.22. In this illustration example, the SPT data are divided into two parts, the first row in the figure is used as input, and the second row is for validation only.

Based on the result of the proposed method, eight SPT boreholes shall be performed with equal spacing of about 50 m within the cross-section A-A'.

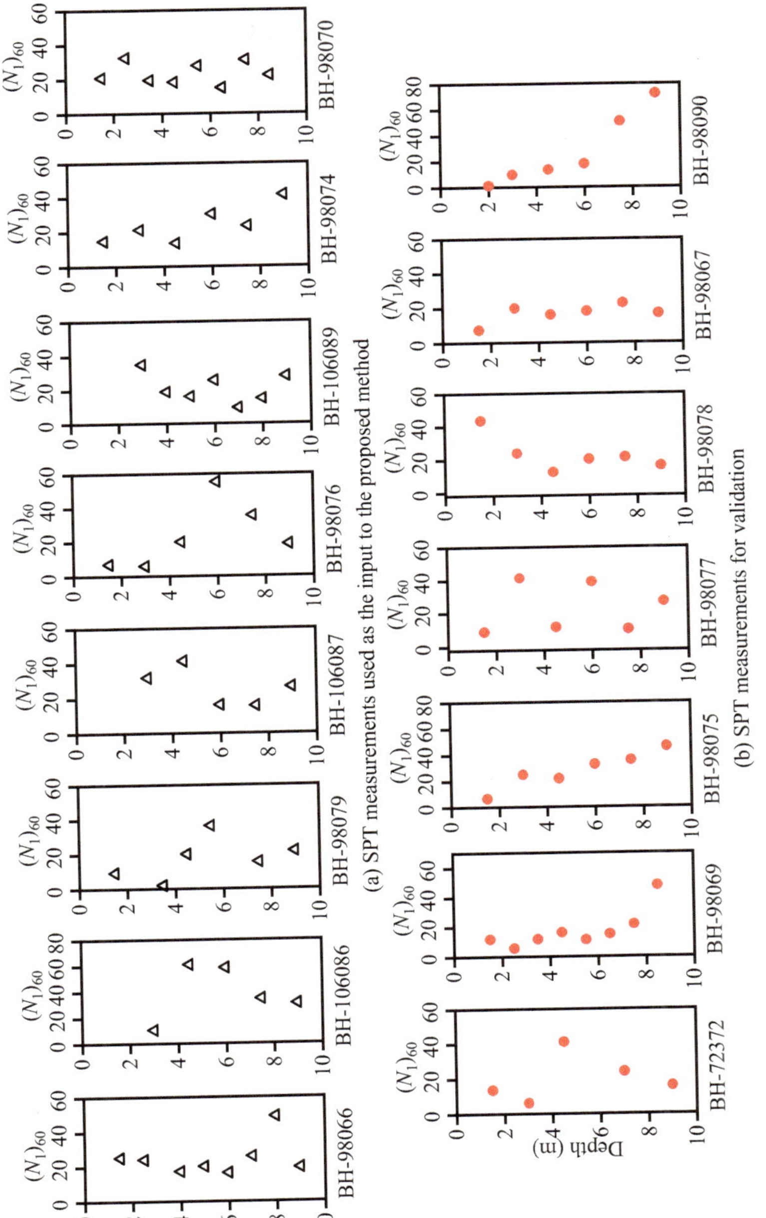

FIGURE 5.22 SPT measurements used as the input to the proposed method and validation.

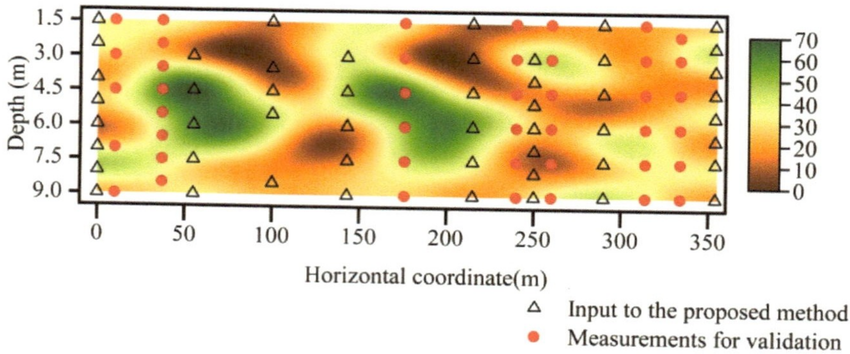

FIGURE 5.23 High-resolution A-A' cross-section of $(N_1)_{60}$ interpreted from eight boreholes with SPTs using 2D BCS (after Guan and Wang 2021a).

Therefore, 51 $(N_1)_{60}$ data points from eight boreholes (i.e., BH-98066, BH-106086, BH-98079, BH-106087, BH-98076, BH-106089, BH-98074, and BH-98070) are taken as measurement data (i.e., input) for the proposed method. As shown in Figure 5.21 by solid circles, these eight boreholes have an approximately equal spacing of 50 m within the cross-section A-A'. In other words, these 51 $(N_1)_{60}$ data points are used as the input to 2D BCS for interpolating high-resolution cross-section A-A' of $(N_1)_{60}$ data. The horizontal and vertical resolutions of cross-section A-A' of $(N_1)_{60}$ data are set as 1 m and 0.1 m, respectively, leading to a dimension of 76 × 355 for the cross-section A-A' of $(N_1)_{60}$. The cross-section A-A' of $(N_1)_{60}$ data interpreted from 51 $(N_1)_{60}$ measurement data using 2D BCS is shown in Figure 5.23 by a contour map.

The original 94 $(N_1)_{60}$ data points from 15 SPT boreholes and those at corresponding locations estimated from 2D BCS are shown in Figure 5.23 by open and solid triangles, respectively. The first row of Figure 5.23 is the input to BCS (i.e., eight boreholes with SPTs), while the second row of Figure 5.23 shows seven SPT boreholes (i.e., BH-72372, BH-98069, BH-98075, BH-98077, BH-98078, BH-98067, and BH-98090) whose data are used for validation only. It is observed from Figure 5.24 that the trend of interpreted $(N_1)_{60}$ values at seven validation boreholes is generally consistent with the true one, although several $(N_1)_{60}$ values are not accurately estimated. The relative error between the original 94 $(N_1)_{60}$ data and those at corresponding locations estimated from 2D BCS is calculated as 44%, which is consistent with the target relative error RE_T = 45%. Such consistent results indicate that the cross-section A-A' of $(N_1)_{60}$ interpreted from eight SPT boreholes, which is quantitatively determined using the statistical chart, satisfies the accuracy requirement.

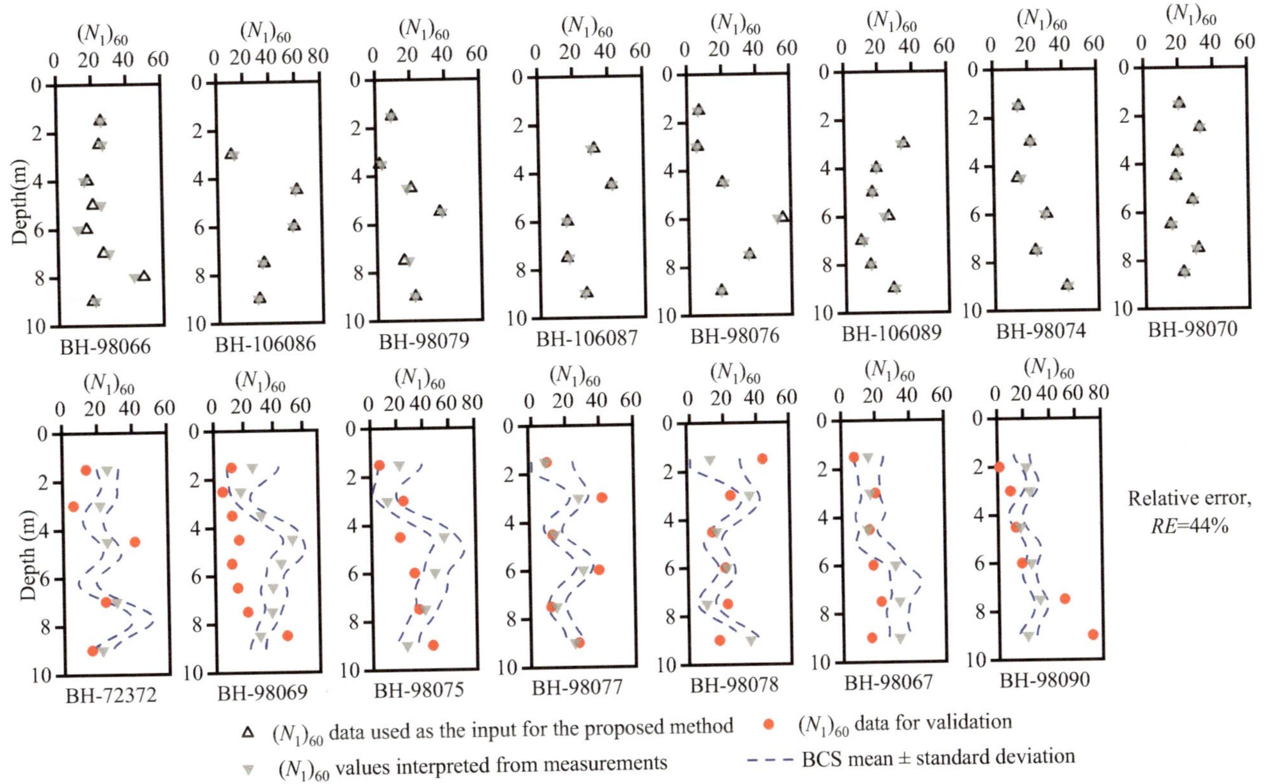

FIGURE 5.24 Comparison between 94 original $(N_1)_{60}$ data points and those at corresponding locations interpreted from eight boreholes with SPTs using BCS (after Guan and Wang 2021a).

REFERENCES

ASHTO. 2018. *Standard specifications for transportation materials and methods of sampling and testing, and AASHTO provisional standards*. Washington, DC: American Association of State Highway and Transportation Officials.

Au, S. K., and Y. Wang. 2014. *Engineering risk assessment with subset simulation*. Singapore: John Wiley & Sons.

Baecher, G. B., and J. T. Christian. 2003. *Reliability and statistics in geotechnical engineering*. Hoboken, NJ: John Wiley & Sons.

Bong, T., and A. W. Stuedlein. 2018. Efficient methodology for probabilistic analysis of consolidation considering spatial variation. *Engineering Geology* 237: 53–63.

Boulanger, R. W., and I. M. Idriss. 2014. *CPT and SPT based liquefaction triggering procedures*. Report No. UCD/CGM-14/01. Davis, CA: Center for Geotechnical Modeling, University of California-Davis.

Cami, B., S. Javankhoshdel, K. K. Phoon, and J. Ching. 2020. Scale of fluctuation for spatially varying soils: Estimation methods and values. *ASCE-ASME Journal of Risk and Uncertainty in Engineering Systems, Part A: Civil Engineering* 6 (4): 03120002.

Candes, E. J., J. K. Romberg, and T. Tao. 2006. Stable signal recovery from incomplete and inaccurate measurements. *Communications on Pure and Applied Mathematics* 59 (8): 1207–1223.

Cao, Z., and Y. Wang. 2014. Bayesian model comparison and selection of spatial correlation functions for soil parameters. *Structural Safety* 49: 10–17.

Cao, Z., Y. Wang, and D. Li. 2016. Quantification of prior knowledge in geotechnical site characterization. *Engineering Geology* 203: 107–116.

CEN (European Committee for Standardization). 2007. *Geotechnical design—Part 2: Ground investigation and testing Eurocode 7-2*. Brussels, Belgium: CEN.

Chin, C., J. Crooks, and Z. Moh. 1994. Geotechnical properties of the cohesive Sungshan deposits, Taipei. *Geotechnical Engineering* 25 (2).

Christian, J. T. 2004. Geotechnical engineering reliability: How well do we know what we are doing? *Journal of Geotechnical and Geoenvironmental Engineering* 130 (10): 985.

Clayton, C. R. 1995. *The standard penetration test SPT: Methods and use*. CIRIA Report 143. London: Construction Industry Research and Information Association.

DeGroot, D. J., and G. B. Baecher. 1993. Estimating autocovariance of in-situ soil properties. *Journal of Geotechnical Engineering* 119 (1): 147–166.

Dietrich, C. R., and G. N. Newsam. 1993. A fast and exact method for multidimensional Gaussian stochastic simulations. *Water Resources Research* 29 (8): 2861–2869.

Donoho, D. L. 2006. Compressed sensing. *IEEE Transactions on Information Theory* 52 (4): 1289–1306.

Fenton, G. A., and D. V. Griffiths. 2008. *Risk Assessment in Geotechnical Engineering*. New York, NY: John Wiley & Sons.

FHWA. 2017. Geotechnical Site Characterization. Geotechnical Engineering Circular No. 5, Publication No. NHI-16-072. Federal Highway Administration, U.S. Department of Transportation, Washington, DC.

Firouzianbandpey, S., D. V. Griffiths, L. B. Ibsen, and L. V. Andersen. 2014. Spatial correlation length of normalized cone data in sand: Case study in the North of Denmark. *Canadian Geotechnical Journal* 51 (8): 844–857.

Gharahbagh, E. A., M. A. Mooney, G. Frank, B. Walter, and M. A. DiPonio. 2013. Periodic inspection of gauge cutter wear on EPB TBMs using cone penetration testing. *Tunnelling and Underground Space Technology* 38: 279–286.

Guan, Z., and Y. Wang. 2020. Statistical charts for determining sample size at various levels of accuracy and confidence in geotechnical site investigation. *Geotechnique* 70 (12): 1145–1159.

Guan, Z., and Y. Wang. 2021a. Selection of standard penetration test number for geotechnical investigation of a vertical cross-section considering spatial variability and correlation in soil properties. *Canadian Geotechnical Journal* 58 (11): 1654–1668.

Guan, Z., and Y. Wang. 2021b. Rational determination of cone penetration test quantity in a two-dimensional vertical cross-section for site investigation. *Tunnelling and Underground Space Technology* 109: 103771.

Guan, Z., and Y. Wang. 2021c. Non-parametric construction of site-specific non-Gaussian multivariate joint probability distribution from sparse measurements. *Structural Safety* 91: 102077.

Guan, Z., and Y. Wang. 2021d. CPT-based probabilistic liquefaction assessment considering soil spatial variability, interpolation uncertainty, and model uncertainty. *Computers and Geotechnics* 141: 104504.

Guan, Z., and Y. Wang. 2023. Data-driven simulation of two-dimensional cross-correlated random fields from limited measurements using joint sparse representation. *Reliability Engineering and System Safety* 238: 109408.

Guan, Z., Y. Wang, and T. Zhao. 2021. Delineating the spatial distribution of soil liquefaction potential in a cross-section from limited cone penetration tests. *Soil Dynamics and Earthquake Engineering* 145: 106710.

Huang, S., S. Quek, and K. K. Phoon. 2001. Convergence study of the truncated Karhunen–Loeve expansion for simulation of stochastic processes. *International Journal for Numerical Methods in Engineering* 52 (9): 1029–1043.

Jaksa, M. B. 1995. *The influence of spatial variability on the geotechnical design properties of a stiff, overconsolidated clay*. PhD diss., The University of Adelaide.

Ji, S., D. Dunson, and L. Carin. 2009. Multitask compressive sensing. *IEEE Transactions on Signal Processing* 57 (1): 92–106.

Ji, S., Y. Xue, and L. Carin. 2008. Bayesian compressive sensing. *IEEE Transactions on Signal Processing* 56 (6): 2346–2356.

Lunne, T., P. K. Robertson, and J. J. M. Powell. 1997. *Cone penetration testing in geotechnical practice*. New York: Chapman & Hall.

Luo, Z., S. Atamturktur, Y. Cai, and C. Juang. 2012. Reliability analysis of basal-heave in a braced excavation in a 2-D random field. *Computers and Geotechnics* 39: 27–37.

Matheron, G. 1963. Principles of geostatistics. *Economic Geology* 58(8): 1246–1266.

Mayne, P. W. 2006. In-situ test calibrations for evaluating soil parameters. In *Characterisation and Engineering Properties of Natural Soils*, Proceedings of the Second International Workshop. London: Taylor & Francis.

McBratney, A. B., and R. Webster. 1986. Choosing functions for semi-variograms of soil properties and fitting them to sampling estimates. *Journal of Soil Science* 37: 617–639.

Michigan Department of Transportation. 2019. *Geotechnical design manual*. Lansing, MI: MDOT.

New York State Department of Transportation. 2013. *Geotechnical design manual*. Albany, NY: NYSDOT.

New Zealand Geotechnical Database (NZGD). 2012. www.nzgd.org.nz (Accessed March 20, 2021).

Oguz, E. A., N. Huvaj, and D. V. Griffiths. 2019. Vertical spatial correlation length based on standard penetration tests. *Marine Georesources and Geotechnology* 37 (1): 45–56.

Ott, W. R. 1994. *Environmental statistics and data analysis*. London: CRC Press.

Phoon, K. K., and F. H. Kulhawy. 1999. Characterization of geotechnical variability. *Canadian Geotechnical Journal* 36(4): 612–624.

Rix, G. J., N. Wainaina, A. Ebrahimi, R. C. Bachus, M. Limas, R. Sancio, B. Fait, and P. W. Mayne. 2018. *Manual on subsurface investigations: NCHRP Project* 21–10. Washington, DC: Transportation Research Board.Robertson, P. K. 1990. Soil classification using the cone penetration test. Can. Geotech., 27(1): 151–158.

Robertson, P. K. 2009. Interpretation of cone penetration tests—A unified approach. *Canadian Geotechnical Journal* 46(11): 1337–1355.

Robertson, P. K. 2016. Cone penetration test (CPT)-based soil behaviour type (SBT) classification system—An update. *Canadian Geotechnical Journal* 53(12): 1910–1927.

Shiau, J., and F. Al-Asadi. 2020. Two-dimensional tunnel heading stability factors Fc, Fs, and Fγ. *Tunnelling and Underground Space Technology* 97: 103293.

Skempton, A. W. 1986. Standard penetration test procedures and the effects in sands of overburden pressure, relative density, particle size, ageing and overconsolidation. *Géotechnique* 36 (3): 425–447.

Stuedlein, A. W., S. L. Kramer, P. Arduino, and R. D. Holtz. 2012. Geotechnical characterization and random field modeling of desiccated clay. *Journal of Geotechnical and Geoenvironmental Engineering* 138(11): 1301–1313.

Tropp, J. A., and A. C. Gilbert. 2007. Signal recovery from random measurements via orthogonal matching pursuit. *IEEE Transactions on Information Theory* 53 (12): 4655–4666.

Uzielli, M., G. Vannucchi, and K. K. Phoon. 2005. Random field characterisation of stress-normalised cone penetration testing parameters. *Geotechnique* 55 (1): 3–20.

Vanmarcke, E. 1977. Probabilistic modeling of soil profiles. *Journal of Geotechnical and Geoenvironmental Engineering* 103 (11): 1227–1246.

Vanmarcke, E. 2010. *Random fields: Analysis and synthesis*. London: World Scientific.

Vanmarcke, E., M. Shinozuka, S. Nakagiri, G. Schueller, and M. Grigoriu. 1986. Random fields and stochastic finite elements. *Structural Safety* 3(3–4): 143–166.

Wang, Y., and T. Zhao. 2016. Interpretation of soil property profile from limited measurement data: A compressive sampling perspective. *Canadian Geotechnical Journal* 53 (9): 1547–1559.

Wang, Y., and T. Zhao. 2017. Statistical interpretation of soil property profiles from sparse data using Bayesian compressive sampling. *Géotechnique* 67 (6): 523–536.

Wang, Y., Z. J. Cao, and S. K. Au. 2010. Efficient Monte Carlo simulation of parameter sensitivity in probabilistic slope stability analysis. *Computers and Geotechnics* 37 (7–8): 1015–1022.

Wang, Y., Z. Guan, and T. Zhao. 2019. Sample size determination in geotechnical site investigation considering spatial variation and correlation. *Canadian Geotechnical Journal* 56 (7): 992–1002.

Wang, Y., T. Zhao, and Z. Cao. 2015. Site-specific probability distribution of geotechnical properties. *Computers and Geotechnics* 70: 159–168.

Wang, Y., T. Zhao, and K. K. Phoon. 2018. Direct simulation of random field samples from sparsely measured geotechnical data with consideration of uncertainty in interpretation. *Canadian Geotechnical Journal* 55 (6): 862–880.

Webster, R., and M. A. Oliver. 2007. *Geostatistics for environmental scientists*. Chichester, England: John Wiley & Sons.

Zhang, J., and B. Ellingwood. 1994. Orthogonal series expansions of random fields in reliability analysis. *Journal of Engineering Mechanics* 120 (12): 2660–2677.

Zhang, L., and J. Chen. 2012. Effect of spatial correlation of standard penetration test (SPT) Data on bearing capacity of driven piles in sand. *Canadian Geotechnical Journal* 49 (4): 394–402.

Zhang, Y., Y. E. Li, and T. Ku. 2019. Geotechnical site investigation for tunneling and underground works by advanced passive surface wave survey. *Tunnelling and Underground Space Technology* 90: 319–329.

Zhao, T., and Y. Wang. 2018. Simulation of cross-correlated random field samples from sparse measurements using Bayesian compressive sensing. *Mechanical Systems and Signal Processing* 112: 384–400.

Zhao, T., Y. Hu, and Y. Wang. 2018. Statistical interpretation of spatially varying 2D geo-data from sparse measurements using Bayesian compressive sampling. *Engineering Geology* 246: 162–175.

6 Adaptive Sampling in Site Characterization

Zheng Guan and Yu Wang

ABSTRACT

Traditional geotechnical site characterization programs often employ an equal sampling spacing strategy with regular grid patterns. While this approach assigns equal importance to all areas of the site, it does not account for spatial variability in subsurface soil, making it potentially inefficient and uneconomical. This chapter introduces an adaptive sampling strategy that continuously monitors interpreted results from already performed exploration points and automatically determines whether additional points are needed, along with their optimal test locations. Initially, exploration points are conducted to establish a basic understanding of subsurface conditions. Then, additional points are introduced sequentially until the interpreted conditions achieve a target reliability level. Each subsequent exploration point is determined based on insights from prior measurements, allowing for efficient decision-making. Even in the absence of sufficient prior knowledge about subsurface conditions, the minimal number of exploration points and their optimal locations can be systematically and automatically determined.

6.1 INTRODUCTION

Site characterization is a multi-stage process, and additional exploration points are often needed in a later stage of site characterization (e.g., Shi et al. 2023). In Chapter 5, methods were developed for determining the sample size of laboratory tests and the number of in-situ tests (i.e., cone penetration test [CPT] and standard penetration test [SPT]) in geotechnical site investigation. Based on the results, an initial number of exploration points (e.g., laboratory tests, borings, and/or in-situ test soundings) may be determined based on the anticipated variability of subsurface conditions. Once the spatial variability of subsurface conditions is estimated based on prior knowledge about the site (e.g., local engineering experience), the minimal number of exploration points can be determined based on a target accuracy level of the interpretation results, as discussed in Chapter 5. However, it might be difficult to accurately estimate the characteristics of soil spatial variability (e.g., coefficient of variation [COV] and correlation length) at the initial stage of site investigation. In addition, the underlying soil properties are unknown, and therefore it is impractical to compute the accuracy of the interpreted results. In other words, the accuracy is unknown.

DOI: 10.1201/9781003381235-6

Since Bayesian compressive sampling (BCS) provides the full probability distribution of the estimated soil property at each location, leading to error bars (e.g., mean ± standard deviation) of the interpreted results, these error bars might be used to give a sense of confidence in the interpreted results. In other words, the reliability or confidence level of the interpreted results can be obtained directly from BCS. In some cases, the preliminary number of exploration points is proven to be sufficient for achieving the target reliability level. However, in other cases, the preliminary number of exploration points determined using the proposed methods might be inadequate, and additional exploration points are needed to obtain a sufficient number of samples or in-situ test measurement data to further improve the reliability of the interpretation results.

An active learning sampling strategy is introduced in this chapter that uses innovative data analytic methods (e.g., BCS and information entropy) for planning the geotechnical investigation of a site in a self-adaptive and data-driven manner (e.g., Guan et al. 2020a, 2022a). The proposed sampling strategy continuously monitors the evolution of the results interpreted from the exploration points that are performed already and automatically decides whether or not additional exploration points are needed and their optimal test locations. In the rest of this chapter, a framework of the proposed active learning strategy is described first, followed by the introduction of an information entropy-based method for determining the optimal sampling locations for one-dimensional (1D) and two-dimensional (2D) problems.

6.2 FRAMEWORK OF THE ACTIVE LEARNING STRATEGY

Geotechnical properties interpreted from a limited number of exploration points unavoidably contain significant statistical uncertainty. In this study, the uncertainty in the interpreted geotechnical properties model is quantified using COV. In general, the statistical uncertainty, i.e., the COV, decreases as the number of exploration points increases. The active learning strategy aims to automatically determine the required number of exploration points and their corresponding optimal locations to obtain interpretation results satisfying a desired target COV, COV_T. The uncertainty in the interpreted geotechnical properties evolves when new measurements from the latest exploration points are obtained and analyzed together with all previous measurements. It is often difficult to deal with such evolution because the site-specific knowledge during the original planning might be insufficient to anticipate all site conditions during the subsequent stages of site investigation. Hence, a self-adaptive strategy is adopted in the decision-making, i.e., deciding whether or not additional exploration points are required.

The basic idea of the proposed sampling strategy is to conduct exploration points in a self-adaptive and data-driven manner until the reliability, i.e., confidence level, of the interpreted results reaches the target level. Note that the target COV_T for a project can be determined based on the risk of geotechnical construction that engineers and owners of the project can tolerate, the anticipated variability of subsurface conditions, and other project-specific considerations. The preliminary number of exploration points is first determined using the methods developed in Chapter 5. After the initial number of exploration points is conducted at their optimal locations, the interpretation results from initial exploration points are used to determine whether a self-adaptive process should be triggered. When the self-adaptive process is triggered and additional

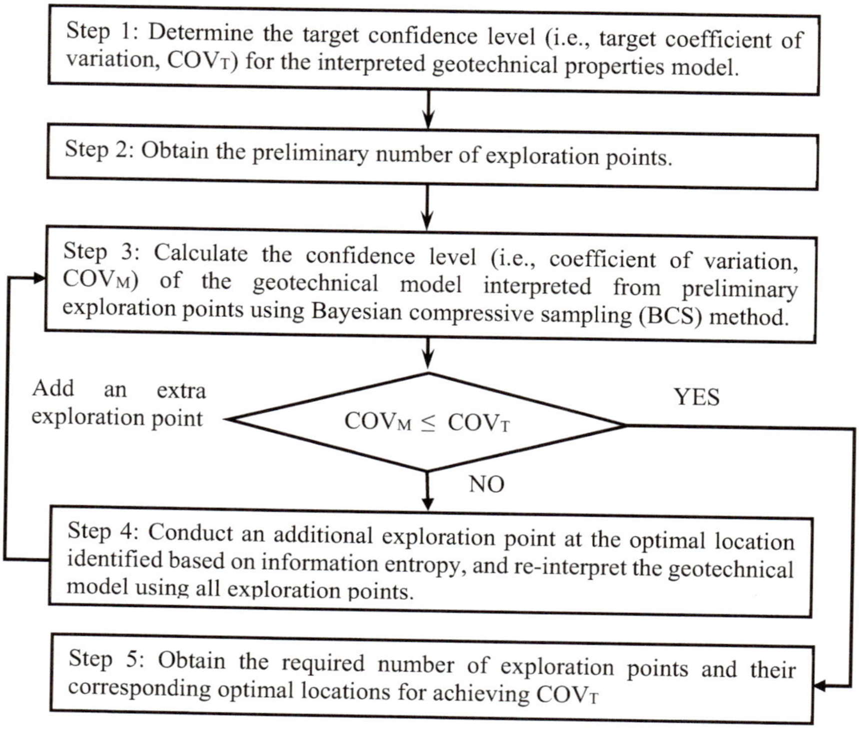

FIGURE 6.1 Framework of proposed active learning sampling strategy (after Guan et al. 2020a).

actions are needed, extra exploration points are conducted one by one following the self-adaptive process. In other words, for each step, only one new exploration point is conducted, and its optimal location is determined automatically based on the geotechnical properties measured from the existing exploration points. Once a new exploration point is conducted at a new location, the measurements from the new exploration point and the previous measurements are used together to update the interpretation results (e.g., the spatial distribution of geotechnical property). The self-adaptive process repeats until a prescribed target COV_T of the interpretation results is achieved.

In summary, a preliminary number of exploration points isfirst determined. Then, the target confidence level is achieved by performing additional exploration points in a self-adaptive and data-driven manner. The framework of the proposed sampling strategy includes five steps, as shown in Figure 6.1. In Step 1, the target confidence level, COV_T, is determined based on the importance and risk of civil structures, the complexity of local geology, previous engineering experience, etc. In Step 2, a preliminary number of exploration points isobtained using the methods presented in Chapter 5. In Step 3, preliminary exploration points are conducted with an equal interval, and their measurements are used to infer the spatial variation of soil properties by BCS. When there is no prevailing prior knowledge about subsurface conditions, exploration points should be uniformly distributed within a specific site to ensure a comprehensive representation of measurement data for the entire volume of geo-material (e.g., FHWA 2017). Furthermore, according to information entropy theory, Zhao and Wang (2019) indicated that in the absence of preexisting measurements, collecting measurement data at equal intervals maximizes the reduction in uncertainty in interpreting results. Therefore, this study adopts an equal spacing strategy for the initial placement of exploration points. If the coefficient of variation of the geotechnical properties model interpreted from the initial exploration points' measurements, defined as COV_M, is larger than the COV_T, the confidence level, or reliability, of the interpreted results is deemed insufficient. Consequently, the self-adaptation process, i.e., the addition of exploration points at their optimal locations, is triggered to reduce COV_M. When $COV_M \leq COV_T$, the adaptation process stops. Otherwise, the self-adaptation continues, and the number of exploration points is updated by adding an extra exploration point measurement at its optimal location (i.e., Step 4). In Step 5, the required number of exploration points is automatically determined once the adaptation terminates, i.e., when $COV_M \leq COV_T$, for achieving the target confidence level of the interpretation results in the site investigation.

In the following two sections, the adaptive sampling strategy is illustrated for 1D and 2D cases, respectively, including (1) the determination of the optimal number and location of specimens to be tested from a borehole (i.e., a 1D case) and (2) the determination of the optimal number and location of in-situ tests in a vertical cross-section (i.e., a 2D case).

6.3 ACTIVE LEARNING FOR THE 1D SITE INVESTIGATION PROBLEM

6.3.1 Determination of Optimal Sampling Location for the 1D Problem

In the adaptive sampling method, the optimal locations of additional exploration points are determined using information entropy (e.g., Guan et al. 2020a). The theory of information entropy has been applied in many fields, including statistics, big data analysis, and graphic detection, for quantifying the uncertainty of a random process (e.g., MacKay 2003). In general, the uncertainty of a random process decreases as the information entropy decreases. When a process is completely known, i.e., it has no uncertainty, the information entropy is zero. In the active learning for site investigation, the spatial variation of soil properties, interpreted from limited measurement data, is considered a random process. Thus, the uncertainty of the interpreted spatial variation of soil properties may be quantified using information entropy. Soil properties are spatially varying and correlated, and therefore the spatial variation of soil properties along depth (i.e., soil property profile in this section) can be considered as a random process (e.g., Vanmarcke 1977; Wang et al. 2010; Zhao and Wang 2019). The information entropy H of an interpreted soil properties profile $\hat{f}$ can be expressed mathematically as follows (MacKay 2003):

$$H(\hat{f}) = -\int p(\hat{f})\ln[p(\hat{f})]\, d\hat{f} \tag{6.1}$$

where $p(\hat{f})$ is the probability density function (PDF) of $\hat{f}$. The basic idea of adaptive sampling is to collect specimens at the optimal locations, which results in the maximum information gained on $\hat{f}$ or reduction in $H(\hat{f})$.

Using the concept of information entropy, the optimal locations for a given number of specimens are those that maximize the reduction from the current information entropy $H(\hat{\boldsymbol{f}})$ to the new entropy $H(\hat{\boldsymbol{f}}_{\text{new}})$ after measurements, $\boldsymbol{y}$, of the geotechnical properties from additional specimens. When no preexisting measurements are available, it is difficult to evaluate $H(\hat{\boldsymbol{f}})$directly. However, the difference between $H(\hat{\boldsymbol{f}})$ and $H(\hat{\boldsymbol{f}}_{\text{new}})$ can be approximately estimated as the information entropy of the measurements $\boldsymbol{y}$, i.e., $\Delta = H(\hat{\boldsymbol{f}}) - H(\hat{\boldsymbol{f}}_{\text{new}}) \approx H(\boldsymbol{y})$. For such a case, Zhao and Wang (2019) showed that the information entropy $H(\boldsymbol{y})$ is maximized when the exploration points are equally spaced. Therefore, the initial exploration points to be tested are equally spaced in the adaptive sampling.

When additional specimens are needed to improve the confidence level of the interpreted geotechnical properties profile, the optimal location of a specimen to be tested at each adaptation stage should be determined first. Different locations of an additional specimen might lead to different $H(\hat{\boldsymbol{f}}_{\text{new}})$ and Δ, and the optimal location is the one that maximizes the difference between $H(\hat{\boldsymbol{f}})$ and $H(\hat{\boldsymbol{f}}_{\text{new}})$. To calculate $H(\hat{\boldsymbol{f}})$ and $H(\hat{\boldsymbol{f}}_{\text{new}})$, the marginal distribution function (e.g., PDF) of the geotechnical properties profile interpreted from different numbers of exploration points should be obtained.

BCS may be used to estimate PDF of the interpreted geotechnical properties profile from sparse measurements. The PDF of $\hat{\boldsymbol{f}}$ with a length of N can be expressed as (Wang and Zhao 2017):

$$p(\boldsymbol{f}) = \frac{\Gamma\left(c_n + \frac{N}{2}\right)}{\Gamma(c_n)\pi^{N/2}\det\left(\frac{d_n}{c_n}\mathbf{B}^1\mathbf{H}\left(\mathbf{B}^1\right)^{\mathrm{T}}\right)^{1/2}\left(2c_n\right)^{N/2}\left[1+\frac{1}{2c_n}(\hat{\boldsymbol{f}}-\boldsymbol{\mu}_{\hat{f}})^{\mathrm{T}}\left(\frac{d_n}{c_n}\mathbf{B}^1\mathbf{H}\left(\mathbf{B}^1\right)^{\mathrm{T}}\right)^{-1}(\hat{\boldsymbol{f}}-\boldsymbol{\mu}_{\hat{f}})\right]^{\left(c_n+\frac{N}{2}\right)}} \tag{6.2}$$

where $\Gamma(\bullet)$ is a Gamma function; $\boldsymbol{\mu}_{\hat{f}}$ is the estimated mean of $\hat{\boldsymbol{f}}$; $\mathbf{B}^1$ is a 1D orthonormal basis function matrix; c_n, d_n are nonnegative parameters of a posterior Gamma distribution; $\det(\bullet)$ represents the determinant of a matrix; $\mathbf{H} = (\mathbf{A}^{\mathrm{T}}\mathbf{A}+\mathbf{D})^{-1}$ where $\mathbf{A}$ is a measurement matrix; and $\mathbf{D}$ is a diagonal matrix. Then, the information entropy of $\hat{\boldsymbol{f}}$ can be derived as (Zhao and Wang 2019):

$$H(\hat{\boldsymbol{f}}) = -\frac{1}{2}\ln\det(\mathbf{A}^{\mathrm{T}}\mathbf{A}+\mathbf{D}) + const1 \tag{6.3}$$

Similarly, the information entropy of $\hat{\boldsymbol{f}}_{\text{new}}$ can be derived as

$$\begin{aligned} H(\hat{\boldsymbol{f}}_{\text{new}}) &= -\frac{1}{2}\ln\det(\mathbf{A}^{\mathrm{T}}\mathbf{A}+\mathbf{D}+\boldsymbol{r}\boldsymbol{r}^{\mathrm{T}}) + const1 \\ &= -\frac{1}{2}\ln(1+\boldsymbol{r}^{\mathrm{T}}(\mathbf{A}^{\mathrm{T}}\mathbf{A}+\mathbf{D})^{-1}\boldsymbol{r}) - \frac{1}{2}\ln\det(\mathbf{A}^{\mathrm{T}}\mathbf{A}+\mathbf{D}) + const1 \\ &= -\frac{1}{2}\ln(1+\boldsymbol{r}^{\mathrm{T}}(\mathbf{A}^{\mathrm{T}}\mathbf{A}+\mathbf{D})^{-1}\boldsymbol{r}) + H(\hat{\boldsymbol{f}}) \end{aligned} \tag{6.4}$$

where $\boldsymbol{r}^{\mathrm{T}}$ represents a new row of **B** to be added to the matrix **A** corresponding to the additional measurement.

Then, the difference between $H(\hat{f})$ and $H(\hat{f}_{\mathrm{new}})$ can be expressed as follows:

$$\Delta = H(\hat{f}) - H(\hat{f}_{\mathrm{new}}) = \frac{1}{2}\ln\left(1 + \boldsymbol{r}^{\mathrm{T}}(\mathbf{A}^{\mathrm{T}}\mathbf{A} + \mathbf{D})^{-1}\boldsymbol{r}\right) \tag{6.5}$$

According to Eq. (6.5), finding the maximum value of Δ is equivalent to maximizing the term $\boldsymbol{r}^{\mathrm{T}}(\mathbf{A}^{\mathrm{T}}\mathbf{A} + \mathbf{D})^{-1}\boldsymbol{r}$, which can be obtained directly from the BCS result. Covariance of the geotechnical properties profile, $\hat{f}$, interpreted from preexisting measurements is expressed as $\mathbf{COV}_{\hat{f}} = \frac{d_n}{c_n - 1}\mathbf{BHB}^{\mathrm{T}}$, and the variance of $\hat{f}$ at the location of the additional specimen is $\sigma^2_{new} = \frac{d_n}{c_n - 1}\boldsymbol{r}^{\mathrm{T}}\mathbf{H}\boldsymbol{r} = \frac{d_n}{c_n - 1}\boldsymbol{r}^{\mathrm{T}}(\mathbf{A}^{\mathrm{T}}\mathbf{A} + \mathbf{D})^{-1}\boldsymbol{r}$. Therefore, the reduction in information entropy can be expressed as $\Delta = \frac{1}{2}\ln\left[1 + \frac{c_n - 1}{d_n}\sigma^2_{new}\right]$. Because both c_n and d_n are constants, the information entropy reduction increases monotonically as the variance σ^2_{new} increases. Thus, finding the maximum value of Δ is equivalent to finding the maximum value of σ^2_{new}. Therefore, when one additional specimen is taken at the location with the maximum variance along $\hat{f}$, the reduction Δ reaches the maximum value.

Using information entropy of BCS results, it is shown that the most efficient location for collecting one additional specimen is the position of the maximum variance along the geotechnical properties profile $\hat{f}$ interpreted from available measurements $\boldsymbol{y}$. In other words, an additional specimen should be collected at the location where the interpreted geotechnical properties profile has the largest uncertainty. This sampling location is consistent with engineering practice, where additional exploration points are usually conducted at locations with the greatest uncertainty.

6.3.2 Real Data Example

To demonstrate the application of the adaptive sampling method to 1D problems, laboratory test results (i.e., natural water content) are used as an illustrative example in this section. Consider, for example, the determination of sample size for characterizing spatial variation of the natural water content, w_n, at St James's Park, London. The site contains a mainly sandy clayey silt layer from a depth of about 10 m to 20 m (Hight et al. 2002).

Suppose that the target relative error RE_{T} is 5% and the target coefficient of variation $\mathrm{COV}_{\mathrm{T}}$ is 4.2%. The COV and λ_v of natural water content in this sandy clayey silt are estimated as 14% and 6.0 m, respectively, based on local experience (e.g., nearby site w_n data). Note that these ranges are also within the typical ranges of 7~46% for COV and 1.6~12.7 m for λ_v, respectively, for natural water content in the sandy clayey silt reported in the literature (e.g., Phoon and Kulhawy 1999). The initial value of the required sample size is determined using the statistical charts shown in Chapter 5. The RE_T / COV = 0.36 is calculated. Using the best-fit curve in the statistical chart, $d_s / \lambda_v = 0.15$ is calculated. Because $\lambda_v = 6.0$ m, the sampling interval $d_s = 0.90$ m is estimated. The thickness of the silty clay layer is 10.0 m, and the number of measurement data points is estimated as $M = 10 / 0.90 + 1 \approx 12$.

In Step 2, 12 w_n shall be measured with an equal sampling interval for interpretation of the w_n profile along depth. Hight et al. (2002) reported that a total number of 52 water content values extended from a depth of 10.0 m to 20.0 m with a measurement interval of about 0.2 m. Figure 6.2 shows the 52 w_n values measured along depth by open triangles. For validating the performance of the adaptive sampling strategy in this illustrative example, 12 w_n values are taken from the 52 w_n values with an approximately equal sampling interval, and they are shown in Figure 6.2 by solid circles. The 12 w_n values obtained are used as the input to BCS for interpolating a high-resolution natural water content profile with 256 data points. Figure 6.2 also shows the mean natural water content profile interpreted from BCS by a solid line and the mean plus or minus one standard deviation profiles by two dashed lines, respectively. It is observed from Figure 6.2 that the solid line follows a trend quite similar to that of the open triangles (i.e., all 52 ω_n values). In Step 3, the COV_M of the interpreted w_n profile is estimated as 4.72%, which is larger than COV_T. Hence, an additional measurement is needed to reduce the uncertainty of the interpreted w_n profile.

In Step 4, an additional measurement data point is taken at the most efficient location to improve interpretation of the w_n profile. As mentioned in Section 6.3.1, the most

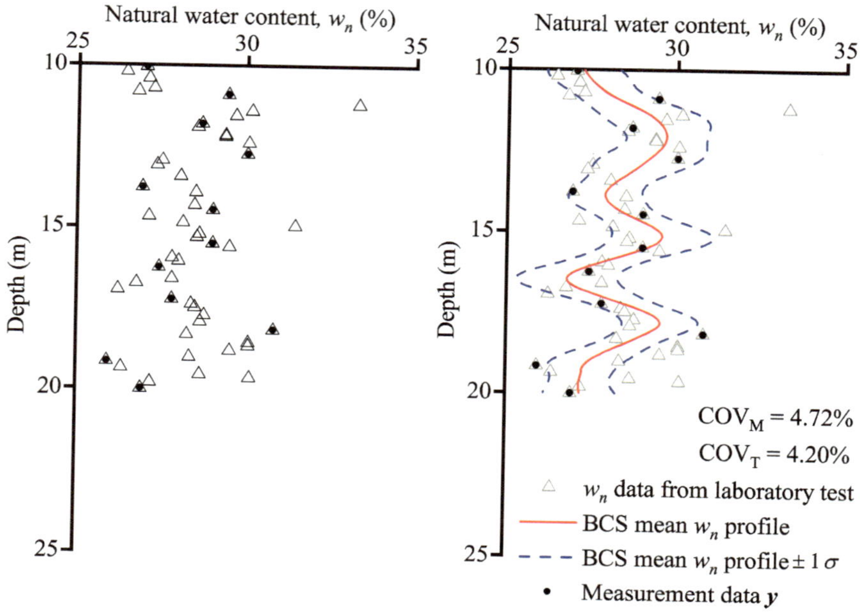

FIGURE 6.2 Fifty-two natural water content values along the depth, measurement data, and BCS result (after Guan et al. 2020b).

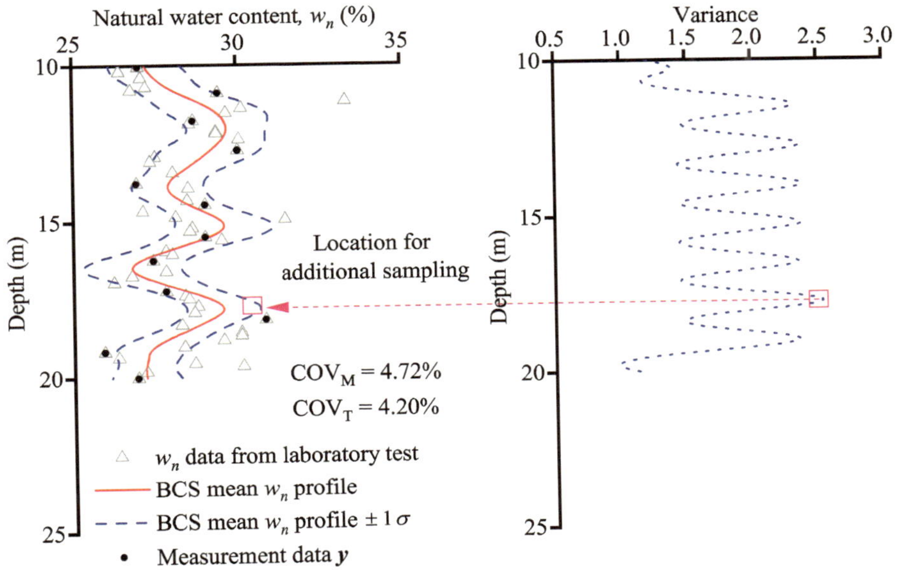

FIGURE 6.3 Efficient sampling location for one additional measurement (after Guan et al. 2020b).

efficient location is the location that has the maximum variance in the interpreted soil property profile. Figure 6.3 shows the spatial variation of variance of the interpreted w_n profile from 12 w_n data along depth. The maximum variance is about 2.5, and the corresponding depth is about 17.5 m. Therefore, an additional measurement data point shall be added at the depth of 17.5 m. Then, one of the 52 w_n data which is closest to the depth of 17.5 m is added to interpret w_n profile. $M = 12 + 1 = 13$ w_n values are used as the input to BCS for interpreting w_n profile. The COV_M of the interpreted w_n profile decreases from 4.72% to 4.57% as the sample size M increases from 12 to 13.

Subsequently, Steps 3 and 4 are performed repeatedly until the $COV_M \leq COV_T$. Figure 6.4 shows the variation of COV_M of the interpreted w_n profile with the number of iterations. It shows that the COV_M of the interpreted w_n profile decreases from 4.72% to 4.18% when the number of iterations increases from 0 to 3. After three iterations (i.e., $M = 12 + 3 = 15$), the $COV_M = 4.18\% < COV_T = 4.2\%$, and the iteration stops. Figure 6.5 shows the 15 w_n data points taken from 52 w_n values by solid circles. Figure 6.5 also shows the mean natural water content profile interpreted from BCS and the mean plus or minus one standard deviation profiles. It can be observed that the area between two dashed lines for $M = 15$ is smaller than that for $M = 12$, which suggests that the confidence level of the interpreted w_n profile improves.

6.4 ACTIVE LEARNING FOR 2D SITE INVESTIGATION PROBLEMS

6.4.1 Determination of Optimal Location for an Additional CPT in a 2D Cross-section

2D vertical cross-sections are frequently developed from line measurements, such as boreholes and CPTs, to represent ground conditions in geotechnical engineering

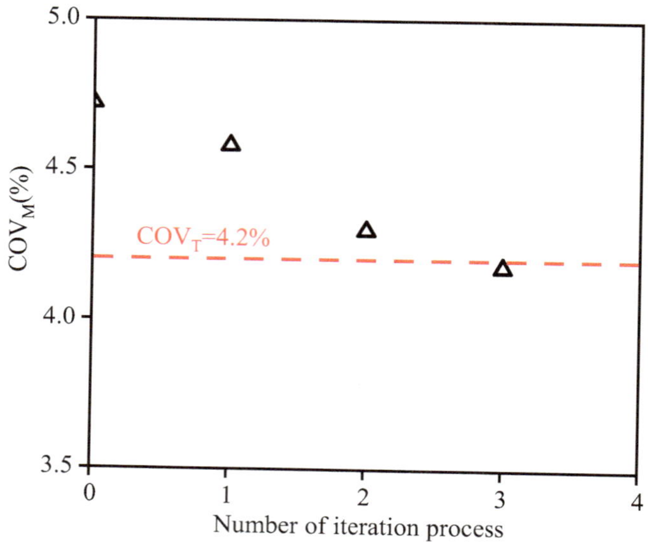

FIGURE 6.4 The variation of COV_M with the number of iteration processes (after Guan et al. 2020b).

practice. When improvement is needed for a 2D geotechnical cross-section interpreted from a preliminary number of CPT soundings, the additional CPT soundings shall be conducted at their optimal locations. In this section, an information entropy-based method is developed for determining the optimal location of an additional CPT sounding. In the context of compressive sensing (e.g., Tipping 2001; Candès et al. 2006), a vertical cross-section of geotechnical properties, **F**, can be represented by a weighted linear combination of $N = N_{x_1} \times N_{x_2}$ 2D basis functions, and it is expressed as $\mathbf{F} = \sum_{i=1}^{N_{x_1}} \sum_{j=1}^{N_{x_2}} \mathbf{B}_{i,j}^{2D} \omega_{i,j}^{2D} = \sum_{t=1}^{N_{x_1} \times N_{x_2}} \mathbf{B}_t^{2D} \omega_t^{2D}$, where $\mathbf{B}_t^{2D}$ is tth 2D basis function and ω_t^{2D} is the corresponding weight coefficient (e.g., Fang et al. 2012; Zhao et al. 2018). Mathematically, **F** can be further expressed as $\mathbf{F} = \mathbf{B}_1^{1D} \boldsymbol{\Omega} (\mathbf{B}_2^{1D})^T = \sum_{i=1}^{N_{x_1}} \sum_{j=1}^{N_{x_2}} \omega_{i,j}^{2D} \boldsymbol{b}_i^1 (\boldsymbol{b}_j^2)^T$, where the $N_{x_1} \times N_{x_1}$ matrix $\mathbf{B}_1^{1D}$ and $N_{x_2} \times N_{x_2}$ matrix $\mathbf{B}_2^{1D}$ are two 1D orthonormal matrices (e.g., DCT matrix) and $\boldsymbol{b}_i^1$ and $\boldsymbol{b}_j^2$ are the ith and jth columns of matrices $\mathbf{B}_1^{1D}$ and $\mathbf{B}_2^{1D}$, respectively. Note that $\boldsymbol{b}_i^1 (\boldsymbol{b}_j^2)^T$ represents 2D basis function $\mathbf{B}_{i,j}^{2D}$; the entry in the ith row and jth column of the matrix $\boldsymbol{\Omega}$ is the corresponding weight coefficient, $\boldsymbol{\omega}_{i,j}^{2D}$. The superscript "T" denotes the transpose operation in linear algebra. Using Kronecker product in linear algebra, **F** or $vec(\mathbf{F}^T)$ can be conveniently expressed as:

$$vec(\mathbf{F}^T) = \mathbf{B}_1^{1D} \otimes \mathbf{B}_2^{1D} vec(\boldsymbol{\Omega}^T) = \mathbf{B}_1^{1D} \otimes \mathbf{B}_2^{1D} \boldsymbol{\omega}^{2D} \tag{6.6}$$

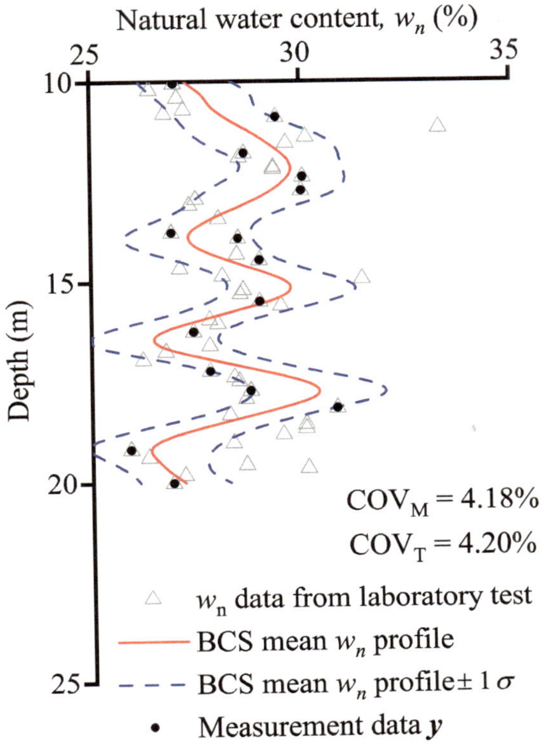

FIGURE 6.5 The BCS interpretation results for $M = 15$ (after Guan et al. 2020b).

where $vec(\mathbf{F}^{\mathrm{T}})$ denotes the vectorized format of the matrix $\mathbf{F}^{\mathrm{T}}$, formed by stacking the columns of $\mathbf{F}^{\mathrm{T}}$ into a single column vector; "⊗" denotes the Kronecker product; and $\boldsymbol{\omega}^{2\mathrm{D}} = vec(\boldsymbol{\Omega}^{\mathrm{T}})$ is a column vector including all weight coefficients.

Mathematically, measurements $\mathbf{Y}$ can be expressed as $\mathbf{Y} = \mathbf{F}\boldsymbol{\Psi}_{x_2} = \mathbf{B}_1^{1\mathrm{D}}\boldsymbol{\Omega}(\mathbf{B}_2^{1\mathrm{D}})^{\mathrm{T}}\boldsymbol{\Psi}_{x_2} = \mathbf{B}_1^{1\mathrm{D}}\boldsymbol{\Omega}(\mathbf{A}_2)^{\mathrm{T}}$, where $\boldsymbol{\Psi}_{x_2}$ indicates the locations of M_c CPT soundings along horizontal direction; $\mathbf{A}_2 = \boldsymbol{\Psi}_{x_2}^{\mathrm{T}}\mathbf{B}_2^{1\mathrm{D}}$. Using $\mathbf{Y} = \mathbf{B}_1^{1\mathrm{D}}\boldsymbol{\Omega}(\mathbf{A}_2)^{\mathrm{T}}$, the estimated weight coefficients from sparse measurements, denoted as $\hat{\boldsymbol{\omega}}^{2\mathrm{D}} = \left[\hat{\boldsymbol{\omega}}_1^{2\mathrm{D}}, \hat{\boldsymbol{\omega}}_2^{2\mathrm{D}}, \ldots, \hat{\boldsymbol{\omega}}_{N_{x_1} \times N_{x_2}}^{2\mathrm{D}}\right]$, can be obtained under a Bayesian framework (e.g., Wang and Zhao 2017). Zhao et al. (2020) showed that the distribution of $\hat{\boldsymbol{\omega}}^{2\mathrm{D}}$ follows a multivariate normal distribution with mean $\boldsymbol{\mu}_{\hat{\boldsymbol{\omega}}^{2\mathrm{D}}}$ and covariance $\mathbf{COV}_{\hat{\boldsymbol{\omega}}^{2\mathrm{D}}}$, which are expressed as follows:

$$\boldsymbol{\mu}_{\hat{\boldsymbol{\omega}}^{2\mathrm{D}}} = \left[\boldsymbol{\mu}_{\hat{\boldsymbol{\omega}}^{2\mathrm{D}}}^{1}, \boldsymbol{\mu}_{\hat{\boldsymbol{\omega}}^{2\mathrm{D}}}^{2}, \ldots, \boldsymbol{\mu}_{\hat{\boldsymbol{\omega}}^{2\mathrm{D}}}^{N_{x_1}}\right]^{\mathrm{T}} \tag{6.7}$$

$$\mathbf{COV}_{\hat{\omega}^{2D}} = \begin{bmatrix} \mathbf{COV}^{1}_{\hat{\omega}^{2D}} & \cdots & 0 \\ \vdots & \ddots & \vdots \\ 0 & \cdots & \mathbf{COV}^{N_{x_1}}_{\hat{\omega}^{2D}} \end{bmatrix} \tag{6.8}$$

where $\boldsymbol{\mu}^{i}_{\hat{\omega}^{2D}} = \tau\sum_{j=1}^{N_1} b_{j,i}\mathbf{COV}^{i}_{\hat{\omega}^{2D}}\mathbf{A}_2^{\mathrm{T}}\boldsymbol{y}^{2D}_j$ and $\mathbf{COV}^{i}_{\hat{\omega}^{2D}} = (\mathbf{A}_2^{\mathrm{T}}\mathbf{A}_2\tau + \mathbf{D}^{\alpha}_i)^{-1}, i = 1, 2, ..., N_{x_1}; b_{j,i}$ indicate the entry in the jth row and ith column of the matrix $\mathbf{B}_1^{1D}$; $\boldsymbol{y}^{2D}_j$ represents the jth M_c consecutive elements of $\boldsymbol{y}^{2D} = vec(\mathbf{Y}^{\mathrm{T}})$; τ represents the reciprocal of the variance of the residuals between y^{2D} and $\mathbf{B}_1^{1D} \otimes \mathbf{A}_2\boldsymbol{\mu}_{\hat{\omega}^{2D}}$; $\mathbf{D}^{\alpha}$ is an $(N_{x_1} \times N_{x_2}) \times (N_{x_1} \times N_{x_2})$ diagonal matrix recording the ith N_{x_2} consecutive diagonal elements of $\mathbf{D}^{\alpha}$; diagonal elements of $\mathbf{D}^{\alpha}$, α_t, t = 1, 2,..., $N_{x_1} \times N_{x_2}$, represents reciprocal of the unknown variance of $\hat{\boldsymbol{\omega}}^{2D}_t$. After $\boldsymbol{\mu}_{\hat{\omega}^{2D}}$ and $\mathbf{COV}_{\hat{\omega}^{2D}}$ are obtained from measurements $\boldsymbol{y}^{2D}$, the reconstructed cross-section of geotechnical properties, $\hat{\mathbf{F}}$, is derived as a multivariate normal distribution. The mean $\boldsymbol{\mu}_{vec(\hat{\mathbf{F}}^{\mathrm{T}})}$ and covariance, $\mathbf{COV}_{vec(\hat{\mathbf{F}}^{\mathrm{T}})}$ of $vec(\hat{\mathbf{F}}^{\mathrm{T}})$, are expressed as follows:

$$\boldsymbol{\mu}_{vec(\hat{\mathbf{F}}^{\mathrm{T}})} = \mathbf{B}_2^{1D} \otimes \mathbf{B}_1^{1D}\boldsymbol{\mu}_{\hat{\omega}^{2D}} \tag{6.9}$$

$$\mathbf{COV}_{vec(\hat{\mathbf{F}}^{\mathrm{T}})} = (\mathbf{B}_1^{1D} \otimes \mathbf{B}_2^{1D})\mathbf{COV}_{\hat{\omega}^{2D}}(\mathbf{B}_1^{1D} \otimes \mathbf{B}_2^{1D})^{\mathrm{T}} \tag{6.10}$$

The information entropy of $\hat{\mathbf{F}}$ or $vec(\hat{\mathbf{F}}^{\mathrm{T}})$, $H(vec(\hat{\mathbf{F}}^{\mathrm{T}}))$, is computed as (e.g., Cover and Thomas 2012; Zhao et al. 2021):

$$H(vec(\hat{\mathbf{F}}^{\mathrm{T}})) = \frac{N_{x_1} \times N_{x_2}}{2}\ln(2\pi e) + \frac{1}{2}\ln(\det(\mathbf{COV}_{vec(\hat{\mathbf{F}}^{\mathrm{T}})})) \tag{6.11}$$

Substituting $\mathbf{COV}_{vec(\hat{\mathbf{F}}^{\mathrm{T}})}$ in Eq. (6.10) into Eq. (6.11) leads to:

$$\begin{aligned} H(vec(\hat{\mathbf{F}}^{\mathrm{T}})) &= \frac{N_{x_1} \times N_{x_2}}{2}\ln(2\pi e) - \frac{1}{2}\sum_{i=1}^{N_{x_1}}\ln\det(\mathbf{A}_2^{\mathrm{T}}\mathbf{A}_2\tau + \mathbf{D}^{\alpha}_i) + \frac{1}{2}\ln\det(\mathbf{B}_1^{1D} \otimes \mathbf{B}_2^{1D}) \\ &\quad + \frac{1}{2}\ln\det(\mathbf{B}_1^{1D} \otimes \mathbf{B}_2^{1D})^{\mathrm{T}} \\ &= \frac{N_{x_1} \times N_{x_2}}{2}\ln(2\pi e) - \frac{1}{2}\sum_{i=1}^{N_{x_1}}\ln\det(\mathbf{A}_2^{\mathrm{T}}\mathbf{A}_2\tau + \mathbf{D}^{\alpha}_i) \end{aligned} \tag{6.12}$$

where $\det(\bullet)$ denotes the determinant of a matrix; $\frac{N_{x_1} \times N_{x_2}}{2}\ln(2\pi e)$ is a constant term that is independent of locations of CPT soundings. Note that $\det(\mathbf{B}_1^{1D} \otimes \mathbf{B}_2^{1D})$= $\det(\mathbf{B}_1^{1D} \otimes \mathbf{B}_2^{1D})^{\mathrm{T}}$=$(\det \mathbf{B}_1^{1D})^{N_{x_2}}(\det \mathbf{B}_2^{1D})^{N_{x_1}}$= 1 because $\mathbf{B}_2^{1D}$ and $\mathbf{B}_1^{1D}$ are two orthonormal matrices, and thus $\frac{1}{2}\ln\det(\mathbf{B}_1^{1D}\otimes\mathbf{B}_2^{1D})+\frac{1}{2}\ln\det(\mathbf{B}_1^{1D}\otimes\mathbf{B}_2^{1D})^{\mathrm{T}}=0$.

When an additional CPT sounding is performed, the information entropy of the re-interpreted cross-section of geotechnical properties, $H(vec(\hat{\mathbf{F}}_{new}^{\mathrm{T}}))$, is expressed as:

$$H(vec(\hat{\mathbf{F}}_{new}^{\mathrm{T}}))=\frac{N_{x_1}\times N_{x_2}}{2}\ln(2\pi e)-\frac{1}{2}\sum_{i=1}^{N_{x_1}}\ln\det\left[(\mathbf{A}_2^{*})^{\mathrm{T}}\mathbf{A}_2^{*}\tau^{*}+\mathbf{D}_i^{*\alpha}\right] \quad (6.13)$$

Note that $\mathbf{A}_2=\mathbf{\Psi}_{x_2}^{\mathrm{T}}\mathbf{B}_2^{1D}$, each row of $\mathbf{B}_2^{1D}$ represents a potential location of an additional CPT sounding. When an additional CPT sounding is conducted, $\mathbf{A}_2^{*}$ is constructed through extracting a new row, $\boldsymbol{r}_2$, from $\mathbf{B}_2^{1D}$ and adding it to $\mathbf{A}_2$, i.e., $\mathbf{A}_2^{*}=\begin{bmatrix}\mathbf{A}_2\\ r_2\end{bmatrix}$; τ^{*} and $\mathbf{D}_i^{*\alpha}$ may be taken as $\tau^{*}=\tau$ and $\mathbf{D}_i^{*\alpha}=\mathbf{D}_i^{\alpha}$ because additional CPT measurements still have not been obtained when determining the optimal location of an additional CPT sounding. Subsequently, $H(vec(\hat{\mathbf{F}}_{new}^{\mathrm{T}}))$ may be rearranged as:

$$H(vec(\hat{\mathbf{F}}_{new}^{\mathrm{T}}))=\frac{N_{x_1}\times N_{x_2}}{2}\ln(2\pi e)-\frac{1}{2}\sum_{i=1}^{N_{x_1}}\ln\det\left[\mathbf{A}_2^{\mathrm{T}}\mathbf{A}_2\tau+\boldsymbol{r}_2^{\mathrm{T}}\boldsymbol{r}_2\tau+\mathbf{D}_i^{\alpha}\right] \quad (6.14)$$

Based on the matrix determinant lemma (e.g., Petersen and Pedersen 2008), $H(vec(\hat{\mathbf{F}}_{new}^{\mathrm{T}}))$ can be further expressed as:

$$\begin{aligned}H(vec(\hat{\mathbf{F}}_{new}^{\mathrm{T}}))=&\frac{N_{x_1}\times N_{x_2}}{2}\ln(2\pi e)-\frac{1}{2}\sum_{i=1}^{N_{x_1}}\ln\det(\mathbf{A}_2^{\mathrm{T}}\mathbf{A}_2\tau+\mathbf{D}_i^{\alpha})\\&-\frac{1}{2}\sum_{i=1}^{N_{x_1}}\ln\left[1+\boldsymbol{r}_2(\mathbf{A}_2^{\mathrm{T}}\mathbf{A}_2\tau+\mathbf{D}_i^{\alpha})^{-1}\boldsymbol{r}_2^{\mathrm{T}}\tau\right]\end{aligned} \quad (6.15)$$

The difference between $H(vec(\hat{\mathbf{F}}^{\mathrm{T}}))$ and $H(vec(\hat{\mathbf{F}}_{new}^{\mathrm{T}}))$ can be expressed as follows:

$$\Delta=H(vec(\hat{\mathbf{F}}^{\mathrm{T}}))-H(vec(\hat{\mathbf{F}}_{new}^{\mathrm{T}}))=\frac{1}{2}\sum_{i=1}^{N_{x_1}}\ln\left[1+\boldsymbol{r}_2\mathbf{COV}_{\hat{\boldsymbol{\omega}}^{2D}}^{i}\boldsymbol{r}_2^{\mathrm{T}}\tau\right] \quad (6.16)$$

Using the concept of information entropy, the optimal CPT location is the one maximizing Δ or the exponential of Δ, $e^{\Delta}=\frac{N_{x_1}}{2}+\frac{\tau}{2}\sum_{i=1}^{N_{x_1}}\boldsymbol{r}_2\mathbf{COV}_{\hat{\boldsymbol{\omega}}^{2D}}^{i}\boldsymbol{r}_2^{\mathrm{T}}$. Note that N_{x_1} and τ is a constant for different $\boldsymbol{r}_2$. Therefore, the problem becomes to find a row of $\mathbf{B}_2^{1D}$ that will maximize the term "$\sum_{i=1}^{N_{x_1}}\boldsymbol{r}_2\mathbf{COV}_{\hat{\boldsymbol{\omega}}^{2D}}^{i}\boldsymbol{r}_2^{\mathrm{T}}$". As shown by Zhao et al. (2021), $\sum_{i=1}^{N_{x_1}}\boldsymbol{r}_2\mathbf{COV}_{\hat{\boldsymbol{\omega}}^{2D}}^{i}\boldsymbol{r}_2^{\mathrm{T}}$ represents a summation of variance of the interpolated soil properties along depth at a given location, which may be obtained directly from BCS results interpreted from previously conducted CPT soundings. In other words, the optimal location of an additional CPT sounding is the one having the maximum summation of interpolation uncertainty along depth. In the following sections, both simulated and real CPT data are used to illustrate the active learning strategy for CPT-based liquefaction assessment.

6.4.2 Procedure of CPT-Based Liquefaction Assessment

Liquefaction triggering analysis is often performed using a simplified procedure (e.g., Seed and Idriss 1971). In the simplified procedure, factor of safety against liquefaction, FS_L, is used to quantify the liquefaction potential of soil, which is computed as (e.g., Youd et al. 2001):

$$FS_L = \frac{\mathrm{CRR}}{\mathrm{CSR}} = \frac{\mathrm{CRR}_{7.5}}{\mathrm{CSR}} \times MSF \tag{6.17}$$

where CRR is the cyclic resistance ratio, which indicates soil liquefaction resistance; CSR is the cyclic stress ratio, which represents the earthquake demand on soil; $\mathrm{CRR}_{7.5}$ is the CRR when the earthquake magnitude $M_w = 7.5$; $MSF = 174 / M_w^{2.56}$ represents the magnitude scaling factor. If the calculated FS_L is smaller than 1, liquefaction is estimated to occur.

Based on case histories, Robertson and Wride (1998) proposed the following equations to estimate $\mathrm{CRR}_{7.5}$ directly from CPT measurement data:

$$\mathrm{CRR}_{7.5} = \begin{cases} 93\left(\dfrac{(q_{c1N})_{cs}}{1000}\right)^3 + 0.08 & \text{if } 50 \le (q_{c1N})_{cs} < 160 \\ 0.833\left(\dfrac{(q_{c1N})_{cs}}{1000}\right) + 0.05 & \text{if } (q_{c1N})_{cs} < 50 \end{cases} \tag{6.18}$$

where $(q_{c1N})_{cs}$ represents the equivalent clean sand cone penetration resistance, as expressed below:

$$(q_{c1N})_{cs} = \mathrm{K}_c q_{c1N} \tag{6.19}$$

where the normalized cone tip resistance $q_{c1N} = (q_c / p_a)C_Q$; q_c= measured cone tip penetration resistance; $C_Q = (p_a / \sigma'_{vo})^n \le 1.7$; σ'_{vo}= effective overburden stress; p_a = 100 kPa; n varies from 0.5 to 1.0 depending on the soil type; and correction factor for grain characteristics, K_c, is computed as (e.g., Robertson and Wride 1998):

$$\mathrm{K}_c = \begin{cases} 1.0 & \text{if } I_c \le 1.64 \\ 33.75 I_c - 21.63 I_c^2 + 5.581 I_c^3 - 0.403 I_c^4 - 17.88 & \text{if } I_c > 1.64 \end{cases} \tag{6.20}$$

where $I_c = [(\log F + 1.22)^2 + (3.47 - \log Q)^2]^{0.5}$; $F = f_s / (q_c - \sigma_{vo}) \times 100\%$; $Q = [(q_c - \sigma_{vo}) / p_a](p_a / \sigma'_{vo})^n$; σ_{vo}= total overburden stress; and f_s= CPT sleeve friction.

CSR is a function of maximum acceleration, a_{max} (e.g., Seed and Idriss 1971):

$$\text{CSR} = 0.65\left(\frac{a_{max}}{g}\right)\left(\frac{\sigma_{vo}}{\sigma'_{vo}}\right) r_d \tag{6.21}$$

where g represents gravitational acceleration constant, and shear stress reduction coefficient, r_d, is expressed as a function of depth, z (e.g., Liao and Whiteman 1986):

$$r_d = \begin{cases} -0.00765z + 1.0 & z < 9.15\text{m} \\ -0.0267z + 1.174 & 9.15\text{m} \le z < 23\text{m} \\ -0.008z + 0.744 & 23\text{m} \le z < 30\text{m} \\ 0.5 & 30\text{m} \le z \end{cases} \tag{6.22}$$

Using the simplified procedure, FS_L can be estimated at the locations with CPT measurements. Proper characterization of the spatial distribution of soil liquefaction potential (e.g., FS_L against liquefaction) is critical for earthquake hazard assessments and mitigation. However, due to constraints in time, cost, and access, only a limited number of CPT soundings are conducted at a specific site, and the liquefaction potential of soil at locations without CPT data has to be interpreted from limited CPT measurements using a proper interpolation method, leading to significant statistical uncertainty in liquefaction assessment results (e.g., Guan and Wang 2022a, 2022b, 2024; Guan et al. 2022a). Such uncertainty might result in a significant risk for structures during an earthquake. In general, conducting a large number of CPT soundings provides increasingly reliable liquefaction assessment results in a cross-section (e.g., a cross-section of FS_L with high spatial resolution), but it also dramatically increases the time and expenditure of site investigation. A fundamental but challenging question in site investigation for liquefaction assessment is how to determine the minimal number of CPTs and their corresponding optimal locations to achieve a reliability level of liquefaction assessment results that decision-makers of a project want to achieve (i.e., the target reliability level).

6.4.3 Simulate Data Example

To illustrate the application of the active learning strategy for liquefaction assessment, a simulated cross-section of $(q_{c1N})_{cs}$ is used in this section. It is widely recognized that soil properties (e.g., $(q_{c1N})_{cs}$) are spatially varying and correlated along both depth and horizontal direction due to a wide variety of physical and chemical processes during the formation of soils (e.g., Vanmarcke 1977; Phoon and Kulhawy 1999; Wang et al. 2016; Guan and Wang 2022c; Guan et al. 2023). Random field is frequently used in geotechnical engineering to model spatial characteristics of geo-materials (e.g., Vanmarcke 1977). Consider, for example, a stationary and normally distributed $(q_{c1N})_{cs}$ 2D random field, with mean $\mu = 100$ and coefficient of variation, COV = 0.2. An anisotropic exponential autocorrelation function is used with $\lambda_v = 1.5$ m and $\lambda_h = 60$ m. After the $(q_{c1N})_{cs}$ random field is specified, a cross-section of $(q_{c1N})_{cs}$ can be generated using a random field generator, such as a circulant embedding algorithm (e.g., Dietrich and Newsam 1993). The generated $(q_{c1N})_{cs}$ cross-section with a depth of 15 m and width of 250 m is shown in Figure 6.6. The vertical and horizontal resolutions of the cross-section are set as 0.1 m and 1.0 m, respectively. In total, the cross-section has 151 × 251 = 37,901 data points.

The adaptive sampling strategy is used to characterize the spatial distribution of FS_L within this cross-section. For illustration purposes, the designed earthquake is taken to have an earthquake magnitude, $M_w = 7.0$, and peak horizontal acceleration, $a_{max} = 0.15g$; the water table is located at the ground surface, and the cross-section consists of liquefiable sand with the saturated unit weight $\gamma_{\text{sat}} = 20$ kN/m^3. In Step 1 of the adaptive sampling, the target reliability level of interpretation results is determined as $\text{COV}_\text{T} = 10\%$. Then, the preliminary number of CPT soundings should be selected first as the starting point of the adaptive sampling strategy. The COV and λ_h of the spatial distribution of FS_L within the cross-section might be estimated as 20% and 40 m, respectively, while the RE_T might be taken as 18%. Then, the preliminary number of CPT sounding is determined using the statistical chart shown in Chapter 5. The RE_T / COV = 0.9 is calculated. Using the lower envelope in the statistical chart, $d_c / \lambda_h = 1.37$, is obtained. Because $\lambda_h = 40$ m, the CPT spacing $d_c = 55$ m is estimated. The width of the cross-section is 250 m, and thus $M_c = 250 / 55 + 1 \approx 6$ CPT soundings should be performed within this cross-section, as shown in Figure 6.6 by dashed lines.

In Step 3, measurement data from six CPT soundings are used to interpret FS_L cross-section. To mimic the process of conducting six CPTs (i.e., C-1, C-2, ..., C-6) in site investigation, 6 × 151 = 906 $(q_{c1N})_{cs}$ data points at the locations of these CPT soundings are extracted from the original simulated $(q_{c1N})_{cs}$ data, as shown in Figure 6.7. Using these $(q_{c1N})_{cs}$ measurement data as input to the 2D BCS method and Markov Chain Monte Carlo (MCMC), $N_{MCMC} = 500$ samples of $(q_{c1N})_{cs}$ cross-section,

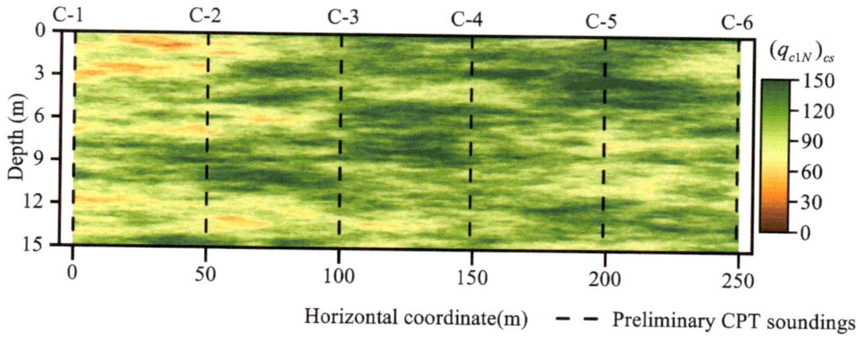

FIGURE 6.6 Simulated equivalent clean sand cone penetration resistance, $(q_{c1N})_{cs}$ in a vertical cross-section (after Guan et al. 2022a).

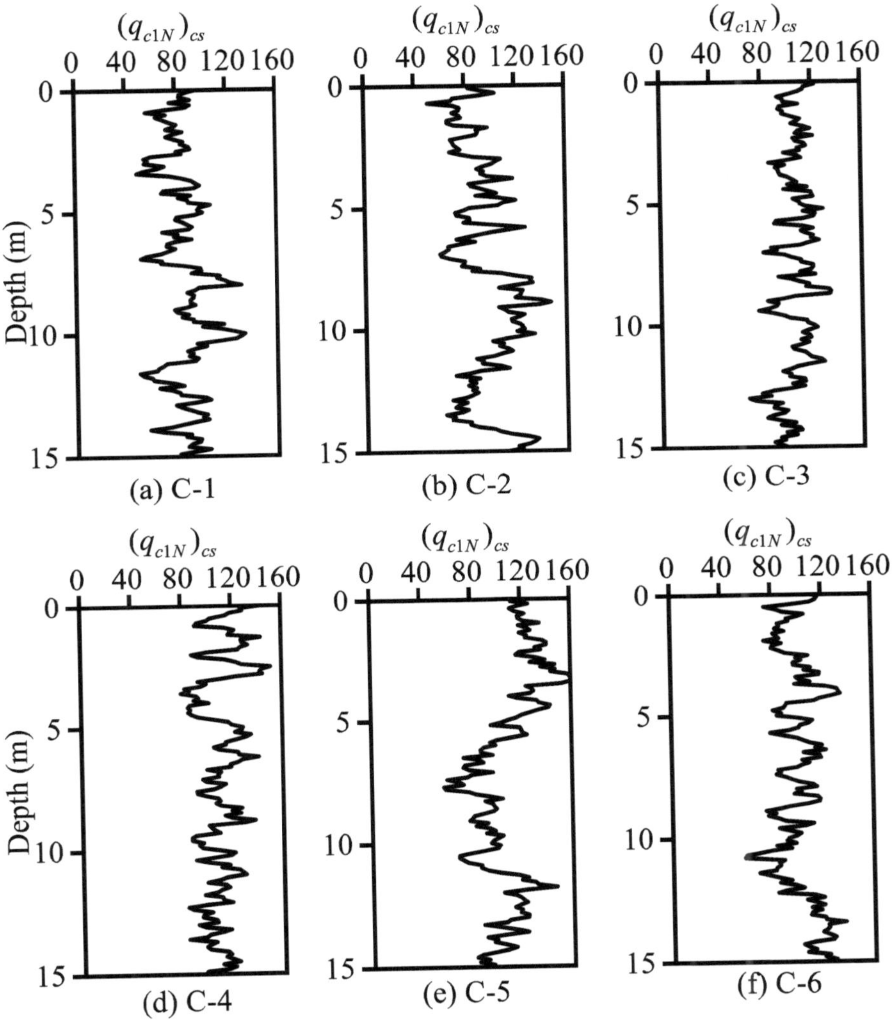

FIGURE 6.7 Equivalent clean sand cone penetration resistance, $(q_{c1N})_{cs}$ measurement data from 6 CPT soundings (after Guan et al. 2022a).

$\hat{\mathbf{F}}_{(q_{c1N})_{cs}}$, with a high spatial resolution are generated. The mean and standard deviation of these generated $(q_{c1N})_{cs}$ samples are illustrated in Figure 6.8. Resolutions of the generated samples of $\hat{\mathbf{F}}_{(q_{c1N})_{cs}}$ are set as the same as the original one.

For each generated $\hat{\mathbf{F}}_{(q_{c1N})_{cs}}$ sample, the CRR is calculated within the cross-section using Robertson and Wride (1998) procedure. The mean and standard deviation of $N_B = 500$ generated samples of CRR cross-section for $M_w = 7.0$ are shown in Figures 6.9(a) and (b), respectively. The cross-section of CSR for the given earthquake is calculated following the simplified procedure (Seed and Idriss 1971), as

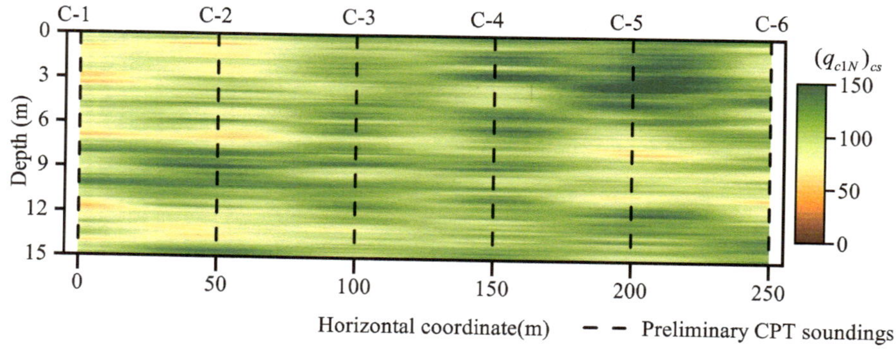

(a) Mean of $(q_{c1N})_{cs}$ cross-section interpreted from 6 preliminary CPT soundings

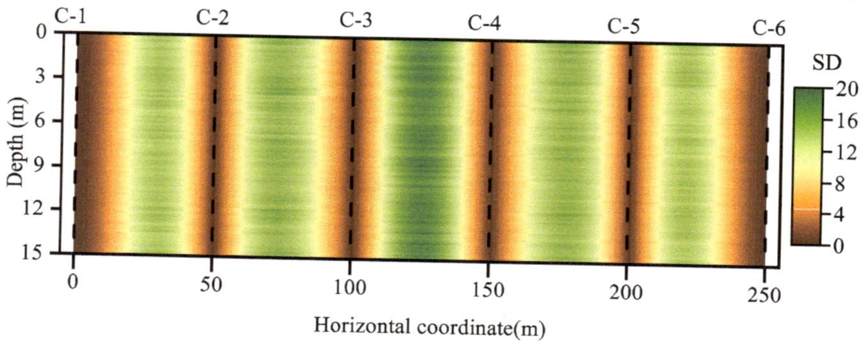

(b) Standard deviation (SD) of $(q_{c1N})_{cs}$ cross-section interpreted from 6 preliminary CPT soundings

FIGURE 6.8 Cross-section of $(q_{c1N})_{cs}$ interpreted from six preliminary CPT sounding (after Guan et al. 2022a).

shown in Figure 6.10. Then, for each sample of CRR cross-section together with the estimated CSR, FS_L over the whole cross-section is obtained by CRR/CSR, leading to N_{MCMC} = 500 samples of FS_L cross-section. The mean and standard deviation of the generated FS_L cross-section samples from six CPT soundings are shown in Figures 6.11(a) and (b), respectively. The reliability level of the interpreted FS_L cross-section is estimated as the mean COV, COV_M = 14.2%, and it is greater than the target reliability level, COV_T = 10.0%. This suggests that the reliability of the interpretation results should be further improved. Hence, the self-adaptive process is triggered to perform additional CPT soundings at their optimal locations.

In Step 4, the target reliability level of the interpreted FS_L cross-section is achieved through adaptive sampling. One additional CPT sounding is conducted at its optimal location for each adaptation based on the information obtained from the previously conducted CPT soundings. The new $(q_{c1N})_{cs}$ measurement from the additional CPT

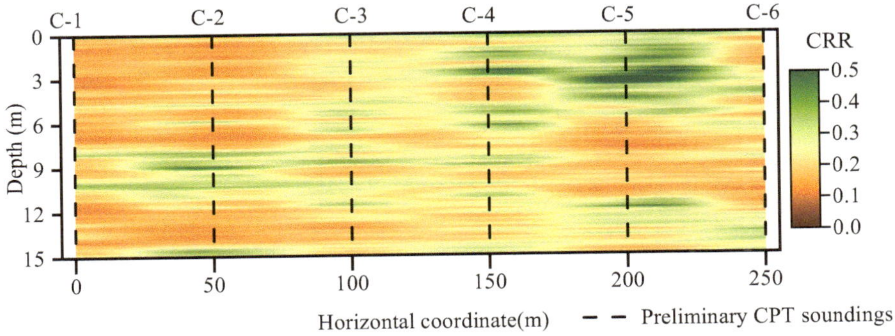

(a) Mean of CRR cross-section interpreted from 6 CPT soundings

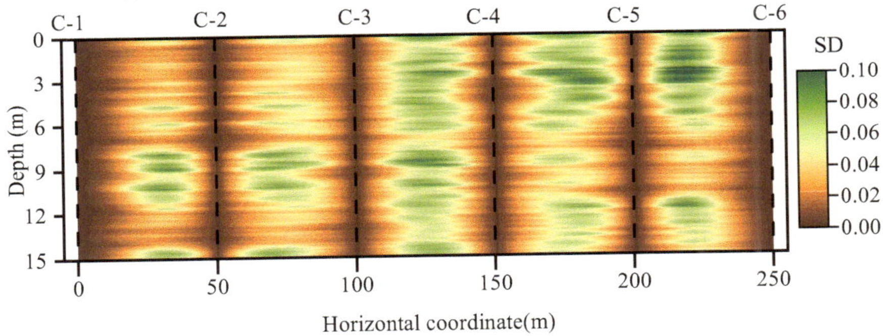

(b) Standard deviation (SD) of CRR cross-section interpreted from 6 CPT soundings

FIGURE 6.9 Cyclic resistance ratio, CRR cross-section interpreted from 6 CPT soundings using Bayesian compressive sampling (BCS) (after Guan et al. 2022a).

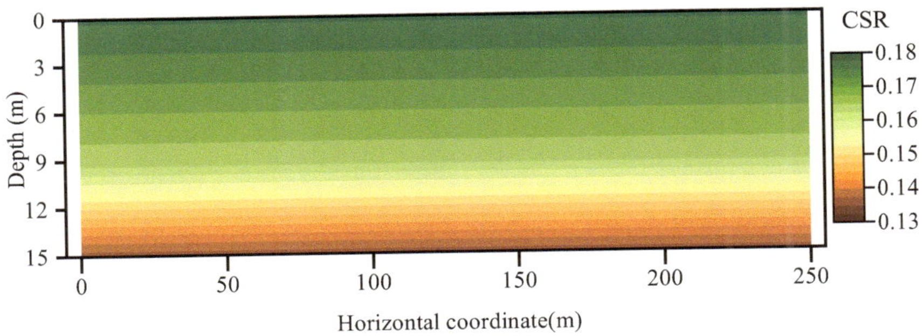

FIGURE 6.10 Cyclic stress ratio, CSR cross-section for peak horizontal acceleration at the ground surface, $a_{max} = 0.15g$ (after Guan et al. 2022a).

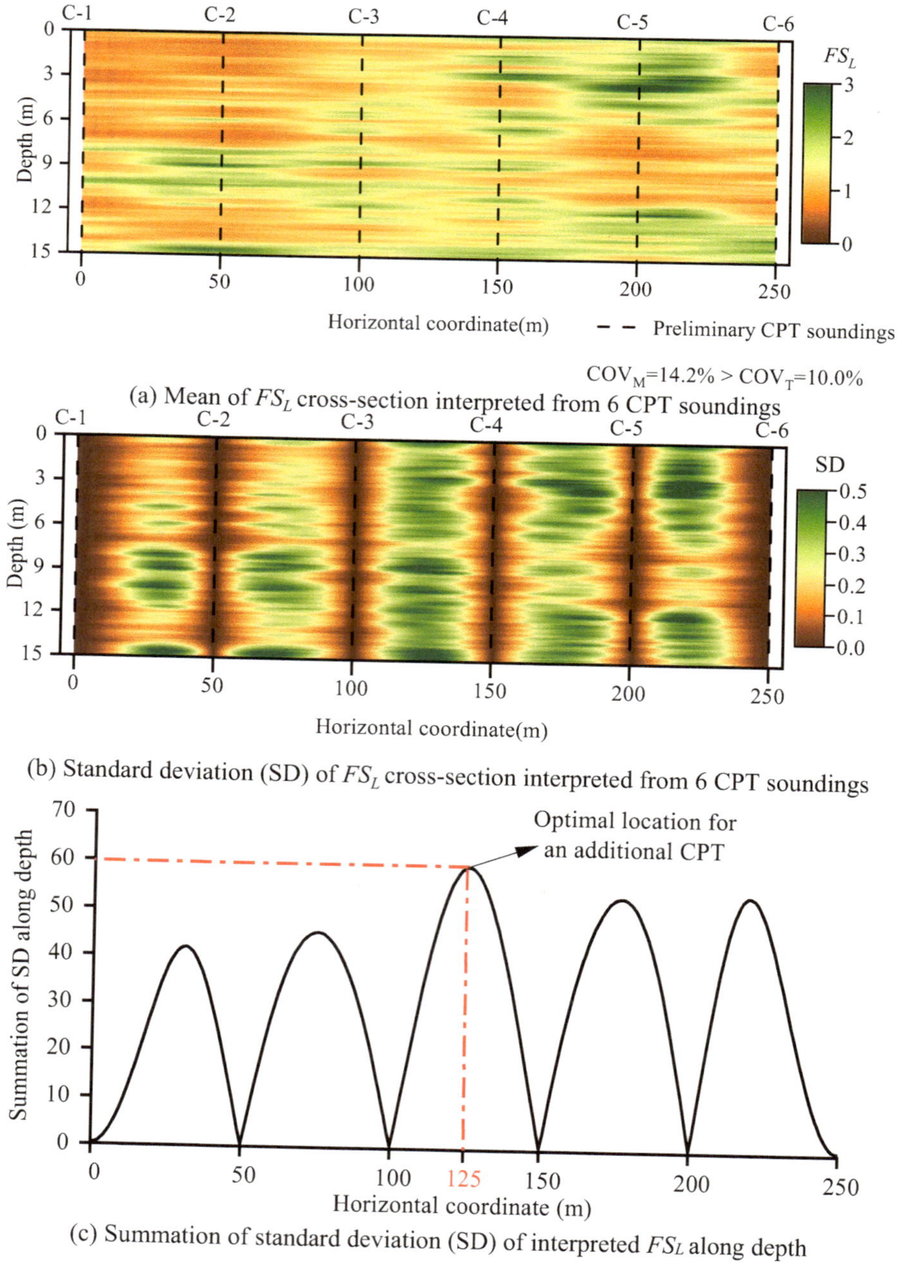

(a) Mean of FS_L cross-section interpreted from 6 CPT soundings

(b) Standard deviation (SD) of FS_L cross-section interpreted from 6 CPT soundings

(c) Summation of standard deviation (SD) of interpreted FS_L along depth

FIGURE 6.11 Cross-section of the factor of safety, FS_L against liquefaction interpreted from six CPT soundings (after Guan et al. 2022a).

sounding together with the previously measured data points are used to re-interpret FS_L cross-section. Once the reliability level of the interpretation results achieves the target one, the self-adaptive process stops. As introduced in section 6.4.1, the optimal CPT location is the one where the summation of variance, or standard deviation, of the interpreted FS_L along depth is the largest. Figure 6.11(c) shows the summation of the standard deviation of interpreted FS_L along depth at different locations. It is observed from Figure 6.11(c) that the maximum summation of standard deviation along depth is 54, which is located at the horizontal coordinate of 125 m. This indicates that an additional CPT sounding is supposed to be conducted at this location for the first adaptation. Based on information entropy, if additional CPT sounding is conducted at this location, the maximum information gain will be obtained. It is also interesting to note that such results are consistent with the first law of geography (Tobler 2004). Based on the first law of geography, the efficient sampling location may be taken as the minimal total Euclidean distance from all unsampled locations to the selected sampling location (e.g., Wang and Li 2021). For this case, when an additional CPT is conducted in the middle it leads to the minimal total Euclidean distance.

To mimic the process of conducting additional CPT, $(q_{c1N})_{cs}$ data points along depth at the horizontal coordinate of 125 m are taken from the simulated $(q_{c1N})_{cs}$ cross-section and used together with the previous 906 $(q_{c1N})_{cs}$ data points to re-interpret FS_L cross-section. In total, $7 \times 151 = 1{,}057$ $(q_{c1N})_{cs}$ measurement data obtained from $M_c = 6 + 1 = 7$ CPT soundings are used to interpret the FS_L cross-section. Figures 6.12(a) and 6.12(b) show the mean and standard deviation of the FS_L cross-section interpreted from seven CPT soundings. Figure 6.12(c) shows the summation of standard deviation of the interpreted FS_L along depth. It can be further observed by comparing Figure 6.11(c) with Figure 6.12(c) that the standard deviation of interpreted FS_L close to the location (i.e., 125 m in the horizontal direction) of the additional performed CPT is significantly reduced. The COV_M of the interpreted FS_L cross-section is calculated as 13.3%. This indicates that the reliability of the interpreted FS_L cross-section is improved after performing an additional CPT.

However, COV_M=13.3% is still greater than COV_T=10.0%, and, therefore, the self-adaptative process continues to further improve the reliability level of interpretation results. It is observed from Figure 6.12(c) that the maximum summation of standard deviation is located at the horizontal coordinate of 29 m, and thus the next CPT sounding is conducted at this location. Figures 6.13(a) and (b) show the mean and standard deviation of the FS_L cross-section interpreted from $M_c = 6 + 2 = 8$ CPT soundings. The COV_M is further reduced to 12.1%.

The self-adaptative process continues until $COV_M \leq COV_T$. The variation of COV_M with the number of adaptations is shown in Figure 6.14. It is found that COV_M decreases from 14.2% to 9.5% as M_c increases from 6 to 12. The reliability level of the interpreted FS_L cross-section is monitored at each adaptation, and the FS_L cross-section is re-interpreted with the newly conducted CPT sounding, which in turn is used for decision on further CPT soundings, if needed. When M_c increases to 12, $COV_M = 9.5\% < COV_T = 10.0\%$. This suggests that the reliability of the interpretation results achieves the target level, and the self-adaptative process stops. Figure 6.15 illustrates the FS_L cross-section interpreted from the 12 CPT soundings.

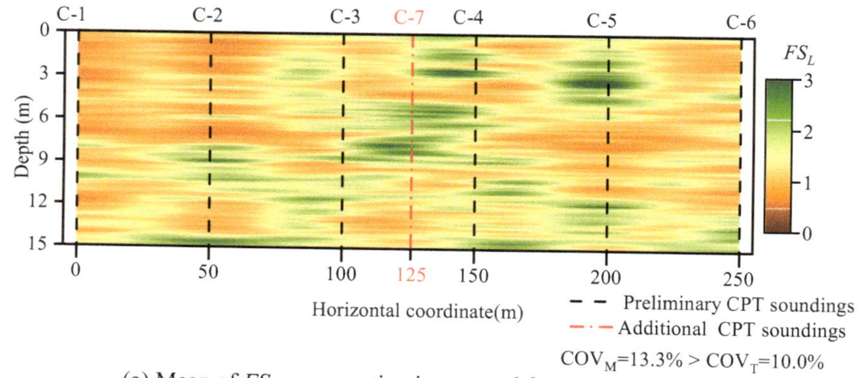

(a) Mean of FS_L cross-section interpreted from 7 CPT soundings

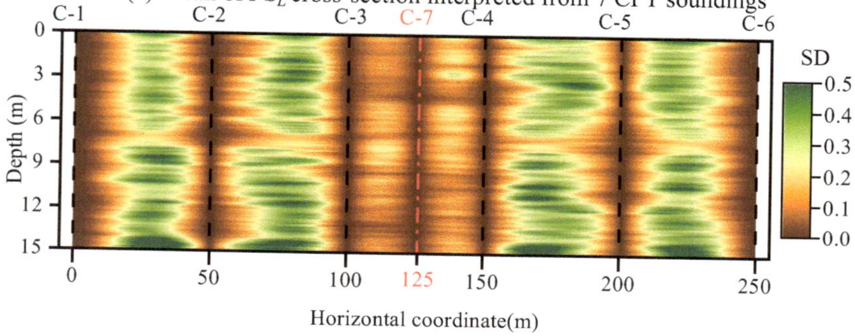

(b) Standard deviation (SD) of FS_L cross-section interpreted from 7 CPT soundings

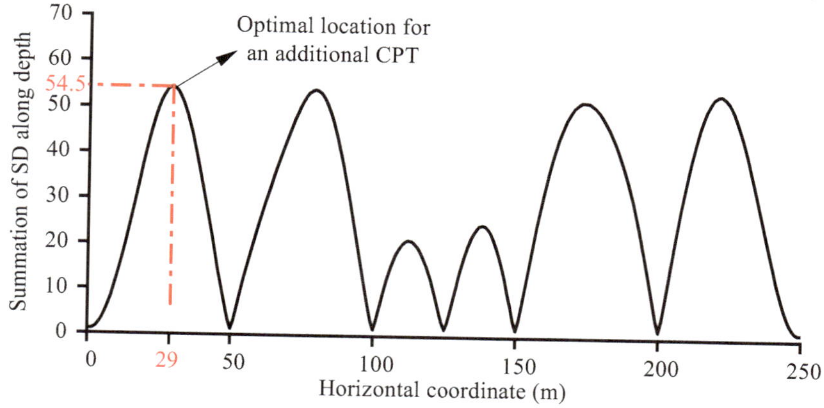

(c) Summation of standard deviation (SD) of interpreted FS_L along depth

FIGURE 6.12 Cross-section of a factor of safety, FS_L against liquefaction interpreted from previously conducted 6 CPT soundings together with one additional CPT sounding (after Guan et al. 2022a).

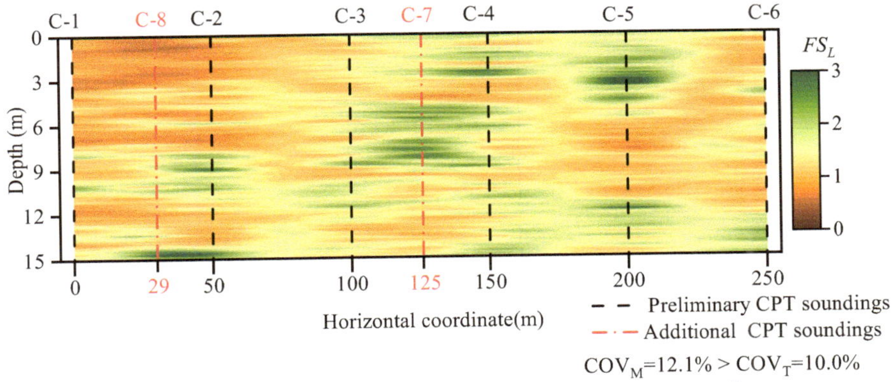

(a) Mean of FS_L cross-section interpreted from 8 CPT soundings

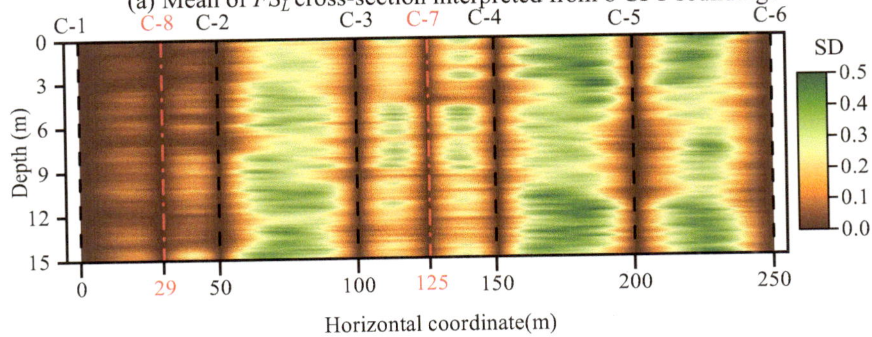

(b) Standard deviation (SD) of FS_L cross-section interpreted from 8 CPT soundings

FIGURE 6.13 Cross-section of the factor of safety, FS_L, against liquefaction interpreted from 8 CPT soundings (after Guan et al. 2022a).

The preliminary CPT soundings and the additional six CPT soundings are illustrated in Figure 6.15 by dashed lines and dash-dotted lines, respectively. Using the adaptive sampling, 12 CPT soundings and their optimal locations are automatically determined.

6.4.4 Real Data Example

In this section, the active learning strategy is used to characterize spatial distribution of liquefaction potential of soil within a cross-section A-A' at a site in Christchurch, New Zealand, as shown in Figure 6.16. The site mainly consists of about 5–6 m thick of clean sand and silty sand near the ground surface (e.g., Cubrinovski 2013). Based on nearby site investigation reports, the saturated unit weight of clean sand and silty sand is about γ_{sat}= 18 kN/m^3. The water table is located at 1.2 m below the ground

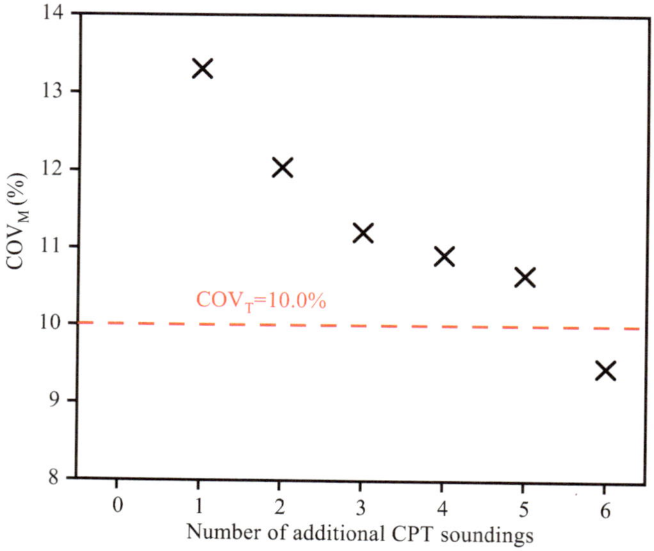

FIGURE 6.14 Variation of COV_M with the number of additional CPT soundings (i.e., number of adaptations) (after Guan et al. 2022a).

surface. The liquefaction hazard at this site under the 2011 Christchurch Earthquake with M_w = 6.2 is evaluated using the proposed method. Based on recordings from a nearby seismic strong motion station, a_{max} = 0.52g is used in this example (e.g., Bradley and Cubrinovski 2011).

The width of cross-section A-A' is about 180 m, while the depth of cross-section A-A' ranges from about 2.0 m to about 5.0 m. In Step 1, a target reliability level, COV_T = 5% is used for illustration purposes. Then, the COV and λ_h of the spatial distribution of $(q_{c1N})_{cs}$ within the cross-section might be estimated as 15% and 60 m, respectively, while the RE_T might be determined as 10%. Then, the preliminary number of CPT soundings is determined using the statistical chart shown in Chapter 5. The RE_T / COV = 0.67 is calculated. Using the lower envelope in the statistical chart, d_c / λ_h = 0.72 is obtained. Because λ_h = 60 m, the CPT spacing d_c = 43.2 m is estimated. The width of the cross-section is 180 m, and thus M_c = 180 / 43.2+1 ≈5 CPT soundings should be performed within this cross-section. Therefore, five CPT soundings (i.e., CPT_385, CPT_55692, CPT_55682, CPT_55684, and CPT_55680) are conducted within this cross-section, as shown in Figure 6.16 by solid triangles. CPT measurement data (q_c and f_s) from these CPT soundings are obtained from the New Zealand Geotechnical Database (NZGD) (NZGD 2012), as shown in Figure 6.17. For each set of q_c and f_s, $(q_{c1N})_{cs}$ is calculated.

In Step 3, measured data points from preliminary CPT soundings are used to interpret the FS_L cross-section. Using measured $(q_{c1N})_{cs}$ data from these five CPT

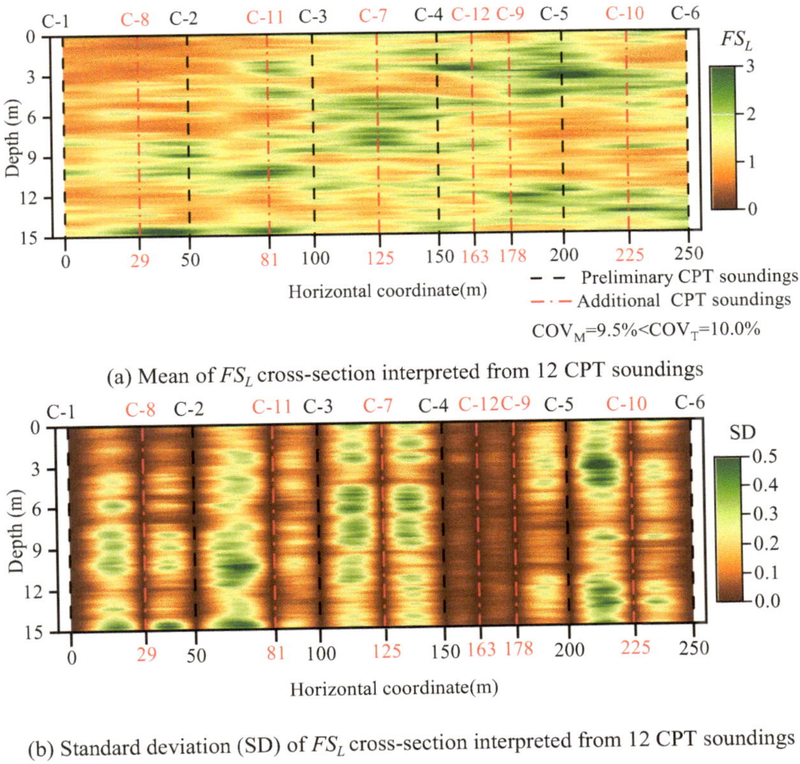

FIGURE 6.15 Cross-section of a factor of safety, FS_L against liquefaction interpreted from 12 CPT soundings (after Guan et al. 2022a).

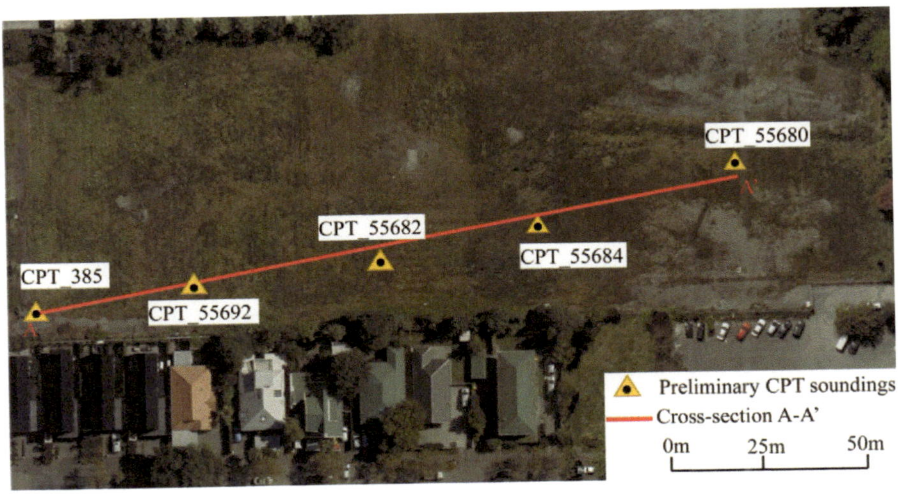

FIGURE 6.16 Layout of five preliminary CPT soundings at a site in Christchurch, New Zealand (after Guan et al. 2022a).

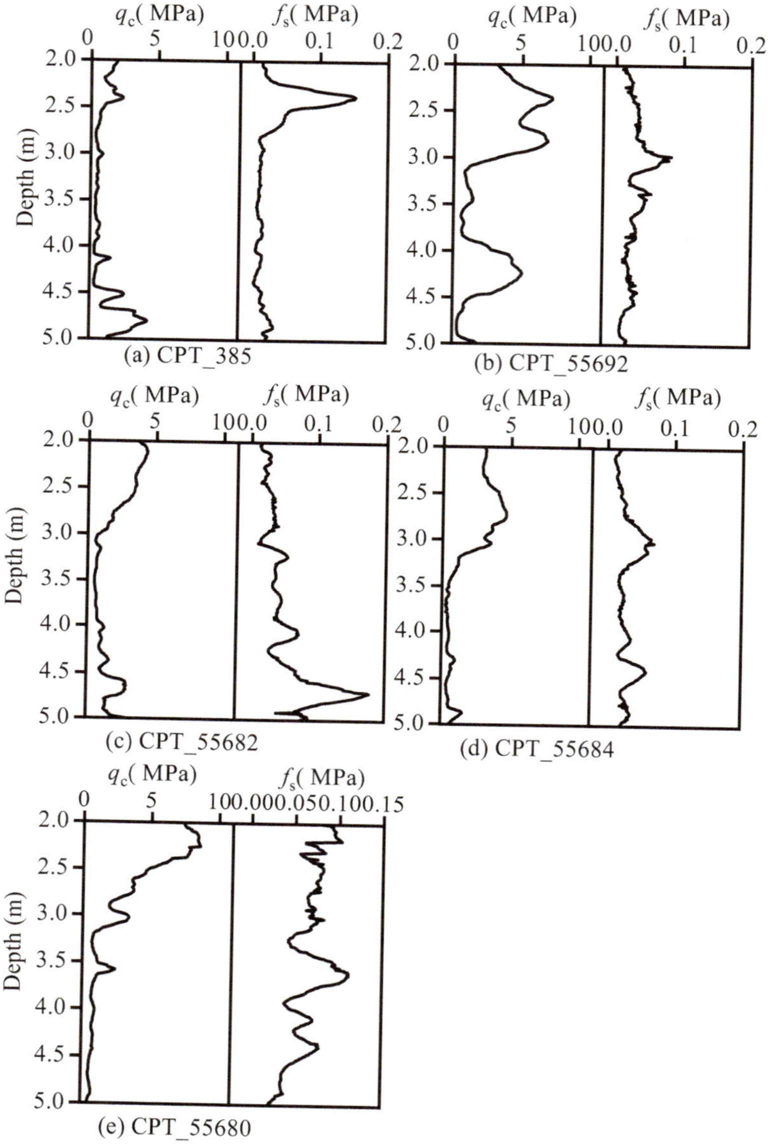

FIGURE 6.17 Measurement data (q_c and f_s) from preliminary CPT soundings at a site in Christchurch, New Zealand (after Guan et al. 2022a).

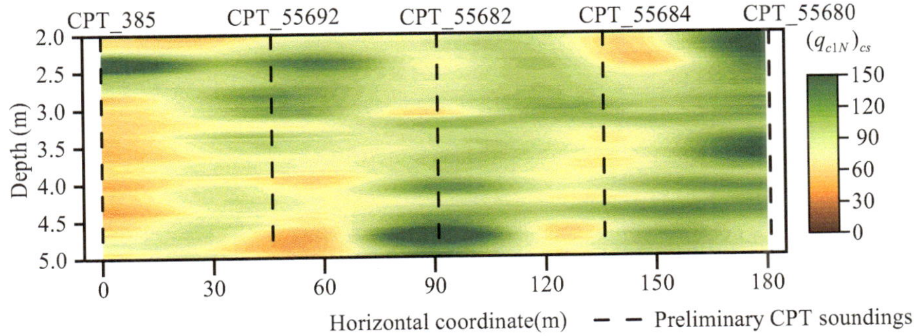

(a) Mean of $(q_{c1N})_{cs}$ cross-section interpreted from 5 CPT soundings

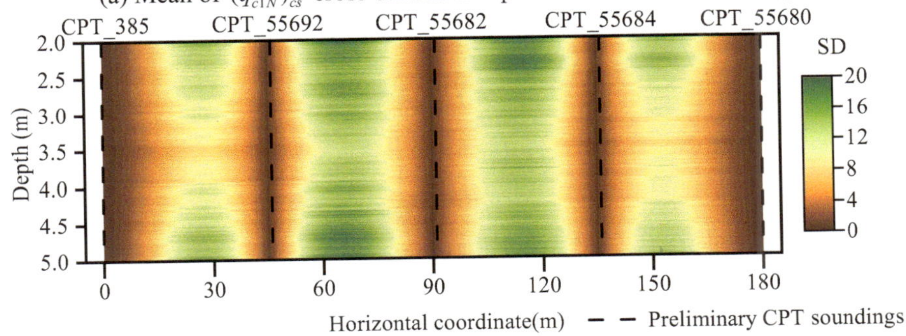

(b) Standard deviation (SD) of $(q_{c1N})_{cs}$ cross-section interpreted from 5 CPT soundings

FIGURE 6.18 Cross-section of $(q_{c1N})_{cs}$ interpreted from preliminary CPT soundings (after Guan et al. 2022a).

soundings, $N_{MCMC} = 500$ samples of $(q_{c1N})_{cs}$ cross-section with high spatial resolution are generated using 2D BCS and MCMC. The mean and standard deviation of the generated $(q_{c1N})_{cs}$ samples are shown in Figure 6.18. The vertical and horizontal resolutions of the samples of $(q_{c1N})_{cs}$ cross-section are 0.01 m and 1 m, respectively. Then, $N_{MCMC} = 500$ samples of CRR cross-section are obtained. The mean and standard deviation of the generated CRR cross-section samples are shown in Figure 6.19. The CSR in cross-section A-A' is computed, as illustrated in Figure 6.20.

Subsequently, using the generated samples of CRR cross-section together with estimated CSR, $N_b = 500$ samples of FS_L cross-section are obtained. The mean and standard deviation of generated FS_L samples are shown in Figure 6.21. The reliability level of the interpreted FS_L cross-section is calculated as $COV_M = 11.4\%$ which is greater than $COV_T = 5.0\%$. Therefore, additional CPT soundings are conducted, one by one, at their optimal locations for improving the reliability of interpretation results in Step 4.

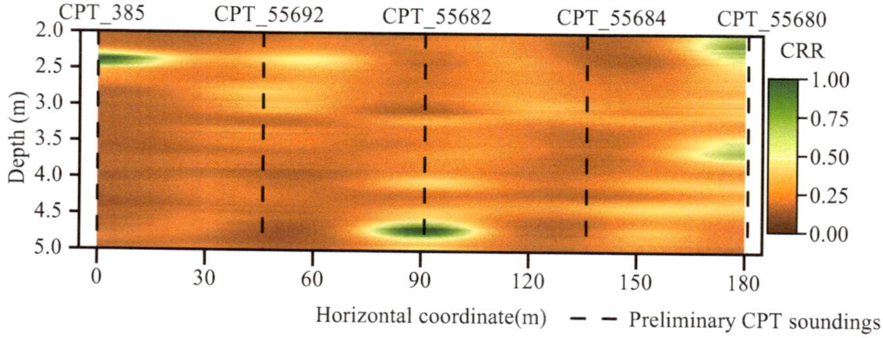

(a) Mean of CRR cross-section interpreted from 5 CPT soundings

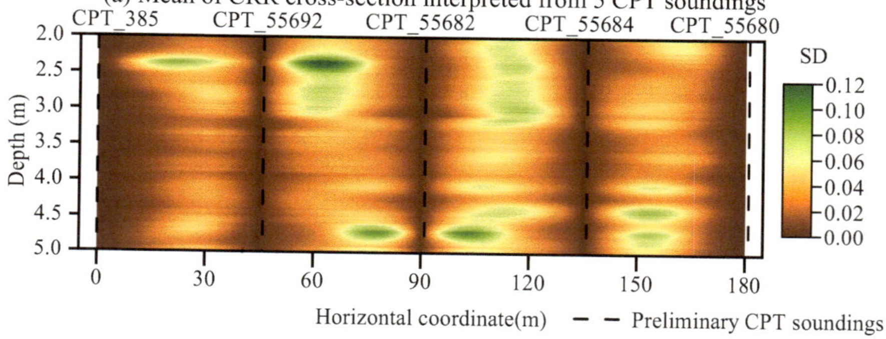

(b) Standard deviation (SD) of CRR cross-section interpreted from 5 CPT soundings

FIGURE 6.19 Cross-section of cyclic resistance ratio, CRR interpreted from preliminary CPT soundings (after Guan et al. 2022a).

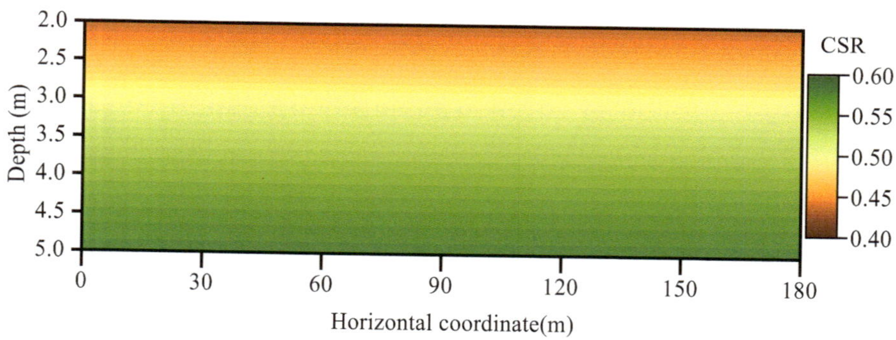

FIGURE 6.20 Estimated cross-section of cyclic resistance ratio, CSR (after Guan et al. 2022a).

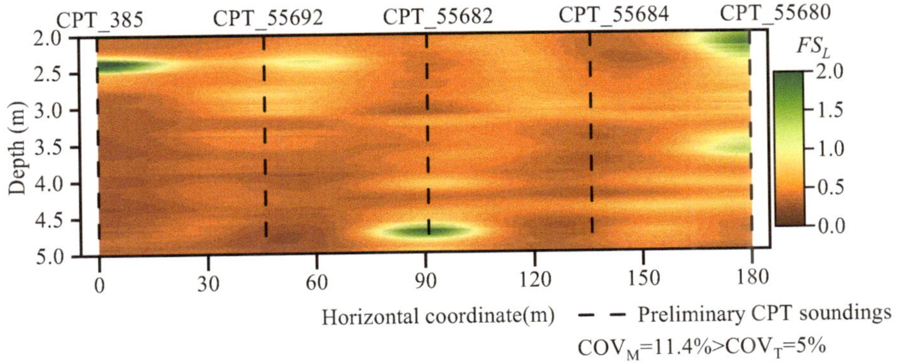

(a) Mean of FS_L cross-section interpreted from 5 CPT soundings

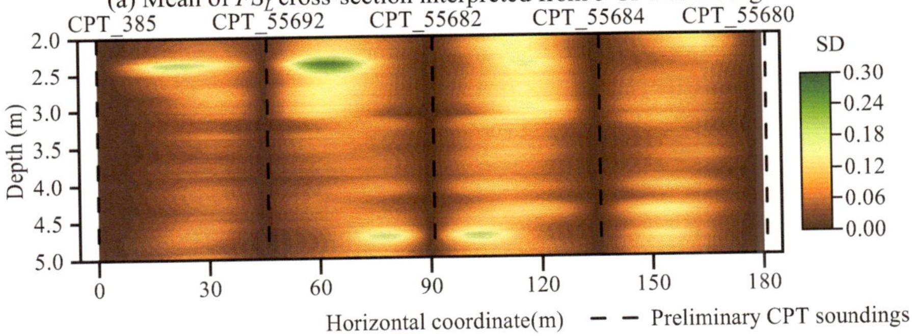

(b) Standard deviation (SD) of FS_L cross-section interpreted from 5 CPT soundings

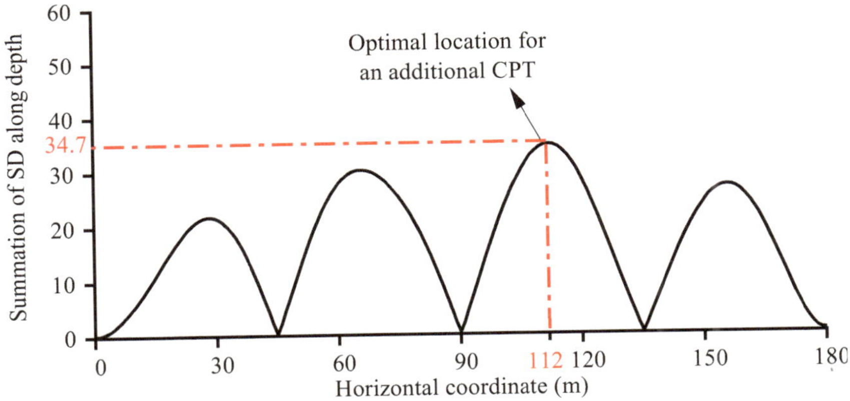

(c) Summation of standard deviation (SD) of interpreted FS_L along depth

FIGURE 6.21 Cross-section of a factor of safety, FS_L against liquefaction interpreted from five preliminary CPT soundings (after Guan et al. 2022a).

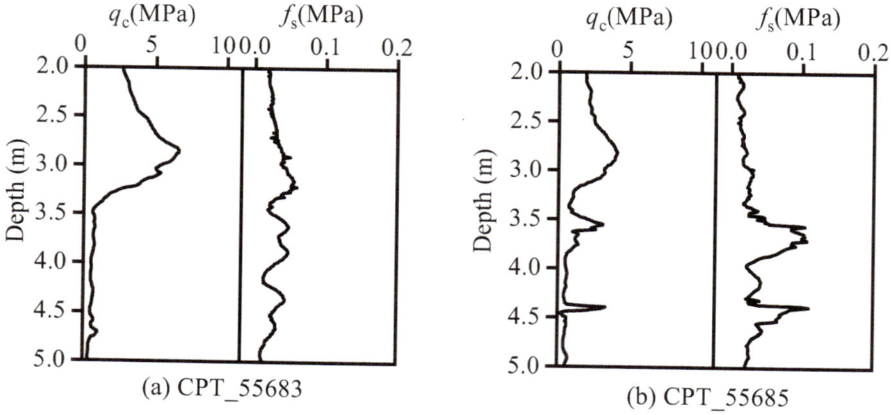

FIGURE 6.22 Measurement data (q_c and f_s) from additional CPT soundings at a site in Christchurch, New Zealand (after Guan et al. 2022a).

Figure 6.21(c) shows the summation of standard deviation of the interpreted FS_L along depth at different locations. It is observed from Figure 6.21(c) that the maximum summation of standard deviation along depth is 34.7 which is located at the horizontal coordinate of 112 m. Then, an additional CPT sounding is conducted at this location. To mimic this procedure, CPT_55683, which was approximately conducted at the horizontal coordinate of 112 m, is used together with the previous five CPT soundings to re-interpret the FS_L cross-section. CPT measurement data (q_c and f_s) from CPT_55683 is shown in Figure 6.22(a). The interpretation results from six CPT soundings are shown in Figure 6.23. The COV_M of the FS_L cross-section interpreted from six CPT is computed as 8.0%, and thus the self-adaptive process continues. As shown in Figure 6.23(c), the optimal location for the next CPT sounding is 158 m in the horizontal direction.

CPT_55685 was approximately conducted at the horizontal coordinate of 158 m, and thus it is used together with the previous six CPT soundings to re-interpret the FS_L cross-section. CPT measurement data (q_c and f_s) from CPT_55685 is shown in Figure 6.22(b). Figure 6.24 shows the FS_L interpreted from seven CPT soundings within cross-section A-A'. It is found from Figure 6.24(b) that the overall standard deviation of interpreted FS_L is significantly reduced after performing two additional CPT soundings. The reliability level of interpretation results is improved from $COV_M = 11.4\%$ to $COV_M = 4.9\%$, and it achieves the target reliability level. The layout of additional CPT soundings is illustrated in Figure 6.25 by solid circles. Using the proposed active learning strategy, seven CPT soundings and their optimal locations are smartly determined for achieving $COV_T = 5.0\%$.

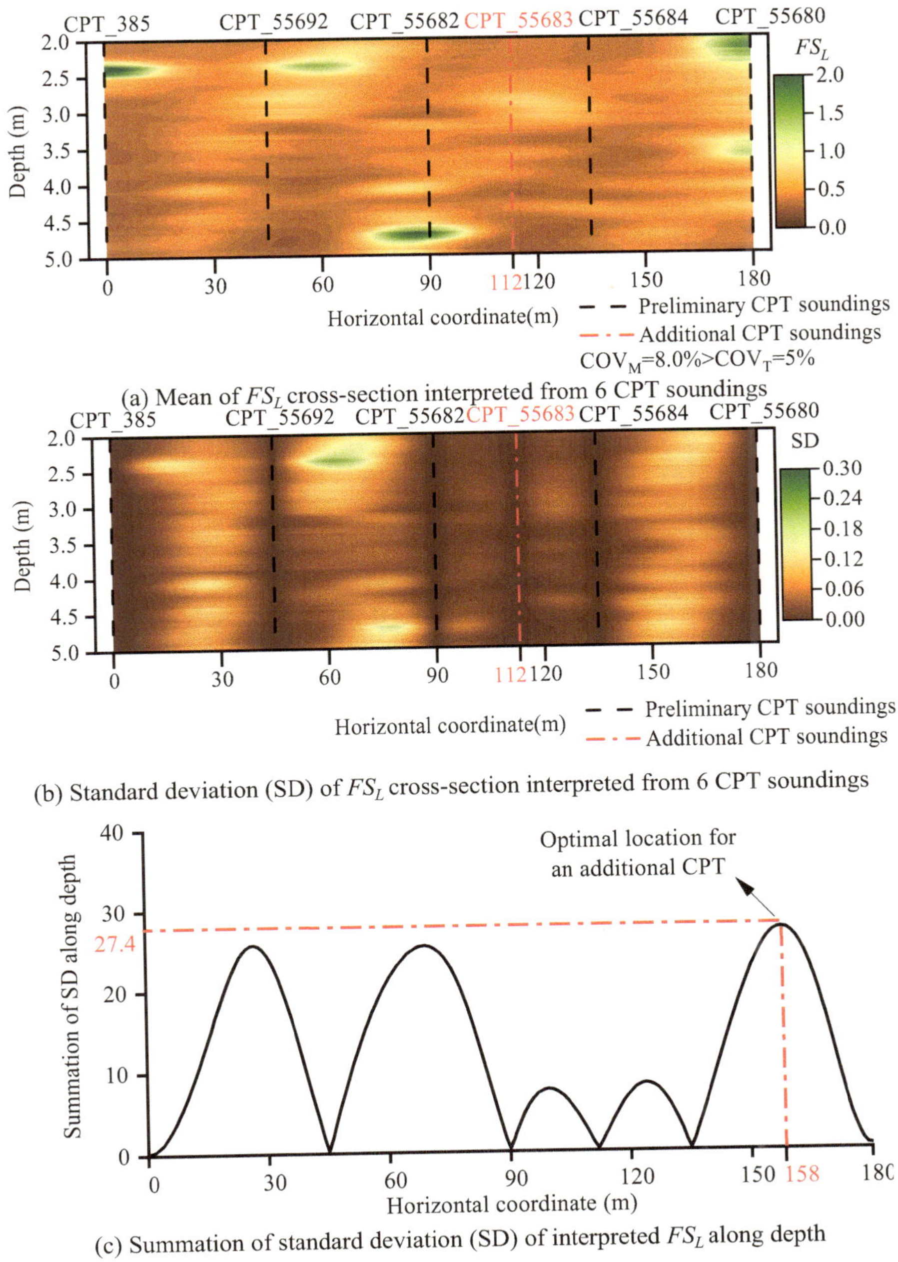

FIGURE 6.23 Cross-section of the factor of safety, FS_L, against liquefaction interpreted from six CPT soundings (after Guan et al. 2022a).

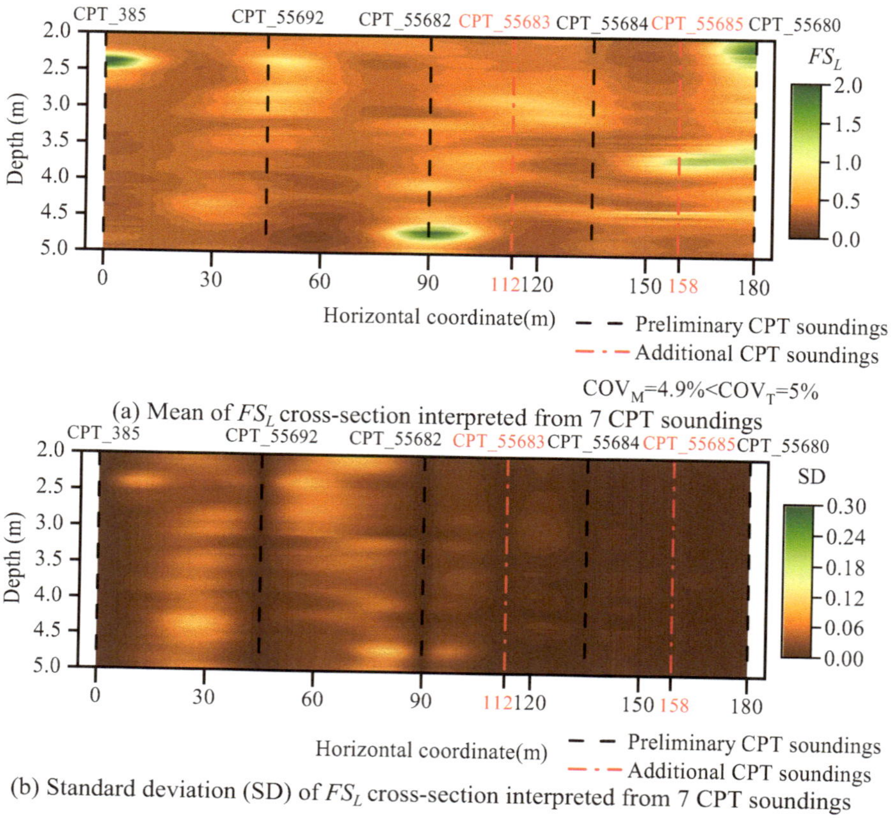

(a) Mean of FS_L cross-section interpreted from 7 CPT soundings

(b) Standard deviation (SD) of FS_L cross-section interpreted from 7 CPT soundings

FIGURE 6.24 Cross-section of the factor of safety, FS_L against liquefaction interpreted from seven CPT soundings using the proposed active learning strategy (after Guan et al. 2022a).

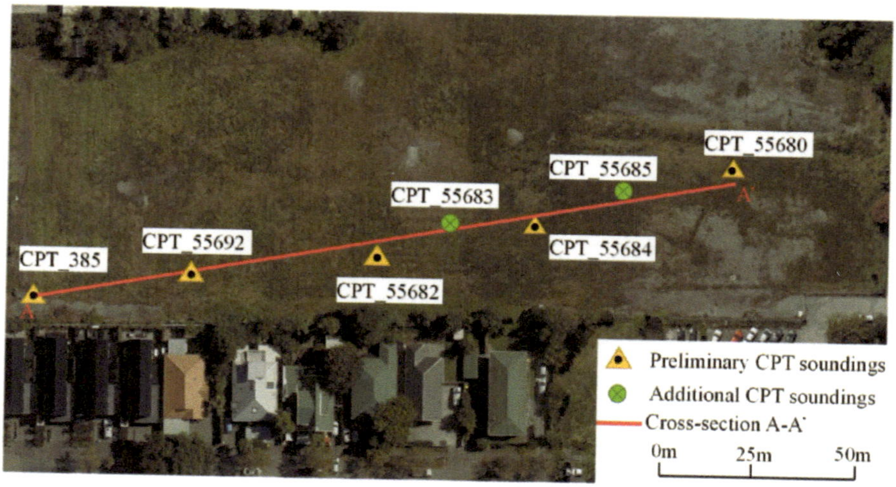

FIGURE 6.25 Layout of seven CPT soundings determined using the proposed sampling strategy at a site in Christchurch, New Zealand (after Guan et al. 2022a).

REFERENCES

Bradley, B. A., and M. Cubrinovski. 2011. Near-source strong ground motions observed in the 22 February 2011 Christchurch earthquake. *Seismological Research Letters* 82 (6): 853–865.

Candes, E. J., J. K. Romberg, and T. Tao. 2006. Stable signal recovery from incomplete and inaccurate measurements. *Communications on Pure and Applied Mathematics* 59 (8): 1207–1223.

Cover, T. M., and J. A. Thomas. 2012. *Elements of information theory*. John Wiley & Sons.

Cubrinovski, M. 2013) Liquefaction-induced damage in the 2010–2011 Christchurch (New Zealand) earthquakes. In S. Prakash (Ed.), Proceedings of the Seventh International Conference on Case Histories in Geotechnical Engineering (Paper No. 1). Missouri University of Science and Technology.

Dietrich, C. R., and G. N. Newsam. 1993. A fast and exact method for multidimensional Gaussian stochastic simulations. *Water Resources Research* 29 (8): 2861–2869.

Fang, Y., J. Wu, and B. Huang. 2012. 2D sparse signal recovery via 2D orthogonal matching pursuit. *Science China Information Sciences* 55 (4): 889–897.

Federal Highway Administration (FHWA). 2017. *Geotechnical site characterization*. Geotechnical Engineering Circular No. 5, Publication No. NHI-16-072. Washington, DC: U.S. Department of Transportation.

Guan, Z., and Y. Wang. 2022a. Assessment of liquefaction-induced differential ground settlement and lateral displacement using standard penetration tests with consideration of soil spatial variability. *Journal of Geotechnical and Geoenvironmental Engineering* 148 (5): 04022018.

Guan, Z., and Y. Wang. 2022b. Quantifying reliability of liquefaction severity map developed from sparse cone penetration tests. *Canadian Geotechnical Journal* 60 (5): 623–641.

Guan, Z., and Y. Wang. 2022c. SPT-based probabilistic evaluation of soil liquefaction potential considering design life of civil infrastructures. *Computers and Geotechnics* 148: 104807.

Guan, Z., and Y. Wang. 2024. Risk-informed adaptive sampling strategy for liquefaction severity mapping. *Georisk: Assessment and Management of Risk for Engineered Systems and Geohazards* 18 (2): 526–539.

Guan, Z., Y. Wang, and K. K. Phoon. 2023. Fusion of sparse non-co-located measurements from multiple sources for geotechnical site investigation. *Canadian Geotechnical Journal* 61 (8): 1574–1592.

Guan, Z., Y. Wang, and T. Zhao. 2022a. Adaptive sampling strategy for characterizing spatial distribution of soil liquefaction potential using cone penetration test. *Journal of Rock Mechanics and Geotechnical Engineering* 14 (4): 1221–1231.

Guan, Z., Y. Wang, Z. Cao, and Y. Hong. 2020a. Smart sampling strategy for investigating spatial distribution of subsurface shallow gas pressure in Hangzhou Bay area of China. *Engineering Geology* 274: 105711.

Guan, Z., Zhao, T., & Wang, Y. 2020b. Smart sampling strategy for geotechnical site characterization. In J. P. Hambleton, R. Makhnenko, & A. S. Budge (Eds.), Geo-Congress 2020: Engineering, Monitoring, and Management of Geotechnical Infrastructure (pp. 728–736). Reston, VA: American Society of Civil Engineers.

Hight, D. W., F. McMillan, J. J. M. Powell, R. J. Jardine, and C. P. Allenou. 2002. Some characteristics of London clay. In *Characterisation and Engineering Properties of Natural Soils*, edited by T. S. Tan, K. K. Phoon, D. W. Hight, and S. Leroueil, 851–908. Rotterdam, the Netherlands: Balkema.

Liao, S., and R. A. Whitman. 1986. *Catalog of liquefaction and non-liquefaction occurrences during earthquakes*. Massachusetts Institute of Technology. Department of Civil Engineering, Cambridge.

MacKay, D. J. 2003. *Information theory, inference and learning algorithms*. Cambridge University Press.

New Zealand Geotechnical Database (NZGD). 2012. www.nzgd.org.nz (accessed March 20, 2023).

Petersen, K. B., and M. S. Pedersen. 2008. *The matrix cookbook*. Technical University of Denmark, Version 20081110.

Phoon, K. K., and F. H. Kulhawy. 1999. Characterization of geotechnical variability. *Canadian Geotechnical Journal* 36 (4): 612–624.

Robertson, P. K., and C. E. Wride. 1998. Evaluating cyclic liquefaction potential using the cone penetration test. *Canadian Geotechnical Journal* 35 (3): 442–459.

Seed, H. B., and I. M. Idriss. 1971. Simplified procedure for evaluating soil liquefaction potential. *Journal of Soil Mechanics and Foundations Division* 97 (9): 1249–1273.

Shi, C., Y. Wang, and V. Kamchoom. 2023. Data-driven multi-stage sampling strategy for a three-dimensional geological domain using weighted centroidal voronoi tessellation and ic-xgboost3d. *Engineering Geology* 325: 107301. https://doi.org/10.1016/j.enggeo.2023.107301

Tipping, M. E. 2001. Sparse Bayesian learning and the relevance vector machine. *Journal of Machine Learning Research* 1: 211–44.

Tobler, W. 2004. On the first law of geography: A reply. *Annals of the Association of American Geographers* 94 (2): 304–310.

Vanmarcke, E. 1977. Probabilistic s *Journal of Geotechnical and Geoenvironmental Engineering* 103 (11): 1227–46.

Wang, Y., and P. Li. 2021. Data-Driven Determination of Sample Number and Efficient Sampling Locations for Geotechnical Site Investigation of a Cross-Section Using Voronoi Diagram and Bayesian Compressive Sampling. *Computers and Geotechnics* 130: 103898.

Wang, Y., and T. Zhao. 2017. Statistical interpretation of soil property profiles from sparse data using Bayesian compressive sampling. *Géotechnique* 67 (6): 523–36.

Wang, Y., Z. J. Cao, and S. K. Au. 2010. Efficient Monte Carlo simulation of parameter sensitivity in probabilistic slope stability analysis. *Computers and Geotechnics* 37 (7–8): 10.

Wang, Y., Z. Cao, and D. Li. 2016. Bayesian Perspective on Geotechnical Variability and Site Characterization. *Engineering Geology* 203: 117–25.

Youd, T. L., I. M. Idriss, R. D. Andrus, I. Arango, G. Castro, J. T. Christian, R. Dobry, W. D. Liam Finn, L. F. Harder Jr., M. E. Hynes, K. Ishihara, J. P. Koester, S. S. C. Laio, W. F. Marcuson III, G. R. Martin, J. K. Mitchell, Y. Moriwaki, M. S. Power, P. K. Robertson, R. B. Seed, K. H. Stokoe. 2001. Liquefaction Resistance of Soils, Summary Report from the 1996 NCEER and 1998 NCEER/NSF Workshops on Evaluation of Liquefaction Resistance of Soils. *J. Geotech. Geoenviron. Eng.* 127 (10): 817–33.

Zhao, T., and Y. Wang. 2019. Determination of efficient sampling locations in geotechnical site characterization using information entropy and Bayesian compressive sampling. *Canadian Geotechnical Journal* 56 (11): 1622–37.

Zhao, T., Y. Hu, and Y. Wang. 2018. Statistical interpretation of spatially varying 2D geo-data from sparse measurements using Bayesian compressive sampling. *Engineering Geology* 246: 162–175.

Zhao, T., Y. Wang, and L. Xu. 2021. Efficient CPT locations for characterizing spatial variability of soil properties within a multilayer vertical cross-section using information entropy and Bayesian compressive sensing. *Computers and Geotechnics* 137: 104260.

Zhao, T., L. Xu, and Y. Wang. 2020. Fast non-parametric simulation of 2D multi-layer cone penetration test (CPT) data without pre-stratification using Markov Chain Monte Carlo simulation. *Engineering Geology* 273: 105670.

7 Software ASSD-BCS

Yue Hu and Yu Wang

ABSTRACT

The background, algorithms, and various applications of Bayesian compressive sampling (BCS) have been explored in previous chapters. This chapter introduces a computer software package for implementing BCS using the Python programming language, providing readers with a practical and user-friendly tool for data analytics and simulation in geotechnical site characterization. The chapter contains three parts: (1) an introduction to Python, one of the most widely used programming languages, followed by the implementation of key BCS algorithms using Python code; (2) a presentation of the software called Analytics of Sparse Spatial Data using Bayesian Compressive Sampling (ASSD-BCS) and its graphical user interface; and (3) demonstrations of ASSD-BCS through various tutorial examples, including some examples from previous chapters. ASSD-BCS is an open-access software and available for free download online, enabling users to gain hands-on experience in applying BCS to geotechnical site characterization.

7.1 PYTHON PROGRAMMING

Python is a popular, interpreted, and object-oriented programming language first released in 1991. It is a general-purpose scripting language that is widely applied in many fields. For example, it has been frequently used in web development, data analysis, geostatistical modeling, and artificial intelligence, among many other fields. Python comes with a large collection of prebuilt and portable functionality known as "standard library", using which many common tasks can be realized, such as accessing web servers, searching text with regular expressions, reading and modifying files, etc. In Python, library is a collection of different packages or modules that are pre-written pieces of codes and can be easily imported into a Python program without having to write everything from scratch. In addition, Python can be extended readily with both locally developed libraries and a vast collection of third-party libraries developed and maintained by the Python community. Examples include the Django used for web development, Pandas, NumPy and SciPy for data analysis and scientific research, TensorFlow and PyTorch for artificial intelligence research, among others. Various reference books about Python are available for users at different levels (e.g., Lutz 2013).

DOI: 10.1201/9781003381235-7

```
>>> def fib(n):
>>>   a, b = 0, 1
>>>   while a < n:
>>>   print(a, end=' ')
>>>   a, b = b, a + b

>>> fib(1000)
>>> 0 1 1 2 3 5 8 13 21 34 55 89 144 233 377 610 987
```

FIGURE 7.1 Example of defining and calling a function in Python.

Python generally focuses more on readability, coherence, and software quality than other scripting languages. Readers with experiences in any other programming languages can follow up Python programming easily. Python is designed with a clean and easily readable syntax. It commonly uses four-space indentation to define code blocks without using curly brackets or "end" statements, making it highly readable and reducing the need for excessive punctuation. Consider, for example, a simple function "fib", as shown in Figure 7.1, for generating a Fibonacci sequence up to a specified number "n". In a Fibonacci sequence, the first two elements are 0 and 1, respectively. Then, the following element is the sum of the two elements that precede it. A clear code structure is shown in Figure 7.1. The first line of the function includes a keyword of definition "`def`" followed by the name of function "`fib`". The input parameter "`n`" is placed in the parenthesis. The body of the "`fib`" function is indented just after the line of definition. The variables "`a`" and "`b`" denote the first two items of Fibonacci sequence (i.e., `a`=0, `b`=1), respectively. The variables "`a`" and "`b`" can be conveniently declared in just one line of code using a single assignment operator "`=`". In the function body of "`fib`", the body of "`while`" loop is again indented just after the "`while`" loop statements. While the variable "`a`" is smaller than the specified number "`n`", the body of the while loop continues: the latest value of "`a`" is shown by calling a built-in function "`print`" and the values of "`a`" and "`b`" are reassigned by the values of "`b`" and "`a + b`". A call of the "`fib`" function for generating a Fibonacci sequence up to 1000 is also shown in the last two lines of Figure 7.1.

7.2 IMPLEMENTATION OF OMP ALGORITHM IN PYTHON

As described in Chapters 2 and 3, the orthogonal matching pursuit (OMP) algorithm plays an important role in solving CS-based formulation for geotechnical site characterization. In this section, a general OMP algorithm described in Chapter 2 is implemented using Python for solving a sparse vector $\boldsymbol{\omega}$, or equivalently $\boldsymbol{f}$, from limited measurement data $\boldsymbol{y}$. In Figure 7.2, a code snippet of a user-defined function "`omp`" for implementing OMP is demonstrated. Each line or block of code is preceded with a line of brief explanation starting with "#". The Python interpreter will skip these lines starting with "#" symbol during execution. First, the library "`numpy`" needs to be imported preliminarily for the "`omp`" function because a series of matrix and vector operations is required in the "`omp`". The library "`numpy`" is therefore declared in the first line and can be abbreviated as "`np`". After importing the library "`numpy`", it can be called as "np" subsequently. The input parameters to this function include the measurement data "y" which is an $M\times1$ vector for one-dimensional (1D) data, the measurement locations "idy" which is also an $M\times1$ vector, the basis function "B" which is an $N\times N$ matrix, and the stopping criteria "stc" of OMP. These input parameters should be correctly configured before implementing this function. The "n" is the length of the target vector $\boldsymbol{f}$ and can also be defined by the number of rows of the basis function matrix "B".

As highlighted in Chapters 2 and 3, in the context of geotechnical data interpolation and simulation, the original vector $\boldsymbol{f}$ is often not available and therefore a specific design of sensing matrix $\boldsymbol{\Psi}$ (e.g., Gaussian random matrix) is not feasible. Therefore, in this "omp" function, the associated sensing matrix "phi" is adapted from an identity matrix "c" by selecting rows from the identity matrix "c" based on the indices given in "idy". Subsequently, the measurement matrix "A" is constructed by multiplying the sensing matrix "phi" with the basis function matrix "B". "A_iter" is a copy of "A" that is subsequently modified during iterations. "As" is an empty matrix which is used to store those selected columns from "A" during the iterations. "pv" stores the coefficient values, and "pvid" stores the indices of the selected columns. "rel" stores the residuals' evolution, tracking how the residual changes over iterations. The regression residual "res" is set to the measurement vector "y" initially. "times" keeps track of the number of iterations. "ss" computes the norm of each column in matrix "A" and is used to normalize the cosine similarity. "t_max" is the maximum number of iterations which is just the number of columns in the basis matrix "B". "con_res" and "con_iter" are two control flags, indicating whether to continue iterations based on the residual and iteration count. The "while" loop continues as long as both control flags are True.

Within the "while" loop, "cos_sim" measures the similarity between the columns of "A_iter" and the residual "res". The "id_descend" holds the indices of the columns sorted by their similarity to the residual. Then, the column of "A_iter" with the highest similarity is added to "As". The selected column in "A_iter" is set to zero to prevent the same column from being selected again. Using the least squares method, the coefficient vector "pv" is computed for the selected columns in "As". Accordingly, the index of the selected column is added to "pvid". The new residual is then computed by subtracting the projection of "y" onto the selected columns from actual measurement data "y". According to the status of "con_res" and "con_iter", the function finally returns the coefficient vector "pv", the indices of the selected columns "pvid", the evolution of the residual "rel", and the number of iterations "times".

Figure 7.2 and the paragraphs above illustrate the "omp" function for solving the underdetermined system in CS as described by Eq. (2.4). The outputs of the "omp" function are a prerequisite for reconstructing the data profile $\boldsymbol{f}$ in CS and subsequent uncertainty quantification in BCS. Many functions, similar to the "omp" function, are needed for the implementation of the CS and BCS algorithms described in the previous chapters. To facilitate practical applications of the CS and BCS for a wide range of audience, particularly those without extensive programming or mathematical backgrounds, a user-friendly software package called Analytics of Sparse Spatial

```
import numpy as np
# OMP algorithm
def omp(y, idy, B, stc):
   # y: M×1 measurement data vector
   # idy: M×1 measurement locations vector
   # B: N×N basis function matrix
   # stc: stopping criteria
   # get the target spatial resolution
   n = B.shape[0]
   # identity matrix
   c = np.identity(n)
   # sensing matrix
   phi = c[idy, :]
   # Matrix A in Equation (2.3)
   A = phi @ B
   A_iter = phi @ B
   # iteration
   # initiate a matrix for storing columns from A
   As = np.empty([A.shape[0], 0])
   # coefficient values and corresponding IDs
   pv, pvid = np.empty([0]), np.empty([0]).astype(int)
   # evolution of residuals
   rel = np.empty([0])
   # initiate a residual as y
   res = y.copy()
   # epoch of iteration
   times = 0
   # norm of A matrix
   ss = np.sqrt(np.sum(A * A, axis=0))
   # the iteration stops when all columns in B are selected
   t_max = B.shape[1]
   # controller
   con_res = True
   con_iter = True
   # iterative algorithm
   while con_res and con_iter:
      cos_sim = (np.abs(A_iter.T @ res) /
                 (np.linalg.norm(res) * ss).T)
      id_descend = cos_sim.argsort()[::-1]
      As = np.append(As, A_iter[:, [id_descend[0]]], axis=1)
      A_iter[:, id_descend[0]] = 0
      pv = np.linalg.inv(As.T @ As) @ As.T @ y
      pvid = np.append(pvid, [id_descend[0]], axis=0)
      res = y - A[:, pvid[:].astype(int)] @ pv
      rel = np.append(rel,
                      [(np.linalg.norm(res)
                       / np.linalg.norm(y))],
```

FIGURE 7.2 Implementation of OMP algorithm using Python.

```
                                axis=0)
        times = times + 1
        con_res = (np.linalg.norm(res) / np.linalg.norm(y)) > stc
        con_iter = times < t_max
    return pv, pvid, rel, times
```

FIGURE 7.2 Implementation of OMP algorithm using Python. (Continued)

Data using Bayesian Compressive Sampling (ASSD-BCS) is developed, together with a graphical user interface (GUI), in this chapter. This software package integrates all necessary functionalities and intermediate steps of the CS and BCS algorithms into an intuitive and easy-to-use GUI. The following section introduces ASSD-BCS in detail.

7.3 INTRODUCTION OF ASSD-BCS

The objective of ASSD-BCS software is to provide a user-friendly software package for researchers and practitioners to implement CS and BCS algorithms for geotechnical data analytics and associated random simulation (Lyu et al. 2023). This section introduces the installation of the software ASSD-BCS and its GUI. The GUI allows users to interact with computers through graphical elements such as windows, icons, and buttons, rather than text-based commands. GUIs are designed to be user-friendly, making it easy for individuals with little programming knowledge to operate software applications. The ASSD-BCS GUI is developed in Python environment and has been compiled to a stand-alone application for Microsoft Windows operating system. The installation file "ASSD-BCS-1.2-win64.msi" of ASSD-BCS (version 1.2) can be downloaded for free from the link: https://sites.google.com/site/yuwangcityu/software-download/bayesian-compressive-samplingsensing-bcs. When one double-clicks the "ASSD-BCS-1.2-win64.msi" file, the installation program starts, and a setup window pops up, as shown in Figure 7.3.

After configuring the local installation path, the installation process initiates after clicking the "Next" button, as shown in Figure 7.4. After the installation and configuration are completed, an executable file of ASSD-BCS software is ready in the specified installation directory and can be opened directly. The operating window of ASSD-BCS is shown in Figure 7.5. The software window has five tabs at the top, i.e., Data-driven Interpolation, BCS Random Field Simulation, Help, About, and References. The Help tab of ASSD-BCS is shown in Figure 7.5. The Help tab provides brief introductions to the ASSD-BCS, the Data-driven Interpolation module with its data format requirements, and the BCS Random Field Simulation module, helping users to get started with the software.

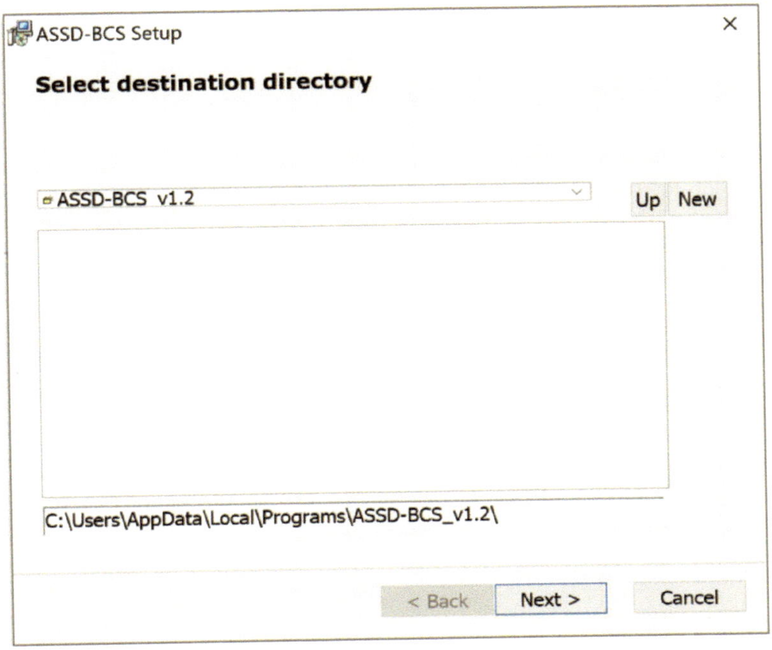

FIGURE 7.3 Windows Installer of ASSD-BCS.

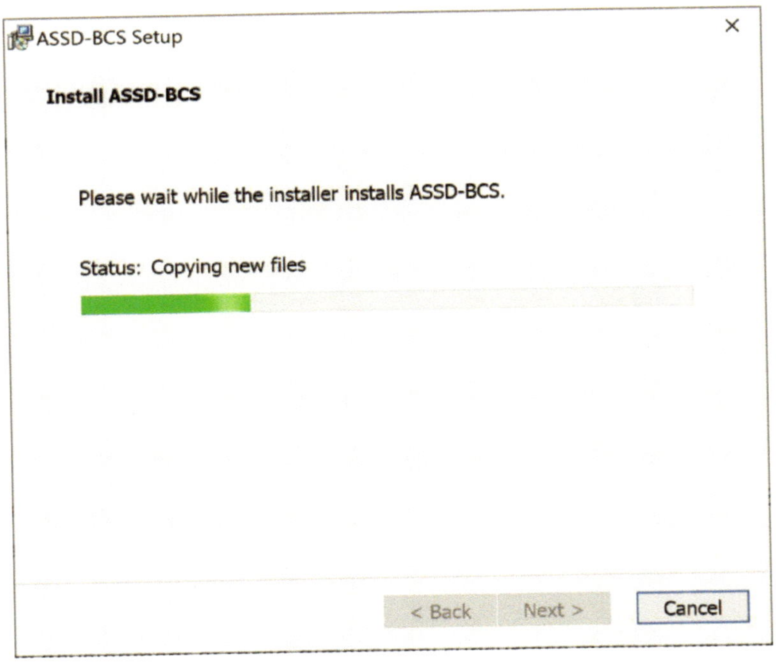

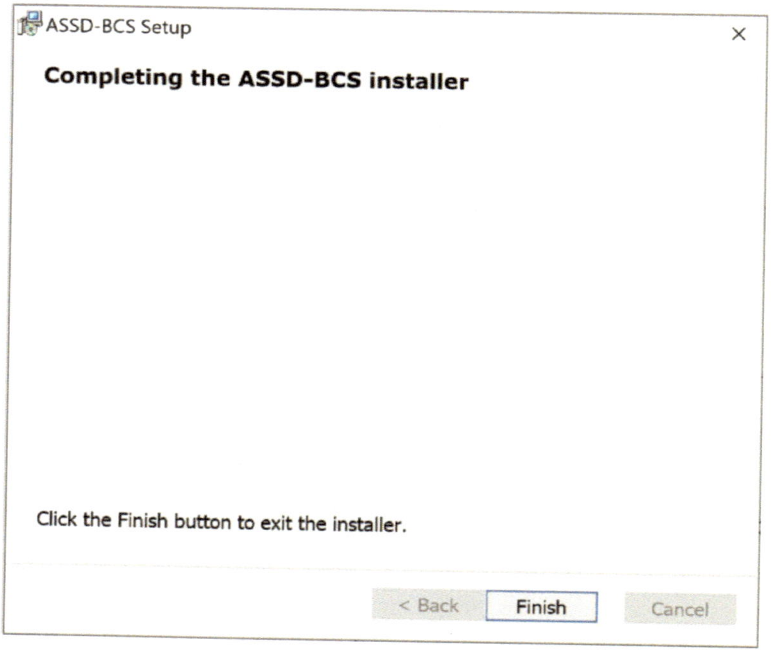

FIGURE 7.4 Installation process of ASSD-BCS.

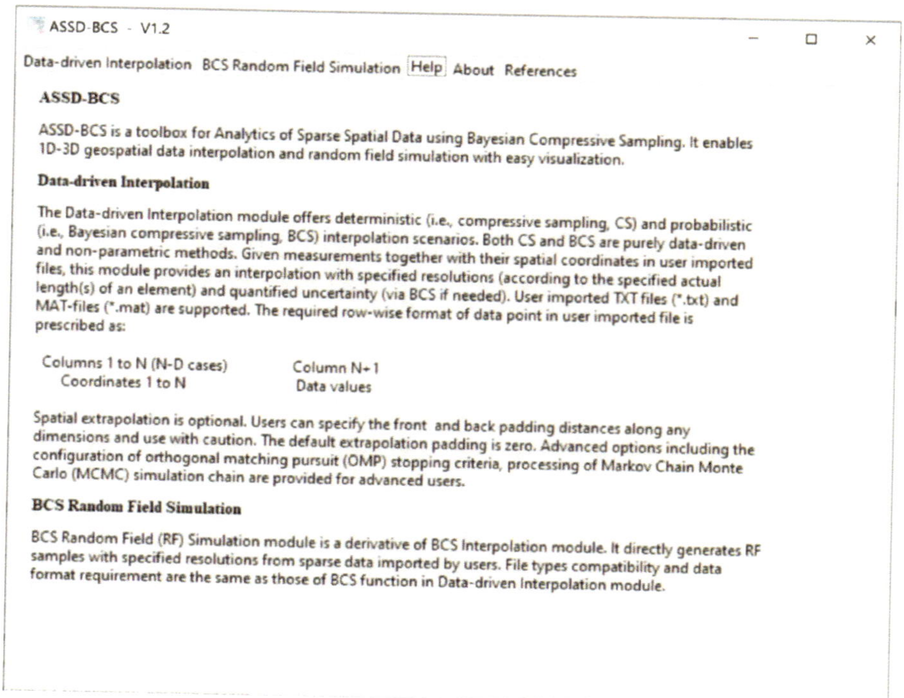

FIGURE 7.5 Graphical user interface of ASSD-BCS.

7.4 ASSD-BCS FUNCTIONS

ASSD-BCS has two basic functions, i.e., Data-driven Interpolation, and BCS Random Field Simulation, as supported by the first two tabs in Figure 7.5. Figure 7.6 shows the Data-driven Interpolation module for interpolating 1D to three-dimensional (3D) sparse spatial data using CS or BCS (e.g., Wang and Zhao 2016&2017; Zhao et al. 2018; Hu et al. 2020; Zhao and Wang 2020&2021). In this tab, there are six available frames to determine the required input, including "Compressive Sampling" task selection, "Bayesian Compressive Sampling" task selection, "Load Input Data File", "Actual Length of Element", "Extrapolation", and "Advanced Options". At the bottom of this frame, there are four buttons, i.e., Run, Save, Plot, and Reset. The first two frames allow users to select the target dimensionality for CS or BCS interpolation. The "Load Input Data File" frame allows users to specify the path of the input data file. The "Actual Length of Element" frame enables users to specify the spatial resolution of interpolated data (i.e., the actual length represented by one interpolated element) along a corresponding dimension. The unit of the spatial resolution is identical to that of the data coordinate in the input data file. The "Extrapolation" frame allows users to extrapolate data outside the range of input measurement data. The "front padding" and the "back padding" enable users to specify the extrapolation extents along both sides of a corresponding dimension. The units of the extrapolation extents are identical to that of the data coordinate in the input data file. The "Advanced

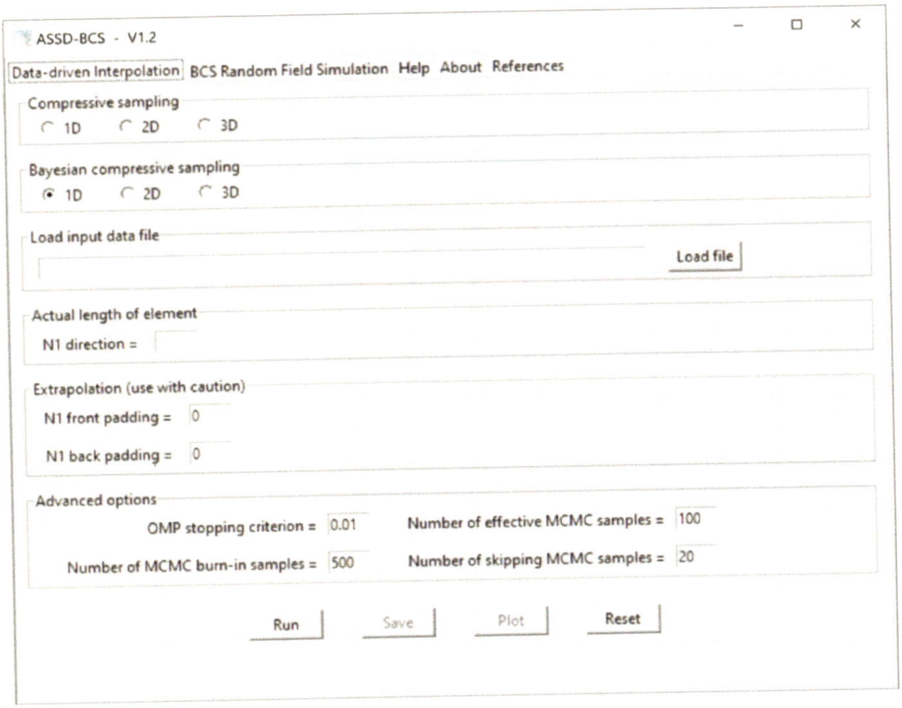

FIGURE 7.6 Data-driven Interpolation module of ASSD-BCS.

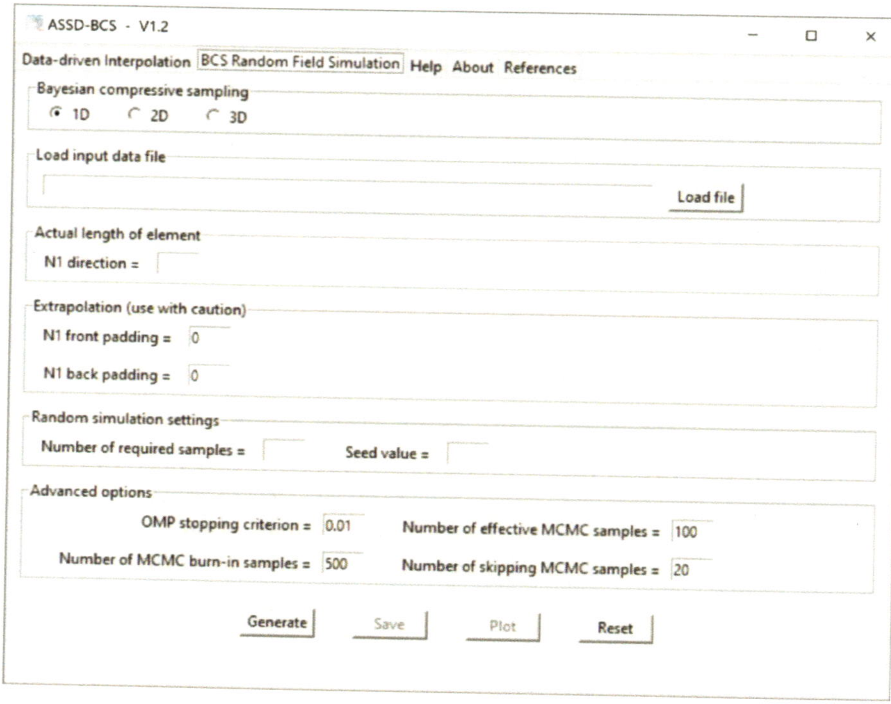

FIGURE 7.7 Random Field Simulation module of ASSD-BCS.

Options" frame provides some advanced settings of CS and BCS algorithms. The "BCS Random Field Simulation" module, as shown in Figure 7.7, is analogous to BCS interpolation but provides another frame "Random Simulation Settings" that enables users to specify the number of required random field samples (RFSs) generated directly from sparse measurements. (e.g., Wang et al. 2018; Hu et al. 2019).

The core of ASSD-BCS is based on discrete cosine transform (DCT) basis function for data compression and reconstruction. The software supports the implementation of three distinct tasks in BCS. The first task is to identify non-trivial DCT weight coefficients of the data using OMP, as illustrated in Figure 7.2. Then, the second task is to estimate the values of identified non-trivial weight coefficients using the least squares method for CS or Markov Chain Monte Carlo (MCMC) simulation for BCS. The third task is to reconstruct the interpolated data or generate RFSs. The details of CS and BCS algorithms are already discussed in Chapters 3 and 4.

7.5 DEFAULT SETTINGS AND USEFUL TIPS

To provide users with a clear understanding of the default configurations, ensuring a smooth and hassle-free experience with the ASSD-BCS software, default settings and some useful tips are introduced here.

1 The format of data points in the input data file should strictly follow the row-wise format prescribed in the Help tab (e.g., see the format of user_1d.txt in

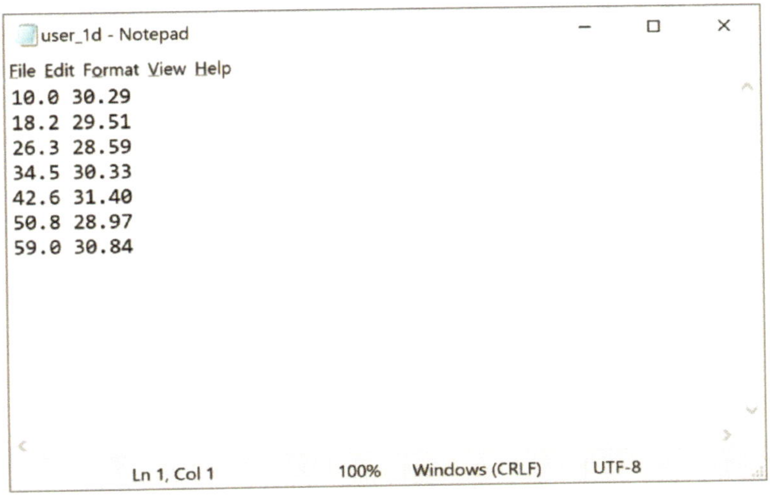
user_1d - Notepad

File Edit Format View Help

```
10.0 30.29
18.2 29.51
26.3 28.59
34.5 30.33
42.6 31.40
50.8 28.97
59.0 30.84
```

Ln 1, Col 1 100% Windows (CRLF) UTF-8

FIGURE 7.8 Input data file of tutorial example user_1d.txt.

Figure 7.8). Both the Text file (*.txt) and the MATLAB™ MAT-file (*.mat) are supported for input data file.

2 The units of both the coordinates and the measured values should not be included in the input data file.
3 Extrapolation should be used with caution. Both the front padding and back padding along each dimension are set to zeros by default.
4 In "Advanced options", the "OMP stopping criterion" is the convergence criterion suggested for OMP algorithm under CS framework and is set to 0.01 by default. The "Number of effective MCMC samples" indicates the number of effective samples in MCMC simulation and is set to 100 by default. The "Number of MCMC burn-in samples" indicates the number of MCMC samples to be discarded at the beginning of MCMC simulation and is set to 500 by default. The "Number of MCMC skipping samples" indicates the thinning of MCMC samples for reducing the autocorrelation of MCMC samples and is set to 20 by default. Details of MCMC simulations are provided in Chapter 4.
5 The array size (i.e., the total number of elements) of the interpolated array should not exceed 10,000,000 to avoid the out of memory error.

7.6 TUTORIAL EXAMPLES: INTERPOLATION OF SPARSE MEASUREMENTS

To facilitate the illustration of the Data-driven Interpolation module, a series of tutorial examples, including 1D, two-dimensional (2D), and 3D spatial data examples, are demonstrated in this section. Step-by-step implementation procedures are described for interpolation of the first 1D tutorial example. Additional tutorial examples, including several simulated data examples presented in previous chapters, are also presented in this section.

7.6.1 1D Data

A 1D tutorial example "user_1d.txt" has been provided to users in the ASSD-BCS installation package. As shown in Figure 7.8, the measurement data of this tutorial example contain seven row-wise data points stored in a Text file (*.txt). The seven data points can be interpreted as sparse measurements of one physical quantity along a line. Consider, for example, soil undrained shear strength in a unit of kPa varying along depth in a borehole in a unit of meter. The first and second columns are measurement coordinates and corresponding measured values, respectively. The implementation of Data-driven Interpolation using ASSD-BCS is straightforward and described as follows:

Step 1: Double-click the executable of the package, i.e., ASSD-BCS.exe, to open the user interface.

Step 2: Determine which algorithm to use and which domain to apply for interpolation, e.g., click the "1D" radio button inside the frame "Bayesian compressive sampling".

Step 3: Load input data file into the software, i.e., click the "Load file" button inside the frame "Load input data file" and then select one local file storing the information of sparse measurements. Consider, for example, loading the tutorial example "user_1d.txt" (see Figure 7.8). After that, the path of the selected file will display on the left.

Step 4: Specify the resolution along each dimension (e.g., N1 direction) inside the frame "Actual length of element". The input value indicates the actual spatial length that each element represents along the corresponding dimension. The unit of specified resolution is the same as the coordinate unit used in the input data file. For example, a target resolution for interpolation can be set to 1.

Step 5: Specify the extent of extrapolation in the frame "Extrapolation". The input values indicate the padding lengths outside spatial range of the input data along a specific dimension. The unit of the specified padding lengths is the same as the coordinate unit used in the input data file. The default values of extrapolation extent are zeros.

Step 6: Confirm the parameters listed in the frame "Advanced options".

Step 7: Click the "Run" button to start interpolation program. After the interpolation is completed, the "Save" and "Plot" buttons will be activated.

Step 8: Click the "Save" button to save the interpolation results in a new Text file under the root folder of the software. Figure 7.9 shows the saved interpolation results, which include three individual Text files. In Figure 7.9a, the interpolated coordinates with a specified resolution of 1 are shown in the Text file "bcs_coordinate.txt". Text files "bcs_best_estimate.txt" and "bcs_standard_deviation.txt" in Figures 7.9b and 7.9c show the best estimate data profile and the standard deviation profile of BCS interpolation, respectively. Click the "Plot" button to plot the interpolation results (e.g., see Figure 7.10). The best estimate of BCS interpolation is denoted by a red solid line. The statistical uncertainty profiles denoted by mean ± one standard deviation are illustrated by two gray dashed

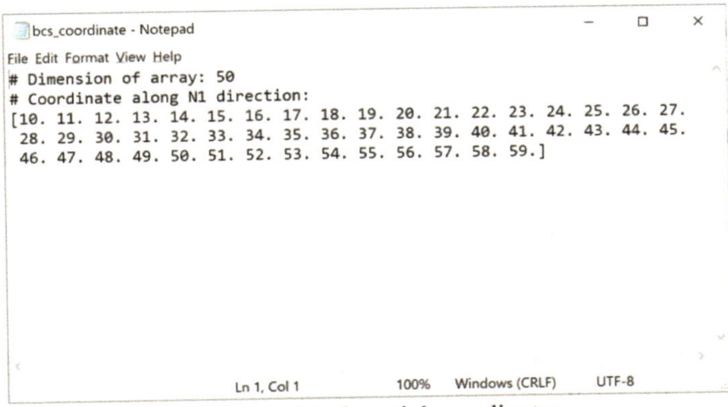

(a) Interpolated spatial coordinates

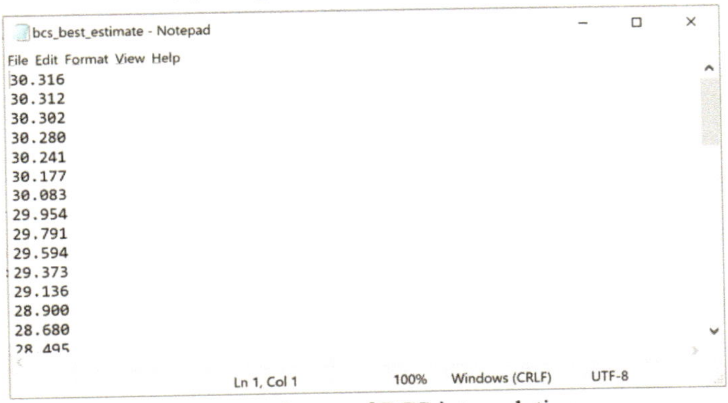

(b) Best estimate of BCS interpolation

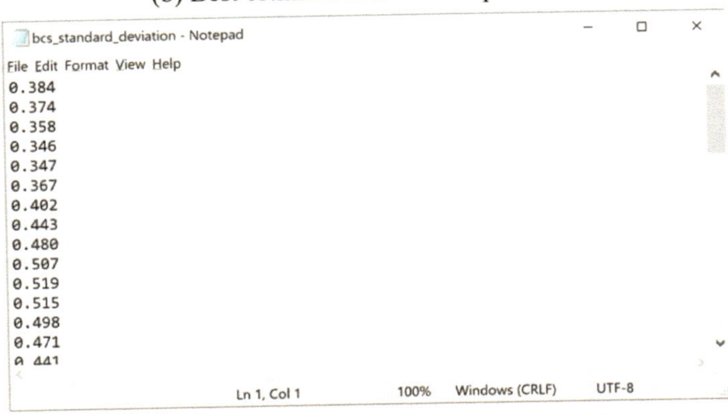

(c) One standard deviation of BCS interpolation

FIGURE 7.9 Saved BCS interpolation results for tutorial example user_1d.txt.

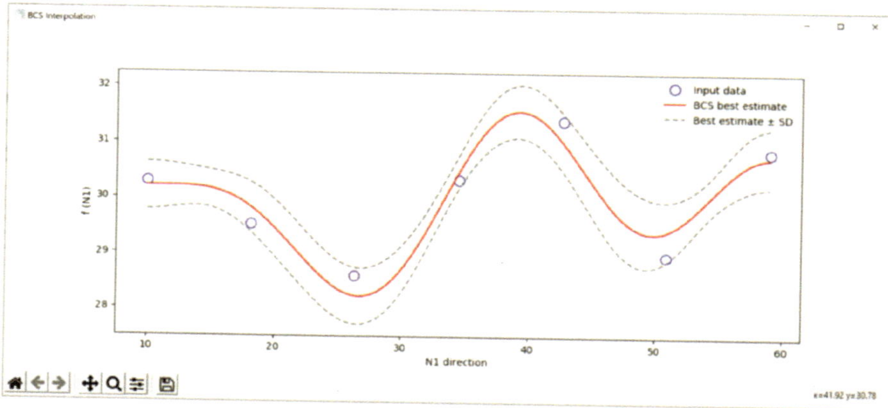

FIGURE 7.10 BCS interpolation results plot of tutorial example user_1d.txt.

lines. The generated plot can also be saved as an image file by clicking the floppy disk button at the bottom of the window.

Step 9: Click the "Reset" button to start another new task, if needed.

7.6.2 2D Data

For 2D spatial data, the input file should include coordinate information for both dimensions of the measurement data. A 2D tutorial example "user_2d.txt" is also provided to users in the ASSD-BCS installation package. As shown in Figure 7.11, the measurement data of this tutorial example contain 25 row-wise data points stored in a Text file (*.txt). The first two columns correspond to the 2D coordinates, while the third column corresponds to the measured values of interest. The implementation procedures are similar to the 1D data example described above, but the interpolation domain is selected as "2D". Consider, for example, click the "2D" radio button inside the frame "Bayesian compressive sampling", load the input data file, specify a target interpolation resolution for both dimensions as "1", keep the other default settings, and click the "Run" button to start interpolation program. The interpolation results can be visualized after clicking the "Plot" button, as shown in Figure 7.12. The sparse measurements in the input data file are represented by color-scaled dots in the left subplot. The best estimate and standard deviation are shown in the middle and right subplots by colormaps, respectively.

7.6.3 3D Data

The ASSD-BCS also supports 3D spatial data interpolation. The simulated 3D data example illustrated in Figure 3.17 of Chapter 3 is demonstrated in this section using the ASSD-BCS. In this 3D data example, ten columns with nonuniform horizontal spacings are sampled from a simulated 3D data array with a dimension of 128×128×16. Ten data points with nonuniform vertical spacings are taken from each

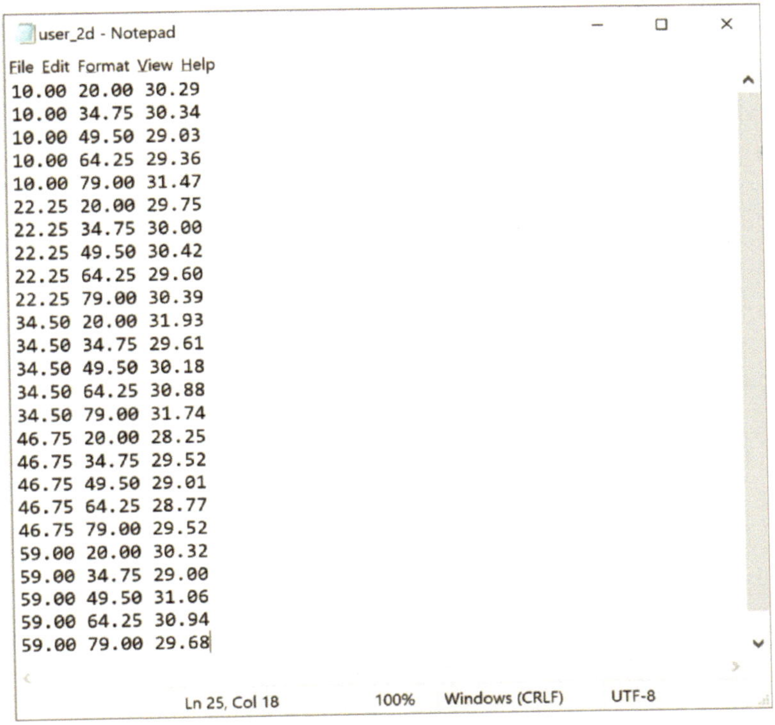

user_2d - Notepad

File Edit Format View Help

```
10.00 20.00 30.29
10.00 34.75 30.34
10.00 49.50 29.03
10.00 64.25 29.36
10.00 79.00 31.47
22.25 20.00 29.75
22.25 34.75 30.00
22.25 49.50 30.42
22.25 64.25 29.60
22.25 79.00 30.39
34.50 20.00 31.93
34.50 34.75 29.61
34.50 49.50 30.18
34.50 64.25 30.88
34.50 79.00 31.74
46.75 20.00 28.25
46.75 34.75 29.52
46.75 49.50 29.01
46.75 64.25 28.77
46.75 79.00 29.52
59.00 20.00 30.32
59.00 34.75 29.00
59.00 49.50 31.06
59.00 64.25 30.94
59.00 79.00 29.68
```

Ln 25, Col 18 100% Windows (CRLF) UTF-8

FIGURE 7.11 Input data file of tutorial example user_2d.txt.

column as measurement data points. This leads to 100 measurement data points in total. The first 15 row-wise data points out of the 100 measurements are illustrated in Figure 7.13 in a Text file "chapter_3_3D.txt". The first three columns represent the coordinates for the three dimensions, respectively. The fourth column is the measured values. Follow the procedures described above by loading the Text file into ASSD-BCS software and specifying a spatial resolution of "1" for all three directions. Then, BCS interpolation with default setting can be performed by clicking the "Interpolate" button. The BCS interpolation results can be visualized instantly after the BCS program is executed. The 100 measurement data points as input, together with 3D best estimate and associated standard deviation as output, are visualized quickly in a pop-up window, as shown in Figure 7.14.

7.7 TUTORIAL EXAMPLES: RANDOM FIELD MODELING FROM SPARSE MEASUREMENTS

To facilitate an illustration of the BCS Random Field Simulation module, two tutorial spatial data examples are demonstrated in this section. Step-by-step implementation procedures are described for the 2D simulated example in Figure 6.6 discussed in Chapter 6, followed by an illustration of another tutorial 3D example.

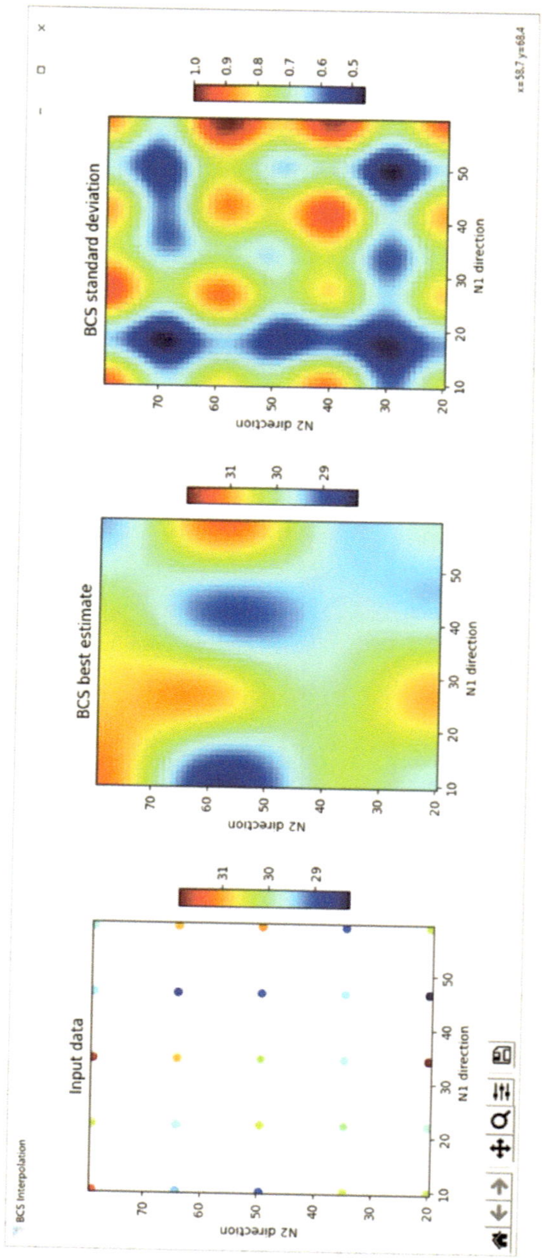

FIGURE 7.12 BCS interpolation results plot of tutorial example user_2d.txt.

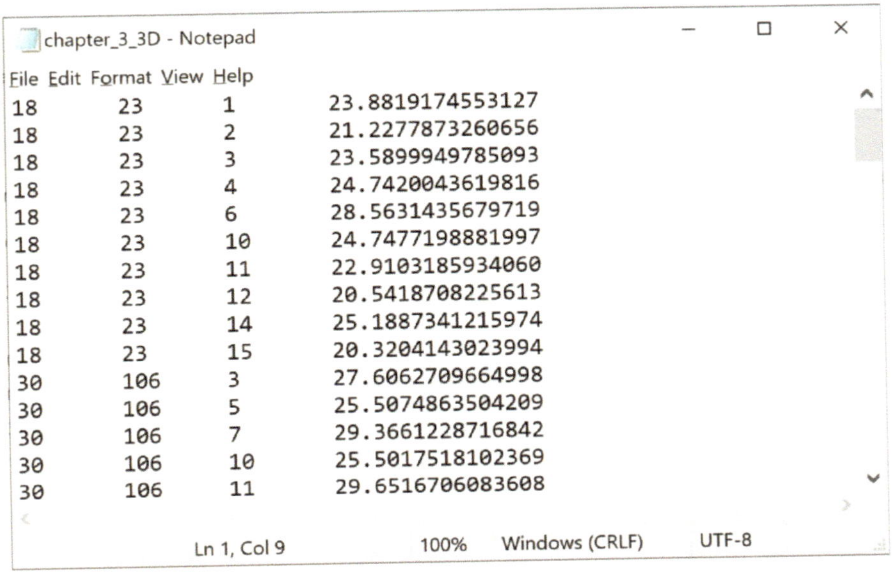

chapter_3_3D - Notepad

File Edit Format View Help

18	23	1	23.8819174553127
18	23	2	21.2277873260656
18	23	3	23.5899949785093
18	23	4	24.7420043619816
18	23	6	28.5631435679719
18	23	10	24.7477198881997
18	23	11	22.9103185934060
18	23	12	20.5418708225613
18	23	14	25.1887341215974
18	23	15	20.3204143023994
30	106	3	27.6062709664998
30	106	5	25.5074863504209
30	106	7	29.3661228716842
30	106	10	25.5017518102369
30	106	11	29.6516706083608

Ln 1, Col 9 100% Windows (CRLF) UTF-8

FIGURE 7.13 The first 15 rows of simulated 3D example in Figure 3.17.

7.7.1 2D Data

The implementation of BCS Random Field Simulation is similar to that of Data-driven Interpolation. However, in BCS Random Field Simulation, a number of RFSs can be generated directly for subsequent stochastic analysis, rather than profiles of statistical characterization as illustrated in Figure 7.10. For illustration, the implementation procedures of BCS Random Field Simulation are described using an example of a simulated CPT data cross-section. The simulated CPT data cross-section of $(q_{c1N})_{cs}$ is taken from Figure 6.6 of Chapter 6. It has a depth of 15 m and a width of 250 m. The vertical and horizontal resolutions of the cross-section are, respectively, 0.1 m and 1 m, leading to a cross-section with a spatial resolution of 151×251. To demonstrate the sparse measurement scenario, measurement data from six CPT soundings are used as input to generate RFSs of the $(q_{c1N})_{cs}$ cross-section. The locations of these six CPT soundings are denoted by the dashed lines shown in Figure 6.6. The measurement process leads to a total number of 6×151=906 measurement data points, and each data point contains depth, horizontal distance, and measured values. The sparse measurements then can be compiled into a Text file with 906 rows and three columns. The first 20 data points are shown in Figure 7.15. The associated BCS random field simulation is implemented as follows.

Step 1: Double-click the executable of the package, i.e., ASSD-BCS.exe, to open the user interface.

Step 2: Determine which algorithm to use and which domain to apply for random field simulation, e.g., click the “2D” radio button inside the frame “Bayesian compressive sampling”.

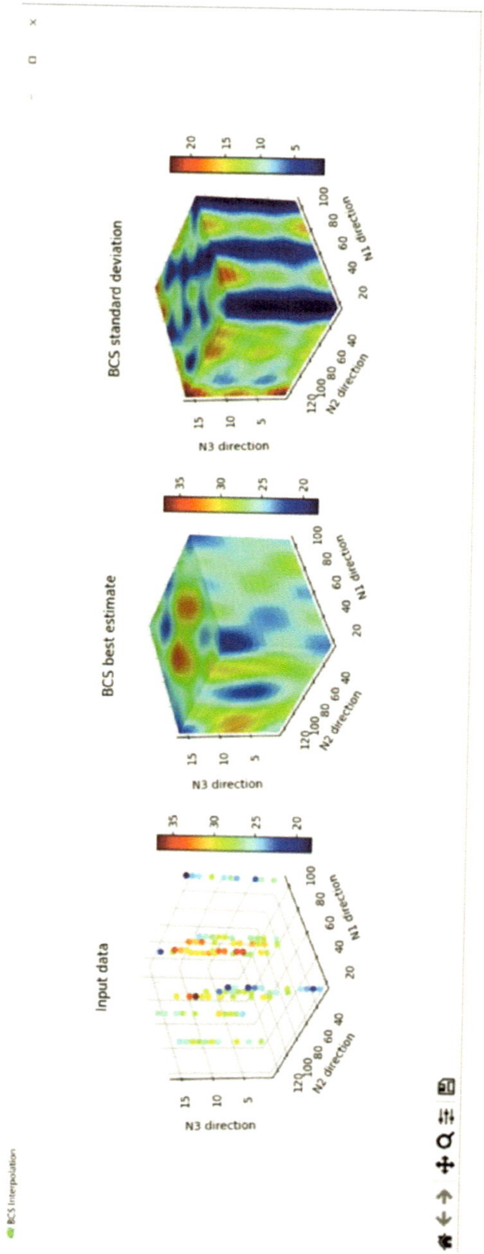

FIGURE 7.14 BCS interpolation results of 3D example shown in Figure 3.17.

Step 3: Load input data file into the software, i.e., click the "Load file" button inside the frame "Load input data file", and then select one local file storing the information of sparse measurements. Consider, for example, load the CPT data example, cpt_1_6_qc_chapter6.txt (see Figure 7.15). After that, the path of the selected file will display on the left.

Step 4: Specify the resolution along each dimension (e.g., depth and horizontal directions) inside the frame "Actual length of element". The input value indicates the actual spatial length that each element represents along the corresponding dimension. For example, specify the resolutions of "1" for N1 (i.e., horizontal) direction, and "0.1" for N2 (i.e., depth) direction, respectively.

Step 5: Specify the extent of extrapolation in the frame "Extrapolation". The default values of zeros are adopted.

Step 6: Confirm the parameters listed in the frame "Random simulation settings", e.g., the number of BCS random field samples required by the user, and the seed value for random simulation. The first entry is set to 100 by default and can be changed by users. The second entry is blank by default and optional.

Step 7: Confirm the parameters listed in the frame "Advanced options".

Step 8: Click the "Generate" button to initiate the simulation. After the simulation is completed, the "Save" and "Plot" buttons will be activated.

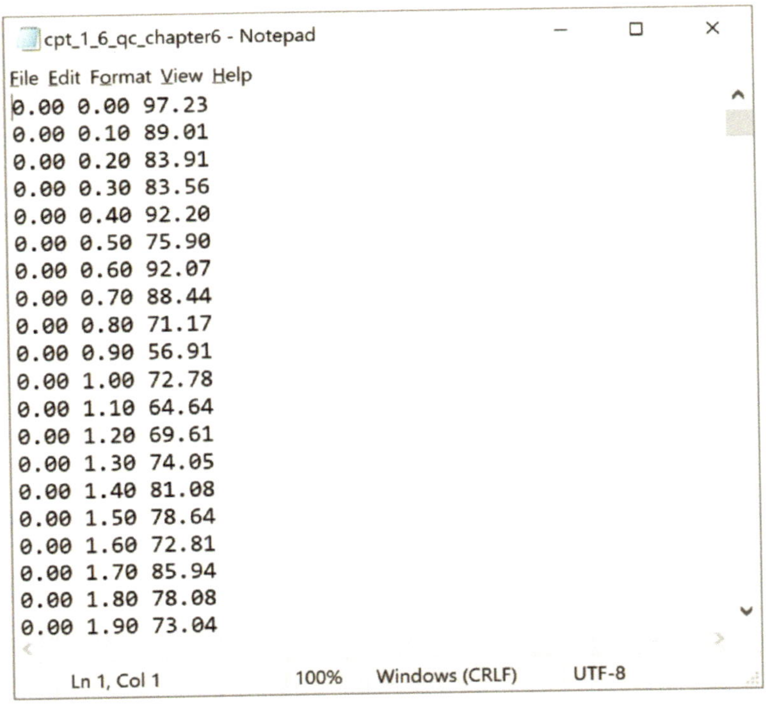

FIGURE 7.15 Measurement data of six CPT profiles in Figure 6.6 of Chapter 6.

Step 9: Click the "Save" button to save the generated RFSs under the root folder of the software. Figure 7.16 shows the saved results, which include two individual Text files. In Figure 7.16a, the interpolated coordinates with specified resolution are shown in Text file "bcs_coordinate.txt". In Figure 7.16b, the 2D arrays of all generated RFSs are stored in Text file "bcs_random.txt". Each generated RFS array is captioned with a line indicating the ID of RFS, e.g., the "# Random field sample # 0:" shown in Figure 7.16b. Click the "Plot" button to plot results of the generated RFSs. The input data and two RFS realizations are visualized quickly in a pop-up window, as shown in Figure 7.17. The generated RFS can be used as input to the subsequent characterization of spatial distribution of soil liquefaction potential in Chapter 6. The pop-up plot window can also be saved as an image file by clicking the floppy disk button at the bottom of the window.

Step 10: Click the "Reset" button to start another new task, if needed.

7.7.2 3D Data

A 3D tutorial example "user_3d.txt" is also provided in the ASSD-BCS installation package. As shown in Figure 7.18, the measurement data of this tutorial example contain 80 row-wise data points stored in a Text file "user_3d.txt". The first three columns store the 3D coordinates, while the fourth column represents the measured values of interest. The implementation procedures are similar to the 2D data example described above, but the spatial domain is selected as "3D". Consider, for example, click the "3D" radio button inside the frame "Bayesian compressive sampling", load the input data file, specify target interpolation resolutions for all dimensions as "1", keep the other default settings, and click the "Run" button to start BCS-based generation of RFSs. The RFS generation results can be visualized after clicking the "Plot" button, as shown in Figure 7.19. The sparse measurements in the input data file are represented by color-scaled dots in the left subplot. Two examples of generated RFSs are shown in the middle and right subplots by color-scaled 3D cubes, respectively.

7.8 APPLICATION EXAMPLES

In the previous two sections, a series of simulated data, including 1D, 2D, and 3D tutorial data examples, have been used to illustrate implementation of ASSD-BCS functions. In this section, ASSD-BCS is further applied to three real data examples.

7.8.1 Interpolation of Soil Strength Data Profile

The example of soil undrained shear strength (s_u) data discussed in Chapter 3 is used in this section for demonstrating the CS interpolation function of ASSD-BCS. As shown in Figure 3.11 of Chapter 3, the soil undrained shear strength data were obtained from vane shear tests at a specific site. Twelve data points of s_u are taken as input data for interpolation and stored in a Text file named "chapter_3_su. txt", as shown in Figure 7.20. The first column represents the depth of s_u measurements in

```
bcs_coordinate - Notepad
File Edit Format View Help
# Dimension of array: [250, 150]
# Coordinate along N1 direction:
[  0.   1.   2.   3.   4.   5.   6.   7.   8.   9.  10.  11.  12.  13.
  14.  15.  16.  17.  18.  19.  20.  21.  22.  23.  24.  25.  26.  27.
  28.  29.  30.  31.  32.  33.  34.  35.  36.  37.  38.  39.  40.  41.
  42.  43.  44.  45.  46.  47.  48.  49.  50.  51.  52.  53.  54.  55.
  56.  57.  58.  59.  60.  61.  62.  63.  64.  65.  66.  67.  68.  69.
  70.  71.  72.  73.  74.  75.  76.  77.  78.  79.  80.  81.  82.  83.
  84.  85.  86.  87.  88.  89.  90.  91.  92.  93.  94.  95.  96.  97.
  98.  99. 100. 101. 102. 103. 104. 105. 106. 107. 108. 109. 110. 111.
 112. 113. 114. 115. 116. 117. 118. 119. 120. 121. 122. 123. 124. 125.
 126. 127. 128. 129. 130. 131. 132. 133. 134. 135. 136. 137. 138. 139.
 140. 141. 142. 143. 144. 145. 146. 147. 148. 149. 150. 151. 152. 153.
 154. 155. 156. 157. 158. 159. 160. 161. 162. 163. 164. 165. 166. 167.
 168. 169. 170. 171. 172. 173. 174. 175. 176. 177. 178. 179. 180. 181.
 182. 183. 184. 185. 186. 187. 188. 189. 190. 191. 192. 193. 194. 195.
Ln 1, Col 1        100%    Windows (CRLF)    UTF-8
```

(a) Spatial coordinates for the generated random field samples

```
bcs_random - Notepad
File Edit Format View Help
# Random field sample # 0:
95.519  94.763  93.260  91.046  88.216  84.944  81.486  78.151
91.967  93.685  96.017  98.064  98.868  97.767  94.654  90.071
95.511  94.753  93.246  91.028  88.193  84.917  81.456  78.121
91.940  93.656  95.986  98.032  98.836  97.735  94.625  90.046
95.494  94.732  93.218  90.990  88.145  84.861  81.396  78.061
91.886  93.599  95.926  97.969  98.771  97.672  94.566  89.996
95.469  94.701  93.175  90.934  88.075  84.778  81.306  77.972
91.805  93.514  95.835  97.874  98.675  97.577  94.478  89.922
95.435  94.659  93.119  90.859  87.981  84.669  81.187  77.853
91.698  93.400  95.715  97.748  98.546  97.451  94.362  89.823
95.393  94.607  93.049  90.766  87.865  84.532  81.039  77.706
91.564  93.259  95.565  97.591  98.386  97.294  94.217  89.700
95.343  94.545  92.966  90.656  87.726  84.370  80.864  77.531
91.405  93.091  95.386  97.403  98.195  97.107  94.044  89.553
95.284  94.474  92.870  90.529  87.566  84.183  80.661  77.329
Ln 1, Col 1        100%    Windows (CRLF)    UTF-8
```

(b) Example of generated random field samples

FIGURE 7.16 Saved BCS random field simulation results for the CPT data example.

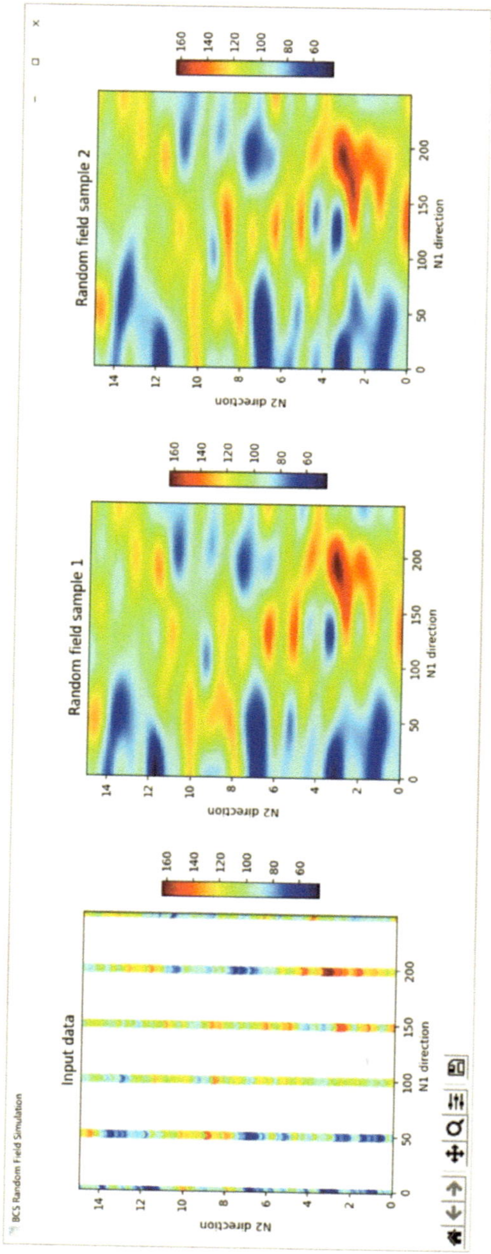

FIGURE 7.17 BCS random field simulation results of $(q_{c1N})_{cs}$ data cross-section from six CPT data profiles.

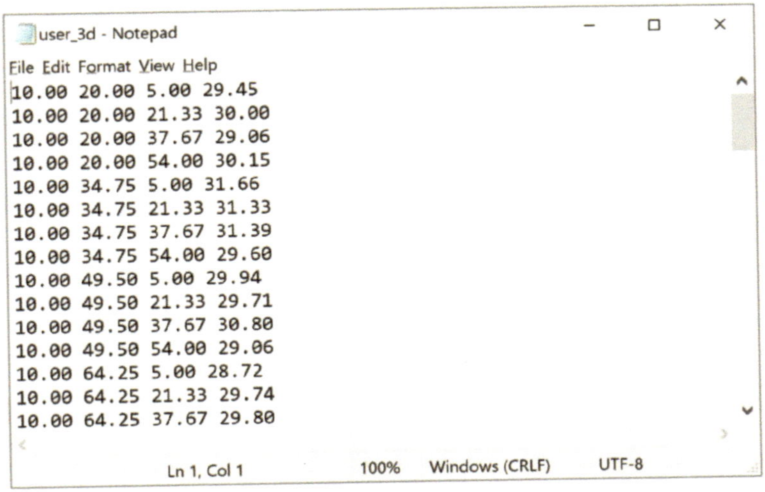

FIGURE 7.18 The first 15 rows of tutorial example user_3d.txt.

meter and the second column represents the s_u values in kPa. To implement CS interpolation, the "1D" radio button inside the frame "Compressive sampling" is selected. Follow the procedure described above by loading the data file and setting a spatial resolution of "0.1" long N1 (i.e., depth) direction. CS interpolation can be performed by clicking the "Run" button. The CS interpolation results and the 12 data points can be visualized in a pop-up window after clicking the "Plot" button, as shown in Figure 7.21. Note that the CS interpolation only provides deterministic interpolation results.

7.8.2 Interpolation of Cone Penetration Test Data Profile

The CPT data profile illustrated in Figure 2.6 of Chapter 2 is re-analyzed in this section for demonstrating the 1D BCS interpolation function of ASSD-BCS. The original q_t data profile has N=256 points and can be treated as a 1D data profile. To illustrate the 1D interpolation function, only M=20 data points are adopted as sparse input measurements for BCS interpolation. These 20 measurement data points are stored in a Text file named "CPT_chapter2.txt", as shown in Figure 7.22. The first column shows the depths of the 20 measurement data points in meter. The second column shows the corresponding measured q_t values in MPa. Follow the procedure described above by clicking the "1D" radio button inside the frame "Bayesian compressive sampling", loading the data file into Data-driven Interpolation module, and setting a spatial resolution of "0.05" long N1 direction. Then, BCS interpolation can be performed by clicking the "Run" button. The BCS interpolation results with a length N=256 can be visualized instantly after an execution of the BCS program. The associated interpolation results, including the measurement data points, the best estimate profile, and a confidence interval of mean ± one standard deviation, are visualized in a pop-up

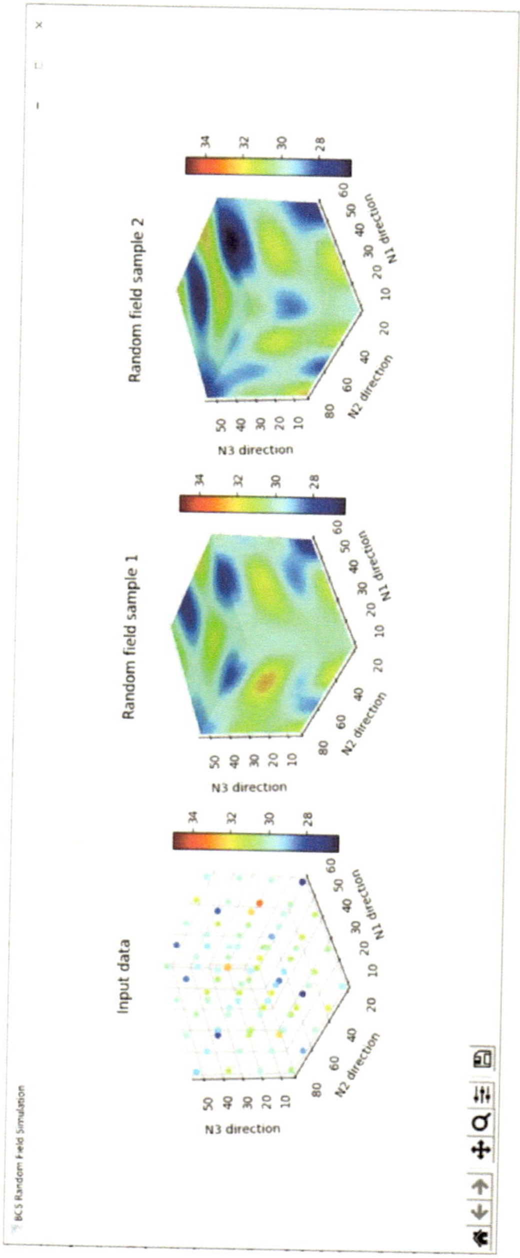

FIGURE 7.19 BCS random field simulation results of tutorial example user_3d.txt.

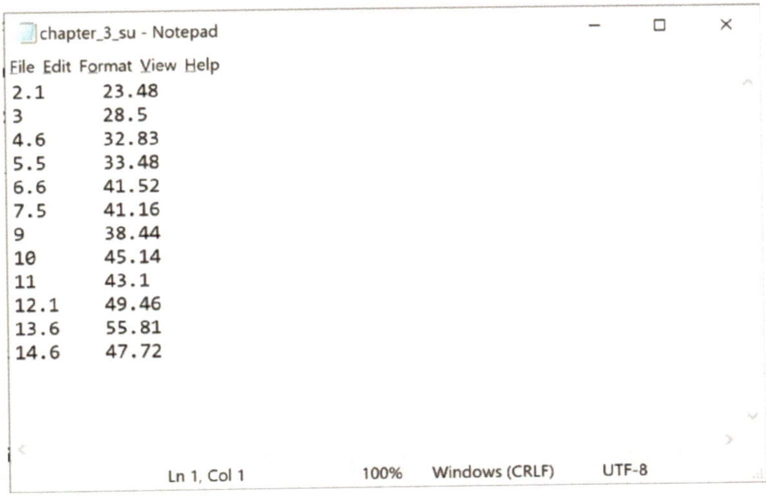
chapter_3_su - Notepad

File Edit Format View Help

2.1	23.48
3	28.5
4.6	32.83
5.5	33.48
6.6	41.52
7.5	41.16
9	38.44
10	45.14
11	43.1
12.1	49.46
13.6	55.81
14.6	47.72

Ln 1, Col 1 100% Windows (CRLF) UTF-8

FIGURE 7.20 Twelve data points of S_u in Figure 3.11.

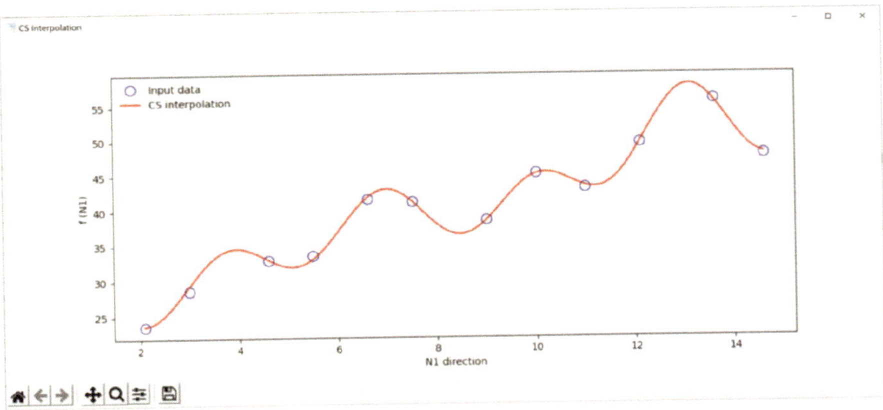

FIGURE 7.21 CS interpolation results of 12 data points of S_u in Figure 3.11.

window, as shown in Figure 7.23. The results in Figure 7.23 are generally consistent with the original q_t data profile shown in Figure 2.6.

7.8.3 2D Interpolation of Cone Penetration Test Data Profiles

The normalized cone resistance Q_t data example illustrated in Figure 5.18 of Chapter 5 is re-analyzed in this section for demonstrating the 2D BCS interpolation function of ASSD-BCS. Q_t data profiles of eight CPT soundings are adopted as sparse input measurement data for interpolating the vertical cross-section denoted in Figure 5.17. Each of the eight CPT soundings has a sounding depth of 2 m with a vertical interval of 0.005 m, leading to 400 data points. The eight CPT soundings consist of a total

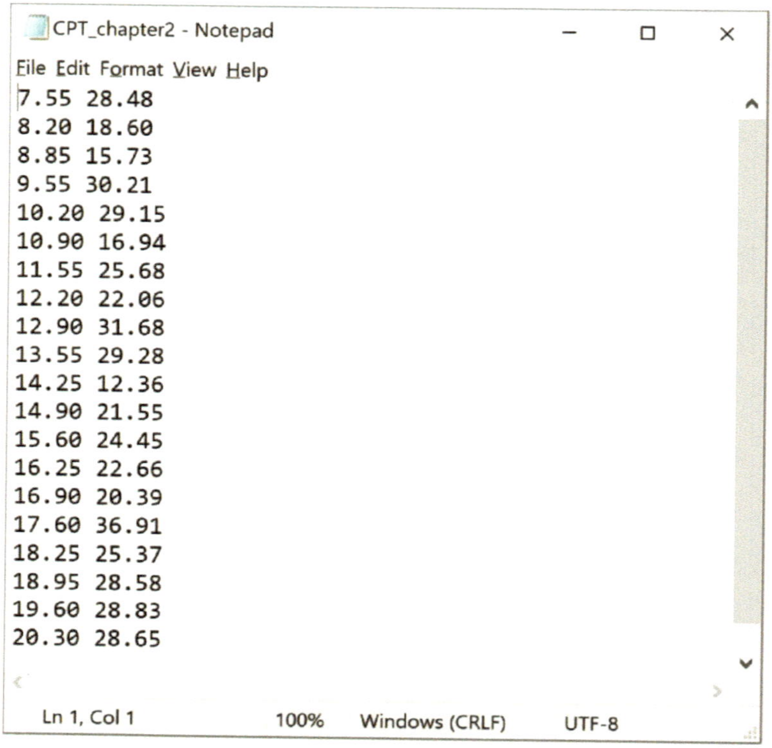
CPT_chapter2 - Notepad
File Edit Format View Help

7.55 28.48
8.20 18.60
8.85 15.73
9.55 30.21
10.20 29.15
10.90 16.94
11.55 25.68
12.20 22.06
12.90 31.68
13.55 29.28
14.25 12.36
14.90 21.55
15.60 24.45
16.25 22.66
16.90 20.39
17.60 36.91
18.25 25.37
18.95 28.58
19.60 28.83
20.30 28.65

Ln 1, Col 1 100% Windows (CRLF) UTF-8

FIGURE 7.22 Sparse measurements of CPT data example in Figure 2.6.

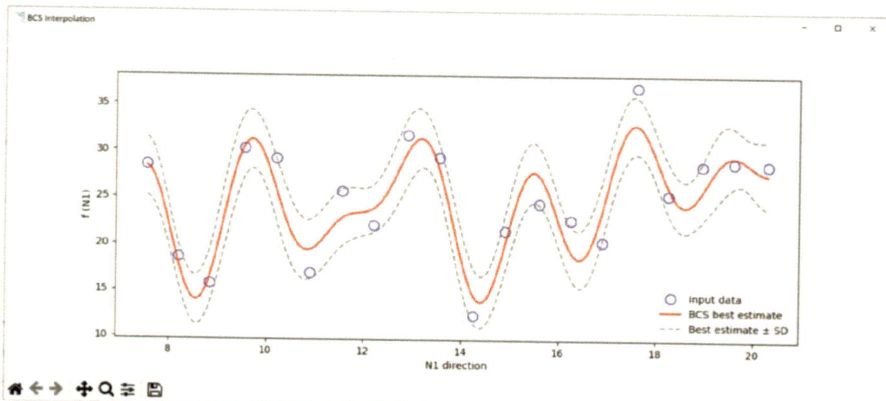

FIGURE 7.23 BCS interpolation results of CPT data example in Figure 2.6.

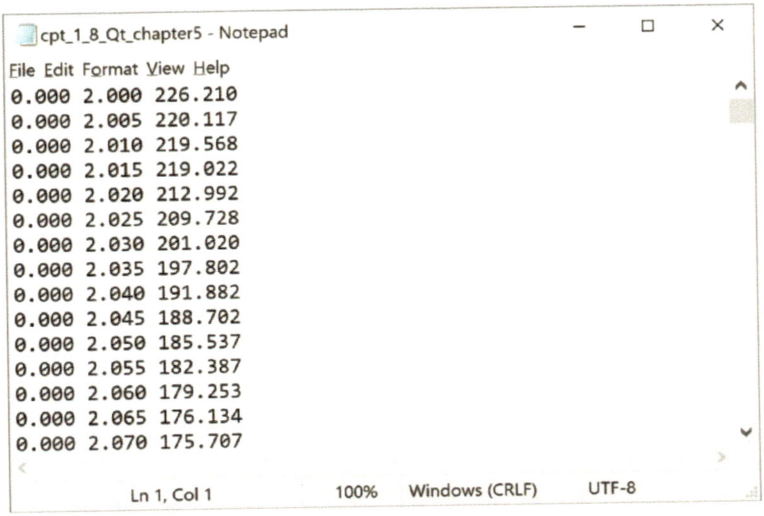

cpt_1_8_Qt_chapter5 - Notepad

File Edit Format View Help

```
0.000 2.000 226.210
0.000 2.005 220.117
0.000 2.010 219.568
0.000 2.015 219.022
0.000 2.020 212.992
0.000 2.025 209.728
0.000 2.030 201.020
0.000 2.035 197.802
0.000 2.040 191.882
0.000 2.045 188.702
0.000 2.050 185.537
0.000 2.055 182.387
0.000 2.060 179.253
0.000 2.065 176.134
0.000 2.070 175.707
```

Ln 1, Col 1 100% Windows (CRLF) UTF-8

FIGURE 7.24 The first 15 rows of normalized cone resistance Q_t data example in Figure 5.18b.

number of 400×8=3,200 measurement data. These 3,200 measurements are stored in a Text file "cpt_1_8_Qt_chapter5.txt", as shown in Figure 7.24. The first two columns represent the horizontal and depth coordinates, respectively. Follow the procedure described above by loading the data file into Data-driven Interpolation module and setting spatial resolutions of "1" and "0.005" for N1 and N2 directions, respectively. Then, 2D BCS interpolation can be performed by clicking the "Run" button. The BCS interpolation results can be visualized after an execution of the BCS program, as shown in Figure 7.25. The input measurement data, best estimate, and one standard deviation of BCS are shown in three subplots, respectively. The best estimate result in Figure 7.25 is generally consistent with the colormap shown in Figure 5.18b. Note that the origin of 2D colormap in Figure 7.25 is at bottom-left corner, as opposed to the top-left corner in Figure 5.18b.

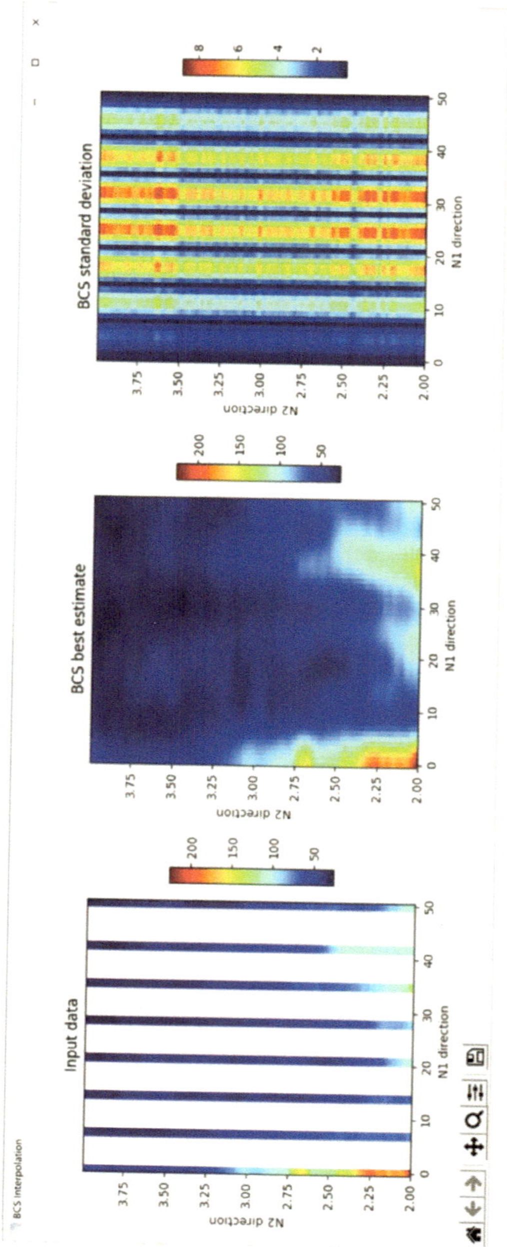

FIGURE 7.25 BCS interpolation results of normalized cone resistance Q_t data cross-sections from eight profiles.

REFERENCES

Hu, Y., Y. Wang, T. Zhao, and K. K. Phoon. 2020. Bayesian supervised learning of site-specific geotechnical spatial variability from sparse measurements. *ASCE-ASME Journal of Risk and Uncertainty in Engineering Systems, Part A: Civil Engineering* 6 (2): 04020019.

Hu, Y., T. Zhao, Y. Wang, C. Choi, and C. W. Ng. 2019. Direct simulation of two-dimensional isotropic or anisotropic random field from sparse measurement using Bayesian compressive sampling. *Stochastic Environmental Research and Risk Assessment* 33 (8): 1477–1496.

Lutz, M. 2013. *Learning Python: Powerful object-oriented programming*. Sebastopol, CA: O'Reilly Media.

Lyu, B., Y. Hu, and Y. Wang. 2023. Data-driven development of three-dimensional subsurface models from sparse measurements using Bayesian compressive sampling: A benchmarking study. *ASCE-ASME Journal of Risk and Uncertainty in Engineering Systems, Part A: Civil Engineering* 9 (2): 04023010.

Wang, Y., and T. Zhao. 2016. Interpretation of soil property profile from limited measurement data: A compressive sampling perspective. *Canadian Geotechnical Journal* 53 (9): 1547–1559.

Wang, Y., and T. Zhao. 2017. Statistical interpretation of soil property profiles from sparse data using Bayesian compressive sampling. *Géotechnique* 67 (6): 523–536.

Wang, Y., T. Zhao, and K. K. Phoon. 2018. Direct simulation of random field samples from sparsely measured geotechnical data with consideration of uncertainty in interpretation. *Canadian Geotechnical Journal* 55 (6): 862–880.

Zhao, T., and Y. Wang. 2020. Non-parametric simulation of non-stationary non-Gaussian 3D random field samples directly from sparse measurements using signal decomposition and Markov Chain Monte Carlo (MCMC) simulation. *Reliability Engineering & System Safety* 203: 107087.

Zhao, T., and Y. Wang. 2021. Statistical interpolation of spatially varying but sparsely measured 3D geo-data using compressive sensing and variational Bayesian inference. *Mathematical Geosciences* 53: 1171–1199.

Zhao, T., Y. Hu, and Y. Wang. 2018. Statistical interpretation of spatially varying 2D geo-data from sparse measurements using Bayesian compressive sampling. *Engineering Geology* 246: 162–175.

8 Summaries and Additional Remarks

Yu Wang and Kok-Kwang Phoon

ABSTRACT

This chapter summarizes the key features of Bayesian compressive sensing or sampling (BCS) discussed in the previous seven chapters. Additionally, it compares BCS with conventional interpolation methods such as Kriging and Gaussian process regression (GPR), both of which require parametric modeling of the autocorrelation structure and parameter estimation. Kriging assumes stationarity and often necessitates detrending, a complex and nontrivial process. While Kriging and GPR perform well when sufficient domain knowledge and extensive measurements are available, they can be unreliable with limited data. Furthermore, Kriging is seldom used for 3D spatial interpolation due to its complexity, whereas BCS is inherently scalable to high-dimensional data and maintains high computational efficiency, making it particularly effective for geotechnical site characterization with sparse measurements. This chapter also includes comments on benchmarking and outlines potential directions for future research.

8.1 SUMMARIES

This book offers a systematic introduction to the essential theories and computational tools for formulating and solving sparse spatial data modeling problems using Bayesian compressive sensing or sampling (BCS), an efficient, data-driven method for spatial data analysis and simulation directly from sparse measurements. In geotechnical site characterization and various geoscience and geoengineering disciplines, measurements are often limited due to time, cost, accessibility, or technical constraints. Phoon (2017) called this the "curse of small sample size". Consequently, the efficient analysis and modeling of sparse measurements have long posed a significant challenge in these fields. BCS addresses this challenge by integrating Bayesian statistics with compressive sensing (CS), a widely used algorithm for data compression in digital signal and image processing.

The background, mathematical theory, and underlying assumptions of CS were reviewed in Chapter 2. For a successful CS application, three key prerequisites must be satisfied: (1) data sparsity, (2) a properly designed sensing matrix, and (3) an appropriate reconstruction algorithm. The first prerequisite concerns the inherent nature of

DOI: 10.1201/9781003381235-8

the signal or image, while the second relates to the measurement modality. The third involves solving an underdetermined system of linear equations. Sparse representation (or sparse approximation) is the foundation of CS, referring to the expression of a signal or image as a linear combination of a small number of basis elements from an overcomplete dictionary. The CS problem can then be formulated as an underdetermined linear system, solvable via optimization techniques, greedy algorithms, or threshold-based methods. In Chapter 2, two widely used and effective reconstruction algorithms were introduced: basis pursuit and orthogonal matching pursuit (OMP).

Geotechnical and geological data inherently exhibit spatial patterns, making them sparse by nature. As a result, they automatically satisfy the CS's prerequisite of data sparsity, allowing CS to be effectively applied in their analysis. Furthermore, geotechnical site characterization typically involves measurements taken at limited locations, resulting in sparse measurements. This presents a unique challenge compared to signal or image processing, where nearly complete measurements of a signal or image are usually available.

The primary objective of using BCS in geotechnical site characterization is to uncover spatial variation patterns from sparse measurements, a fundamentally different goal from traditional CS applications in signal or image processing, which focus on data compression and full signal recovery for storage and transmission.

To address the additional challenges posed by sparse measurements in geotechnical applications, Chapter 3 introduces careful modifications to the sensing matrix and algorithms, including:

1 Adaptation of the OMP algorithm to identify nontrivial coefficients and their corresponding basis functions from sparse measurements.
2 Uncertainty quantification of the identified nontrivial coefficients under a Bayesian framework, accounting for sparse measurements.

Variational Bayesian inference (VBI) is then employed in Chapter 3 to approximate the joint posterior probability density function (PDF) of the nontrivial coefficients with a tractable distribution. The resulting method, termed OMP-VBI, is scalable from one-dimensional (1D) to multi-dimensional (2D, 3D, or higher) geospatial data. Chapter 3 demonstrates and validates OMP-VBI using both simulated and real-world data.

Chapter 4 extends BCS to random field simulation for spatial and temporal data using sparse measurements. Two novel random field generators are presented with systematic derivations: (1) the BCS-KL generator, based on Karhunen–Loève (KL) expansion, and (2) the BCS-MCMC generator, which integrates BCS with Markov Chain Monte Carlo (MCMC) simulation. These generators offer significant advantages for geotechnical applications:

1 They produce random field samples (RFSs) directly from sparse measurements.
2 They eliminate the need to predetermine autocorrelation functions.
3 They require no parameter estimation (e.g., autocorrelation length) for the preselected autocorrelation models.

This capability is particularly valuable in geotechnical site characterization, where conventional approaches typically demand extensive measurements to establish autocorrelation functions and estimate their parameters. Such data are often unavailable. The BCS-based generators demonstrate unique advantages. Their data-driven nature enables simulation of nonstationary, non-Gaussian RFSs directly from sparse measurements, maintaining consistency with actual site conditions. They are also directly applicable for both 2D and 3D simulations. Chapter 4 validates these generators using field data from South Parklands, Australia. The results confirm their effectiveness in subsurface stratification, characterization of 3D soil variability, and reliable simulations directly from limited measurement data.

Chapters 5 and 6 apply the developed BCS framework to the planning and optimization of geotechnical site investigations. These chapters present an uncertainty-based adaptive sampling strategy that continuously monitors the results interpreted from already performed exploration points and automatically decides whether additional exploration points are needed and, if so, their optimal test locations. The adaptive sampling approach leverages two key outputs from BCS, interpretation results of existing measurements and quantified uncertainty estimates

In the first stage of the adaptive sampling, initial exploration points are carried out to obtain a basic understanding of subsurface conditions of a site. Then, additional exploration points are conducted sequentially until the interpreted subsurface conditions reach a target reliability level. Each additional exploration point is determined according to the information obtained from previously conducted exploration points. Thus, even without sufficient prior knowledge about subsurface conditions at a specific site, the minimal number of exploration points and their optimal locations can be efficiently and automatically determined.

To determine the number of initial exploration points in the first stage of adaptive sampling, an extensive parametric study is conducted using BCS in Chapter 5. This study develops quantitative methods and statistical charts for sample size determination, explicitly considering the spatial variation and correlation of geotechnical properties. These methods and charts are also valuable in addressing a common question in geotechnical practice: How many samples (such as test specimens, borings, or in-situ test soundings) are needed for geotechnical investigation of a site?

Existing geotechnical guidelines and design codes only provide conceptual principles for site investigation, often recommending general rules of thumb for the minimum number of exploration points. These typically assume equal sampling spacing and regular grid patterns, treating all areas of the site as equally important. However, such an approach fails to account for the spatial variability of subsurface soil, potentially resulting in an inefficient and uneconomical investigation.

Adaptive sampling is a data-driven approach that incrementally adds exploration points in areas of high uncertainty or variability. By accounting for site-specific spatial variability in subsurface soils or rocks, it optimizes both the number and locations of exploration or sampling points. Chapters 5 and 6 demonstrate and validate this strategy using both simulated data and real-world datasets from Wellington and Christchurch, New Zealand, as well as St James's Park, London.

Chapter 7 introduces a computer software package called Analytics of Sparse Spatial Data using Bayesian Compressive Sampling (ASSD-BCS), providing readers with a convenient tool for hands-on experience with BCS. ASSD-BCS features a user-friendly graphical interface and is openly accessible, available for free online. Additionally, various tutorial examples are provided to guide users in its application.

8.2 COMPARISON WITH KRIGING OR GAUSSIAN PROCESS REGRESSION (GPR)

BCS is a data-driven approach, which is different from conventional parametric or semi-parametric interpolation methods, such as Kriging (e.g., Webster and Oliver 2007) or GPR (e.g., Rasmussen and Williams 2006). Before using Kriging or GPR for interpolation, it requires parametric modeling of the autocorrelation structure (e.g., semi-variogram for Kriging), as well as estimation of its parameters (e.g., range and sill for Kriging), by fitting the experimental semi-variogram to a predetermined function form (e.g., an exponential semi-variogram model). Furthermore, Kriging generally assumes that the spatially varying quantity of interest is stationary (e.g., constant mean and variance over spatial directions). Detrending is often performed for the measurements before calculation of semi-variogram and estimation of parameters such that the residuals are hopefully stationary (e.g., Phoon et al. 2003). Selection of the most appropriate trend function form (e.g., linear or high order polynomial function) for detrending, however, is tricky and nontrivial (e.g., Ching and Phoon 2017; Wang et al. 2017).

When domain knowledge or extensive measurements are available to determine the appropriate functional forms of both the trend function and semi-variogram, Kriging or GPR performs effectively for 1D and 2D data. However, in the absence of domain knowledge to guide the selection of the trend function type or when measurements are limited, the experimental semi-variogram derived from the data may be unreliable, leading to inaccurate parameter estimation. In such cases, interpolation using Kriging or GPR based on inaccurate parameters can be misleading.

For instance, a minimum of 50–100 measurement data points are typically required for reliable 1D Kriging, while 100–150 measurements are recommended for 2D Kriging under isotropic spatial variability conditions (e.g., Webster and Oliver, 2007). If the quantity of interest exhibits anisotropic spatial distribution, at least 250 data points are necessary to reliably estimate autocorrelation for 2D Kriging (e.g., Webster and Oliver, 2007).

When measurements are limited, a common challenge in geotechnical site characterization, BCS provides a data-driven and effective approach to uncovering spatial variation patterns at a specific site, with quantified uncertainty. Notably, Kriging is rarely used for interpolating 3D spatial data due to the complexity of defining a 3D trend function and constructing a semi-variogram for anisotropic spatial distributions. In contrast, BCS is inherently scalable to high-dimensional data and maintains high computational efficiency, even in complex, high-dimensional applications.

8.3 BENCHMARKING AND MODEL COMPARISON IN DATA-CENTRIC GEOTECHNICS

Data-centric geotechnics is an emerging field that advocates a "digital first" agenda. One aspect of this agenda is the development of methods that make sense of actual data. Perhaps more importantly, it calls for the industry to develop methods without imposing thresholds on quantity, quality, and/or other "ugly" data attributes as prerequisites for exploring data-informed decision-making. In short, the industry should engage in the spirit of offering the data an opportunity to speak for itself.

Research on data-driven site characterization has been active in recent years, with different methods being developed (e.g., Ching 2025). Because site-specific data are incomplete and sparse, these data-driven site characterization methods are mainly probabilistic, with Bayesian machine learning (ML) being the most popular. For example, BCS is also known as Bayesian sparse learning in machine learning (e.g., Hu et al. 2020). It is not completely known which method is more appropriate and/or more efficient for a given circumstance. The answer must largely depend on the attributes of the training data and the desired outcomes of site characterization. To promote the development of data-driven site characterization methods that are applicable to routine projects in a more purposeful way, one important step is to create reference geotechnical data sets and standard performance metrics to train and measure different data-driven site characterization methods on a uniform basis.

This benchmark testing, or benchmarking, is widely used in machine learning to foster unbiased and competitive evaluation of different machine learning methods, as it provides a standardized way to evaluate and compare different algorithms, models, and methodologies (Phoon 2025). It enables researchers and practitioners to assess performance objectively, ensuring that improvements in accuracy, efficiency, or robustness are measurable and replicable. By using well-established datasets and metrics, benchmarking facilitates fair comparisons across various approaches, helping identify the most effective solutions for specific problems. Additionally, it drives innovation by highlighting strengths and weaknesses in existing techniques, encouraging the development of more advanced and efficient models. Benchmarking also plays a vital role in practical applications, aiding industries in selecting the best models for deployment, reducing risks associated with untested solutions. In academic and research settings, benchmarks serve as reference points for progress, allowing for consistent evaluation over time. Without benchmarking, there would be no reliable means to track advancements in machine learning, making it difficult to distinguish between genuinely improved models and those that perform well due to chance or overfitting. Furthermore, it promotes transparency and reproducibility, ensuring that findings and results can be validated by others in the field. Overall, benchmarking is fundamental to fostering growth, reliability, and trust in machine learning technologies.

For data-centric geotechnics or other geo-disciplines, benchmark examples should be constructed based on the following guidelines: (1) configurations must be 3D and realistic in scale, stratigraphy, and properties; (2) range of ground conditions should be sufficiently representative; (3) training datasets should be realistic, i.e., should be restricted to only data that an engineer has at his/her disposal in practice; and (4) validation datasets and performance metrics should be selected to demonstrate the value of machine learning to decision-making in practice.

The first benchmark testing in data-driven site characterization was presented in Phoon et al. (2022). Each benchmark example is defined by a combination of a virtual ground ("reality") and a training dataset (measured "reality" or what is known). Four virtual grounds and two training datasets were proposed, resulting in eight benchmark examples. These virtual grounds (S-VG1, S-VG2, S-VG3, and S-VG4) consist of three soil types (a top layer of sand, a middle layer of clay, and a base layer of silty-sand) that are horizontal, inclined, or discontinuous. The distinctive features of these examples are: (1) a realistic 3D ground volume 20 m long × 20 m wide × 10 m deep and (2) realistic MUSIC-3X (Multivariate, Uncertain and Unique, Sparse, Incomplete, and potentially Corrupted with "3X" denoting 3D spatial variability) data commonly produced in a site investigation program. A more realistic virtual ground S-VG5 containing more complex intermixing of soil types was created using a coupled Markov chain simulation (Shuku and Phoon 2023).

A new article type called "GEOAI benchmark examples" was included as part of the journal Geodata and AI to fill this critical gap in the literature. These benchmark examples are labelled based on the template: BM/challenge/GEO AI volume/year of publication. The first example is BM/FootingDesign/1/2024 published in Volume 1 (Cai and Phoon 2024). The machine learning challenge in BM/FootingDesign/1/2024 is to optimize the characterization of a location-specific probability distribution for the design parameter (effective stress friction angle) based on site-specific data and a generic database alone. The second example published in Volume 2 is named BM/AirportSoilProperties/2/2025 (Otake et al. 2025a). BM/AirportSoilProperties/2/2025 is built on actual geotechnical data collected from a large-scale offshore airport in Japan, comprising 2,922 boreholes and 67,760 records across the entire site, with 51 boreholes and 1,001 records concentrated in a 300 m × 300 m verification site. This extensive dataset reflects the complex characteristics of soft ground unique to reclaimed land, providing a solid foundation for realistic field evaluations that capture complexities beyond conventional theoretical models. In contrast, the stratification in S-VG5 is synthetic (generated from a coupled Markov chain). To the author's best knowledge, BM/AirportSoilProperties/2/2025 is the first benchmark example based on a large-scale real site characterization dataset.

Geodata and AI encourages the submission of benchmark examples to foster greater competition among geo ML/AI models in the literature, track the progress of data-centric developments in geo-disciplines, and showcase the value of geo ML/AI to practice. A special virtual collection called "GEOAI Benchmark Examples" was established to provide a one-stop shop for researchers and practitioners to access a relevant benchmark example for measuring the performance of, tracking progress in, and identifying weaknesses in different models developed for a specific task meaningful to decision-making in a geo-discipline (www.sciencedirect.com/special-issue/10M34CDR6DT).

8.4 FUTURE RESEARCH DIRECTIONS

8.4.1 Multivariate Data

The BCS discussed in this book focuses on spatial variation of a single variable. However, site-specific data obtained during geotechnical site characterization often includes multiple soil or rock properties that are not only spatially autocorrelated but also cross-correlated. Phoon et al. (2019) introduced the mnemonic MUSIC-X (Multivariate, Uncertain and Unique, Sparse, Incomplete, and potentially Corrupted, with "X" denoting the spatial/temporal dimension) to describe site characterization data. This acronym succinctly captures the seven common attributes encountered in geotechnical site characterizations.

Different types of data from various sources are often available in a site characterization program, providing complementary information. This raises an important question for future research on BCS in geotechnical site characterization: How can multiple sources of sparse data be integrated within the BCS framework to effectively profile different geotechnical properties? Guan et al. (2024b) developed a joint sparse representation framework for the fusion of sparse, non-co-located measurements from multiple sources for geotechnical site characterization. Further studies in this direction are warranted.

8.4.2 Fusion of Geotechnical and Geophysical Data

In a site characterization program, geophysical surveys, such as seismic refraction, electrical resistivity tomography, and ground penetrating radar, are sometimes conducted to rapidly and nondestructively assess subsurface conditions over large areas. The abundant geophysical data obtained provides valuable complementary information for developing a reliable 2D or 3D spatial variability model of subsurface soil or rock properties. A promising future research direction is the development of practical and robust methods for the quantitative fusion of geotechnical and geophysical data in a 3D space.

Guan and Wang (2024) extended the joint sparse representation framework to efficiently integrate limited geotechnical data with high-resolution geophysical data in a 3D space. This approach leverages the auto- and cross-correlation of 3D geotechnical and geophysical data using joint sparse representation, enabling efficient information exchange in a data-driven manner. The integration of geotechnical and geophysical data has led to significant improvements in constructing 3D models of geotechnical properties. This is an active research area with many ongoing developments with exciting potential.

8.4.3 Basis Function

The basis function is a fundamental component of sparse representation, and this book mainly utilizes a DCT basis function. Other suitable basis functions may also be employed, such as discrete wavelet transformation basis functions or those derived from domain expertise or prior knowledge. For example, during the desk study stage of site characterization, geotechnical data from neighboring sites or locations with similar geological conditions are often collected and used as valuable prior knowledge in geotechnical engineering practice. This prior knowledge or existing database can also serve as a foundation for developing suitable basis functions for geotechnical site characterization.

Guan et al. (2024a) made the first attempt to use basis functions developed from existing data from neighboring sites or sites with similar geological settings for characterization of soil property spatial variability in a data-driven manner. The method begins by constructing a suite of basis functions, or a dictionary, that captures the dominant spatial variation patterns of a property measured at geologically similar sites within a dictionary learning framework. By leveraging the developed dictionary, Bayesian learning is applied to interpret the spatial variability of a property based on sparse measurements obtained from a specific site. Further research in this direction is also warranted.

REFERENCES

Cai, Y. and K. K. Phoon. 2024. GEOAI benchmark example BM/FootingDesign/1/2025. *Geodata and AI*, 1, 100002.

Ching, J. and K. K. Phoon. 2017. Characterizing uncertain site-specific trend function by sparse Bayesian learning. *Journal of Engineering Mechanics* 143 (7): 04017028.

Ching, J. 2025. *Bayesian machine learning in geotechnical site characterization*. Boca Raton, FL: CRC Press.

Guan, Z. and Y. Wang. 2024. Fusion of three-dimensional geotechnical and geophysical data for developing digital twin of underground space. *Soils and Foundations* 64: 101528.

Guan, Z., Y. Wang, and K. K. Phoon. 2024a. Dictionary learning of spatial variability at a specific site using data from other sites. *Journal of Geotechnical and Geoenvironmental Engineering* 150(9): 04024072.

Guan, Z., Y. Wang, and K. K. Phoon. 2024b. Fusion of sparse non-co-located measurements from multiple sources for geotechnical site investigation. *Canadian Geotechnical Journal* 61(8): 1574–1592.

Hu, Y., Y. Wang, T. Zhao, and K. K. Phoon. 2020. Bayesian supervised learning of site-specific geotechnical spatial variability from sparse measurements. *ASCE-ASME Journal of Risk and Uncertainty in Engineering Systems, Part A: Civil Engineering* 6(2): 04020019.

Otake, Y., J. Ching, T. Saito, and K. Asano. 2025a. GEOAI benchmark problems BM/AirportSoilProperties/2/2025. *Geodata and AI* 2: 00011.

Phoon, K. K. 2017. Role of reliability calculations in geotechnical design. *Georisk: Assessment and Management of Risk for Engineered Systems and Geohazards* 11(1): 4–21.

Phoon, K. K., J. Ching, and Y. Wang. 2019. Managing risk in geotechnical engineering–from data to digitalization. In *Proceedings, 7th International Symposium on Geotechnical Safety and Risk (ISGSR 2019)*: Taipei, Taiwan, 13–34.

Phoon, K. K., S. T. Quek, and P. An. 2003. Identification of statistically homogeneous soil layers using modified Bartlett statistics. *Journal of Geotechnical and Geoenvironmental Engineering* 129 (7): 649–659.

Phoon, K. K., T. Shuku, J. Ching, and I. Yoshida. 2022. Benchmark examples for data-driven site characterization. *Georisk: Assessment and Management of Risk for Engineered Systems and Geohazards* 16(4): 599–621.

Phoon, K. K. 2025. GEOAI Editorial Volume 3: Where are the benchmarking tests in geo-disciplines? *Geodata and AI* 3: 100017.

Rasmussen, C. E., and C. K. Williams. 2006. *Gaussian processes for machine learning*. Cambridge, Massachusetts: MIT Press.

Shuku, T. and K.K. Phoon. 2023. Comparison of data-driven site characterization methods through benchmarking: methodological and application aspects. *ASCE-ASME Journal of Risk and Uncertainty in Engineering Systems, Part A: Civil Engineering* 9(2): 04023006.

Wang, Y., O. V. Akeju, and T. Zhao. 2017. Interpolation of spatially varying but sparsely measured geo-data: A comparative study. *Engineering Geology* 231: 200–217.

Webster, R. and M. A. Oliver. 2007. *Geostatistics for environmental scientists*. Hoboken, NY: John Wiley & Sons.

Index

N

O

P

R

S

T

U

V

Y

Z